SOTHIS PRESS

Sothis Press aims to completely reimagine the public domain book with an emphasis on universal access to captivatingly designed print editions. We believe that technology will enable every single book ever published to be perpetually accessible in beautiful print form. Sothis is beauty in ubiquity.

www.sothis.press
1 Waterhouse Square, London, ECIN 2ST

Methods in Plant Histology

Charles J. Chamberlain

SOTHIS PRESS

ISBN 978-1-0365-1694-9

This edition is published by
Sothis Press © 2023. The company
is registered in England and Wales.
The Company Number is 14900389.

THE UNIVERSITY OF CHICAGO PRESS
CHICAGO, ILLINOIS

THE BAKER & TAYLOR COMPANY
NEW YORK

THE CAMBRIDGE UNIVERSITY PRESS
LONDON

THE MARUZEN-KABUSHIKI-KAISHA
TOKYO, OSAKA, KYOTO, FUKUOKA, SENDAI

THE MISSION BOOK COMPANY
SHANGHAI

METHODS IN
PLANT HISTOLOGY

By

CHARLES J. CHAMBERLAIN, PhD., Sc.D.
Professor of Morphology and Cytology in the University of Chicago

FOURTH REVISED EDITION

THE UNIVERSITY OF CHICAGO PRESS
CHICAGO · ILLINOIS

Published June 1901
Second Edition October 1905
Third Edition May 1915
Second Impression December 1916
Third Impression July 1920
Fourth Impression January 1923
Fourth Edition October 1924

Composed and Printed By
The University of Chicago Press
Chicago, Illinois, U.S.A.

PREFACE TO THE FIRST EDITION

This book has grown out of a course in histological technic conducted by the author at the University of Chicago. The course has also been taken by non-resident students through the Extension Division of the University. The *Methods* were published over a year ago as a series of articles in the *Journal of Applied Microscopy*, and have called out numerous letters of commendation, criticism, suggestion, and inquiry. The work has been thoroughly revised and enlarged by about one-half. It is hoped that the criticism and suggestion, and also the experience gained by contact with both resident and non-resident students, have made the directions so definite that they may be followed, not only by those who work in a class under the supervision of an instructor, but also by those who must work in their own homes without any such assistance.

More space has been devoted to the paraffin method than to any other, because it has been proved to be better adapted to the needs of the botanist. The celloidin method, the glycerin method, and freehand sectioning are also described, and their advantages and disadvantages are pointed out.

The first part of the book deals with the principles of fixing and staining, and the various other processes of microtechnic, while in the later chapters these principles are applied to specific cases. This occasions some repetition, but the mere presentation of general principles will not enable the beginner to make good mounts.

The illustrations and notes in the later chapters are not intended to afford a study of general morphology, but they merely indicate to students with a limited knowledge of plant structures the principal features which the preparations should show. The photomicrographs were made from the author's preparations by Dr. W. H. Knap, and Figures 52, 57, and 59 (Figs. 61, 66, and 68 of second edition) were drawn by Miss Eleanor Tarrant; all other figures of plant structures were made from the author's drawings.

Corrections and suggestions will be heartily appreciated.

CHARLES J. CHAMBERLAIN

CHICAGO
June 1, 1901

PREFACE TO THE SECOND EDITION

It is gratifying to the author to learn that the kindly reception accorded to *Methods in Plant Histology* has exhausted the edition. Since the first edition appeared, a little more than four years ago, laboratory methods have been greatly improved, and systematic experiments have made it possible to give much more definite directions for making preparations.

In the present edition much more attention has been given to collecting material. Professor Kleb's methods for securing various reproductive phases in the algae and fungi have been outlined in a practical way. Methods for growing other laboratory material are more complete than in the earlier edition.

The paraffin method has been much improved, and the glycerin method has been almost entirely replaced by the Venetian turpentine method, to which a whole chapter is devoted. Other new chapters deal with michrochemical tests, freehand sections, special methods, and the use of the microscope.

The author is deeply indebted to his colleague, Dr. W. J. G. Land, for numerous suggestions and improvements in methods.

Corrections and suggestions will be heartily appreciated.

CHARLES J. CHAMBERLAIN

CHICAGO
July 1, 1905

PREFACE TO THE THIRD EDITION

The continued appreciation accorded to *Methods in Plant Histology* has exhausted the second edition. Since that edition appeared, methods have become more and more exact, so that the present volume is practically a new book. The general arrangement of the subject-matter, and directions for collecting material and for securing reproductive phases in the algae and fungi have been retained, and a chapter on "Photomicrographs and Lantern Slides" (chap. xii) has been added.

Great improvements have been made in the paraffin method, so that sections are easily cut which were impossible ten years ago, while ten years of added experience with the Venetian turpentine method have made it possible to describe it so definitely that even the beginner should find no serious difficulty.

The author is deeply indebted to his colleague, Dr. W. J. G. Land, for numerous suggestions and improvements covering the whole field of microtechnic. He is also greatly indebted to Dr. S. Yamanouchi for many improvements in the methods applicable to algae and mitotic figures.

Corrections and suggestions will be heartily appreciated.

CHARLES J. CHAMBERLAIN

CHICAGO
May, 1915

PREFACE TO THE FOURTH EDITION

It is gratifying to the author that the appreciation accorded to *Methods in Plant Histology*, when it first appeared as a series of articles in the *Journal of Applied Microscopy*, has continued through three editions of the book. While the chapter headings and general arrangement remain about the same as before, the book has been almost entirely re-written.

Directions for collecting material have been amplified and the preparation of the most familiar laboratory types has received particular attention. While no radical changes have been made in the paraffin method, the process has been shortened and improved; the Venetian turpentine method, introduced in the second edition and improved in the third, has come into such general use that the experience of many laboratories has been added to that of our own, and the directions have become so definite that there is little excuse for failures. The cellulose acetate method, which may do as much for woody structures as the Venetian turpentine method has done for its class of mounts, is outlined in a tentative way, and the chapter on "Photomicrographs and Lantern Slides" has been extended and improved.

The introduction of American stains, which are becoming very accurately standardized, has occasioned some modifications throughout.

The author is even more deeply indebted than before to his colleague, Dr. W. J. G. Land, for suggestions and improvements covering the whole field of microtechnic and photography. He is also indebted to Dr. S. Yamanouchi for improvements applicable to algae and mitotic figures. Dr. Paul J. Sedgwick is responsible for much of the improvement in photomicrography and for many of the photomicrographic illustrations. To Miss Ethel Thomas, who assisted me for many years, I am indebted for improvements, criticisms, and suggestions covering the whole range of the book. Besides, I must thank a host of colleagues and students all over the world for help in all phases of the subject.

Corrections and suggestions will be heartily appreciated.

CHARLES J. CHAMBERLAIN

CHICAGO
October, 1924

CONTENTS

PART I

PART II

PART I

INTRODUCTION

The pollen grain of a lily, placed on a dark background, is barely visible to the naked eye; but with modern technic, such a pollen grain can be cut into fifty sections, the sections can be mounted and stained without getting them out of order, a photomicrograph can be made from the preparation and a lantern slide from the photomicrograph, and finally there appears upon the screen a pollen grain 10 feet long, with nuclei a foot in diameter, nucleoli like baseballs, and starch grains as large as walnuts. With such preparations, botanists are now showing clearly the nature of structures which, only a few years ago, were good subjects for philosophical speculation.

To become a good technician, the student should follow carefully every detail of the various schedules and, when the routine becomes familiar, he should try to understand the reason for every step. In some cases, a dozen different schedules might have been given. When fundamentals have been grasped, the student will be able to make such variations as individual cases may require.

Those who regard the making of mounts as mere mechanical drudgery which can be done by an assistant are likely to become armchair investigators, drawing false conclusions or becoming scholastic grafters, according as the assistant is mediocre or talented. Besides, there is always the danger that a talented but underpaid assistant may "hold out" something. Some time-honored theories would have been abandoned long ago if certain prominent investigators had not relied upon comparatively untrained assistants for their mounts. Benjamin Franklin's advice, "If you would have your business done, go; if not, send," applies very well to the case in hand.

We strongly advise the student to collect his own material in the field, for such collecting is a valuable part of a botanical education. There are details of habitat and behavior which are never described in books. One learns gradually, by experience, that certain kinds of plants grow in certain kinds of places; and further, that not only the season, but even the weather may be an important factor. A heavy rain may cause some algae to disappear; while the same rain, followed by a few days of sunshine, will bring ideal conditions for collecting Myxomycetes and many other fungi. One learns that while *Lycogala* is pink, it is in the free nuclear condition, and that *Stemonitis* is in that condition as long as it is white; that *Volvox* may be abundant at the bottom of a pond when there is scarcely any in suspension. The successful investigator learns how a flower bud should look, if it is to yield floral development, how the flower looks when the embryo sac is mature; and how it looks after fertilization.

Such studies add immensely to the value of a preparation. A considerable part of a botanical education can be gained by collecting material, making and studying preparations, reading what is available, and *thinking*.

Finally, do not imagine that you must always make an elaborate permanent mount before you examine anything with the microscope. Look at living material whenever possible; make freehand sections or tease with needles, and thus make that preliminary survey which should always precede the study of permanent mounts.

CHAPTER I

APPARATUS

The amount of apparatus required for histological work varies, temporary mounts, glycerin mounts, and freehand sections requiring only a razor and a microscope, while the paraffin method, which represents the highest development of technic, brings into use nearly all the equipment of the histological laboratory. The following list includes only the apparatus necessary for making preparations: a microscope; a microtome; a razor; a hone and a good razor strop; a paraffin bath; a turntable; a scalpel; a pair of needles; a pair of scissors; a pair of forceps; staining-dishes; solid watch glasses; bottles; a graduate (50 or 100 c.c.); pipettes; slides, 1×3 inches; round covers, 18 mm. or $\frac{3}{4}$ inch in diameter; and square covers, $\frac{7}{8}$ inch. Longer covers will be needed for some of the serial sections.

A microscope should have a rack and pinion coarse adjustment, a fine adjustment, two eyepieces magnifying about four and eight diameters, a low-power objective of about 16-mm. focus, and a high-power objective of about 4-mm. focus, a double nosepiece, an iris diaphragm, and an Abbé condenser. A cheap and practical form is shown in Figure 1, and similar instruments are for sale by all the leading companies.

Since the chemicals used in histological technic are likely to damage the stage and substage of the microscope, it is well to place upon the stage a piece of glass 3 or 4 inches square. A lantern-slide cover is just right for this purpose. It is not necessary to fasten it to the stage, since it is merely for protection while examining slides which are wet with reagents. In our own laboratory we use for examining wet slides a cheap microscope with only a single low-power objective and a single ocular.

Some knowledge of the structure and optics of the microscope is necessary if one is to use it effectively. Why are there so many diaphragms? Why is there an arrangement for raising and lowering the condenser? Why does the mirror bar swing? Why is one side of the mirror plane and the other concave? Everyone who uses even a cheap microscope should know the answers to questions like these. All the leading manufacturers furnish, free of charge, booklets, explaining the construction of the microscope and giving practical directions for its care and use.

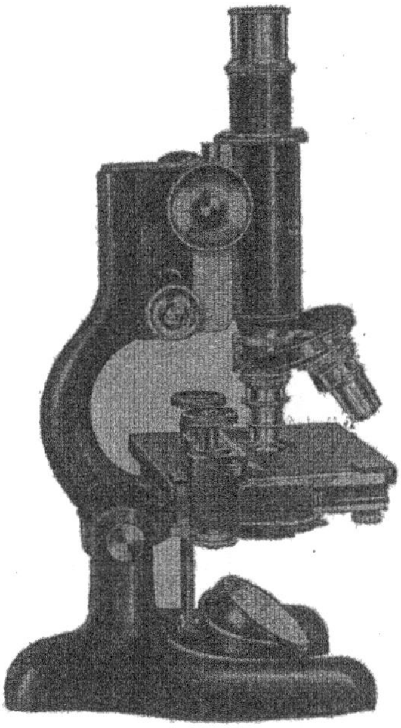

FIG. 1.—An efficient microscope of moderate price. The leading optical companies put the same objectives and oculars upon such instruments as upon their most expensive stands.

Aside from the microscope itself, the microtome is the most important piece of apparatus in the laboratory. In recent years there has been considerable improvement in microtomes, but we still have two general types—the sliding and the rotary.

The cheapest microtomes which have proved to be efficient for general work are simple forms of the sliding microtome like the one shown in Figure 2. It should be provided with a clamp which will hold any kind of a knife (Fig. 3). For large or hard objects the weakness of these small instruments is evident from the figures, but such an instrument is very useful; it is small and is easily carried around; it does not get out of order; it cuts celloidin well and is good for wood sections, except the hardest; and it will do for general paraffin work.

Where expense is not too great an objection, a larger microtome should be secured. There is great difference of opinion as to the relative merits of the sliding and rotary types. As far as convenience and rapidity are concerned, the rotary microtome is unquestionably superior; further, it will produce good sections with less care and skill, because the movements are automatic. The fact that a ribbon carrier is so easily used with the rotary is another great advantage. But the sliding microtome also has its advantages. Obviously, for sections of stems and general celloidin work, where the knife is used in a

FIG. 2.—The student's microtome

very oblique position, it is not only superior, but it is the only type which has proved to be efficient. Attempts to place the knife in an oblique position in rotary microtomes have not been encouraging. For very thin paraffin sections the advantages of the sliding microtome are such as appeal only to the expert. With both rotary and sliding types, a little of the paraffin is sure to stick to the side of the knife next the object after every section. Unless this be wiped off, the face of the block is dragged

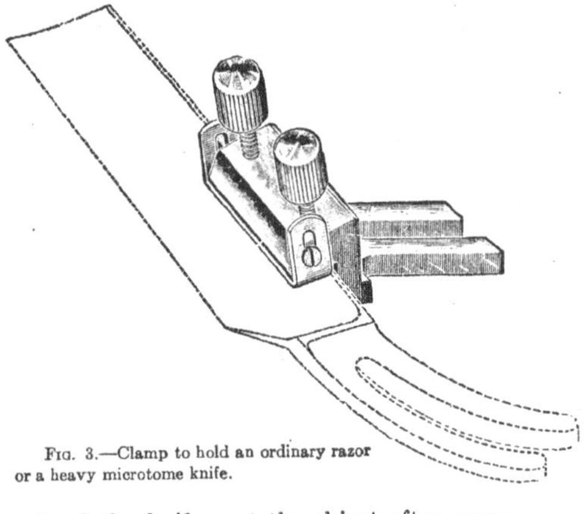

FIG. 3.—Clamp to hold an ordinary razor or a heavy microtome knife.

across it and the next section is damaged even before it is cut. The side of the knife next the object should be wiped with the finger, theoretically after every section. It is very inconvenient to wipe the knife in a rotary microtome. Another advantage of the sliding type is easy to feel but difficult to describe: In the rotary microtome the stroke is so automatic that there is little room for skill, but in the sliding microtome, with one's hand on the sliding block, little variations in the stroke, variations which become instinctive, give the expert a control not yet attained in the rotary forms.

Amateurs, and even professional botanists who have little aptitude in the use of machines, had better rely upon the rotary microtome. However, no better comment on the comparative merits of the two forms could be given than the practice of an expert technician in our own laboratory, who uses a rotary microtome when making sections for ordinary class work, but who turns to a sliding microtome of the Jung-Thoma pattern when cutting sections for his own research.

A much-improved rotary microtome, devised by Dr. H. N. Ott, president of the Spencer Lens Company, combines the precision and stability of the sliding microtome with the convenience of the rotary. Its effectiveness depends, in large measure, upon the simple but rigid clamp for holding the object. The same firm has also produced a very rigid sliding microtome, embodying features suggested by Professor R. B. Thomson, of the University of Toronto. The stability, which makes such a microtome desirable for cutting sections of hard, woody structures, is an advantage in cutting very thin sections of material imbedded in paraffin.

Microtome knives are available everywhere and, when perfectly sharpened, are unsurpassed. Those who sharpen knives for surgeons can grind out nicks, but they do not know how to sharpen a microtome knife and they cannot be taught; they sharpen knives for Dr. Carver and Dr. Cutterout.

In recent years several clamps have been devised to hold the blade of the Gillette safety razor, the hard, even edge of which is very satisfactory for microtome sections. After dealers had ignored our suggestions, Mr. A. W. Strickler, at our request, devised the form of holder[1] shown in Figure 4. It is made of brass and can be used in either rotary or sliding microtomes. The sectional view shows that

[1] At 1320 East Fifty-seventh Street, Chicago, Illinois.

the two pieces of the holder are curved, a feature which insures great rigidity. It is neither necessary nor desirable to have pins fitting the three holes in the blade, since they add nothing to the rigidity and even interfere with the insertion and adjustment of the knife. The knife should not project more than a millimeter beyond the holder. The clamp for holding Gillette blades in the Spencer rotary

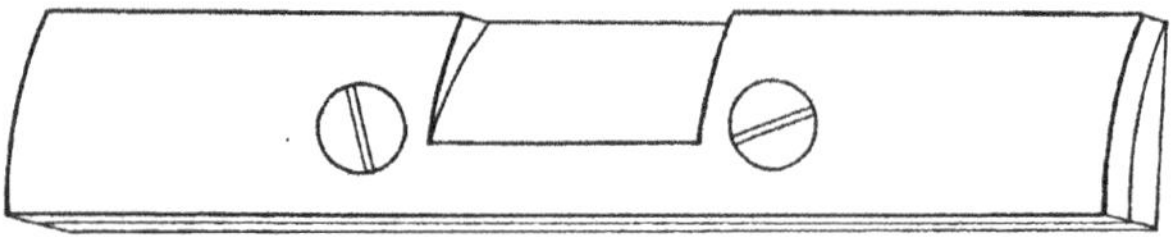

FIG. 4.—Strickler's clamp for holding Gillette blades in the Bausch and Lomb rotary microtome

microtome is necessarily much heavier (Fig. 5). With the Gillette blade in these holders we have cut smooth sections, 2 and 3 μ in thickness, and have cut large sections 2 cm. in diameter and 15 μ in thickness, even such refractory objects as the strobili of *Isoetes* and *Selaginella* cutting as smoothly as with a first-class microtome knife.

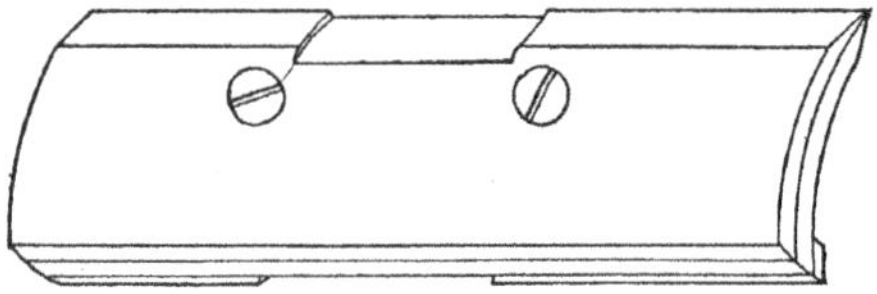

FIG. 5.—Clamp for holding Gillette blades in the Spencer rotary microtome

When the success of the holder—or rather, its sale—became evident, two prominent optical companies, without any apologies or reference to Mr. Strickler, began to manufacture it and advertised it in their catalogues. Their holders are inferior to Mr. Strickler's, doubtless because they overlooked a very important, but very obscure, detail.

Mr. Ralph B. Larsen, 7126 Woodlawn Avenue, Chicago, Illinois, makes these holders, selling the holder for the Bausch and Lomb rotary microtome at \$5.00, and the heavier holder for the Spencer rotary microtome at \$7.50.

Many have trouble with the holders, even as made by Mr. Larsen, because they fail to remember that the blade is not straight, like a microtome knife, but is bent into a curve. Consequently, the holder should not be inclined toward the paraffin as far as the microtome knife is inclined. The exact angle cannot be given, since it will vary with the amount the blade is allowed to project beyond the holder. But, just as in case of the microtome knife, which, when sharpened with a back as it always should be, has a double bevel like that of the safety-razor blades, the angle should be as near the vertical as possible. With either a microtome knife or blade in a holder, the paraffin block will hit the shoulder of the bevel after the section is cut if the position be too nearly vertical. A study of Figure 6 should enable anyone to get the proper orientation.

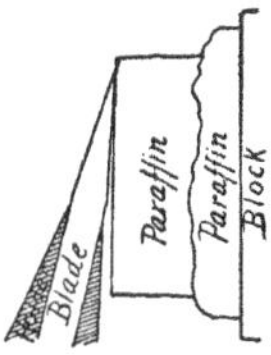

FIG. 6.—Proper relative positions of holder, Gillette blade, and paraffin.

When the Gillette blade begins to lose a little of its effectiveness for microtome work it will make two or three scalpels. With a pair of stout shears, cut the blade into pieces, as indicated in Figure 7. Take a small steel nail and flatten the head and upper part by laying it upon a piece of iron and hitting it with a hammer, or by squeezing it in a vise;

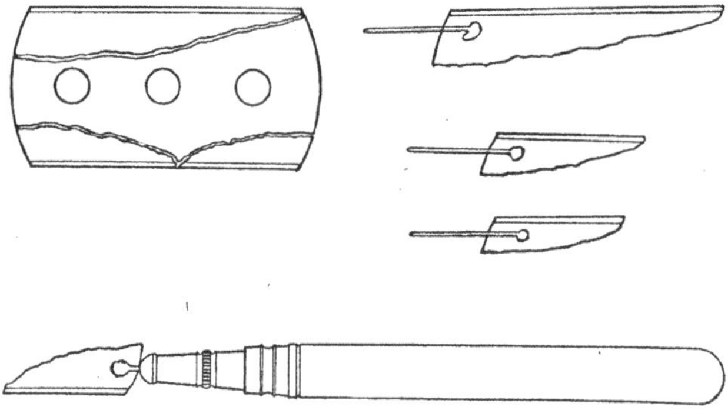

FIG. 7.—Scalpels made from Gillette blades, showing a blade which has been cut into pieces with shears, three of the pieces soldered to nails with flattened heads, and one of the pieces used in an ordinary needle-holder.

then solder the blade to the nail, and use the scalpel in an ordinary needle-holder, or drive the nail into any wooden holder. A dozen of these scalpels can be made in ten minutes.

The stout razors our grandfathers used to shave with are excellent for freehand sectioning and even for cutting sections on the microtome. The blade should be flat on one side (Fig. 8*A*). Modern razors (Fig. 8*B*), with delicate blades ground hollow on both sides, are worthless for cutting sections of plants.

There should be two good hones: a fine carborundum hone for the preliminary sharpening, and a yellow Belgian hone for finishing. About $10 \times 2\frac{1}{2}$ inches is a good size. If the second hone be quite hard and the finishing skilfully done, little or no stropping may be necessary. The best strops used by barbers are satisfactory for microtome knives.

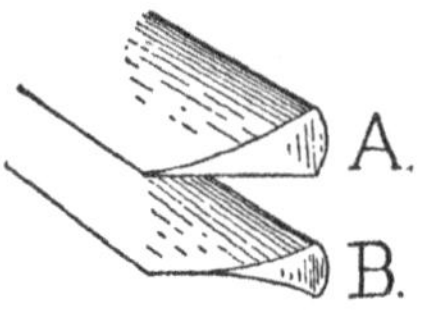

Fig. 8.—The type shown in *A* is good for microtome work; that shown in *B* is worthless for microtome work, but can be used for freehand sections of leaves.

There are numerous forms of the paraffin bath. Those with a water-jacket, a thermometer, and a thermostat to maintain an even temperature are the most convenient.

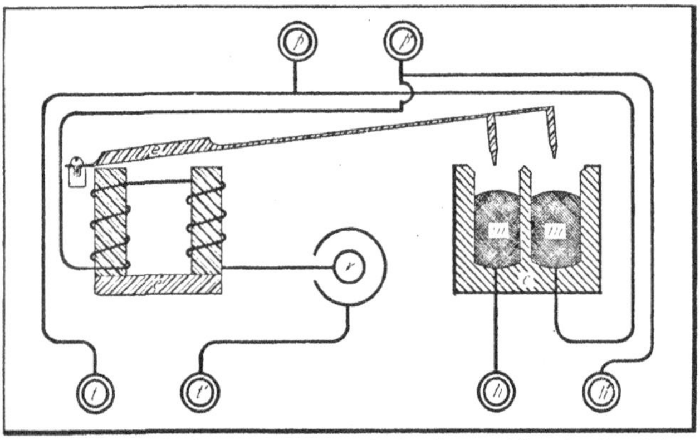
Fig. 9.—Land's electric constant apparatus, showing diagram of the automatic switch, as described in the *Botanical Gazette*, November, 1911.

A bath with a water-jacket and the electric thermostat devised by Dr. Land maintains an even temperature, in spite of any variations in the current. The appearance and principal features of this thermostat are shown in Figures 9 and 10. A detailed description of this thermostat is given in the *Botanical Gazette* for November, 1911. One familiar with tools and electricity could make this thermostat at a very moderate outlay. Unless the coil in the heater is perfectly protected, there will be a short circuit. This danger can be obviated, in large measure, by using oil instead of water in the jacket.

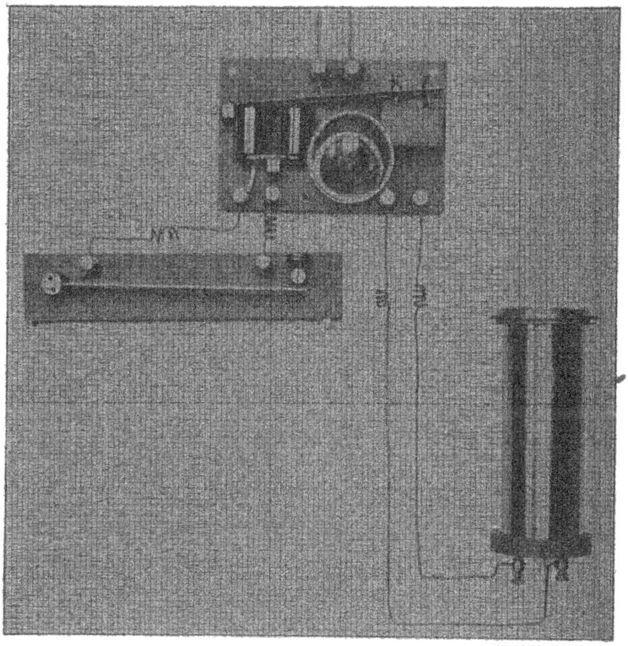

FIG. 10.—Thermostat, heater, and switch of Land's electrical constant apparatus

Another form of heater, using long electric bulbs instead of a coil, can be made with less skill. From a sheet of transite $\frac{1}{8}$ inch thick, make a box just the size of the bottom of the bath and about 4 inches deep. In this box put three long electric bulbs. Two of the bulbs, about 250 watt, giving a temperature of about 40° C., should be connected with the regular outlets and should be on all the time.

The third bulb, which should be capable of raising the temperature 15° or 20° C., should be connected with a thermostat. An efficient thermostat, capable of keeping the variation in temperature within about 1° can be made in a short time. Take a piece of soft steel $\frac{3}{4}$ inch wide and $\frac{1}{16}$ inch thick, and 17 inches long; lay upon it three strips of aluminium of the same width and length, but only $\frac{1}{32}$ inch thick. These four pieces are fastened together by drilling numerous holes, not more than $\frac{1}{16}$ inch in diameter and riveting with pieces of brass wire. After fastening the four pieces together for about 7 inches, begin to bend all four pieces, fastening them together as the bending continues, so that finally there will be a horseshoe shape with the parallel sides about $1\frac{1}{2}$ inches apart. One end of

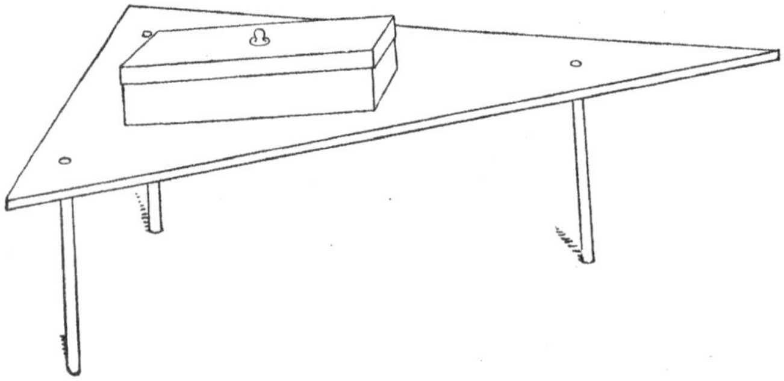

FIG. 11.—Paraffin bath

the horseshoe is fastened to a block of transite or other non-conductor, while the other end moves freely. On this free end is fastened a good platinum contact, which can be bought at any automobile supply store. This contact should be on a screw, so that it may be adjusted. A similar contact is fastened to a post, so that the two contacts are about $\frac{1}{8}$ inch apart. Anyone familiar with electricity can make the connections. Various temperatures are secured by changing the distance between the platinum points. To prevent sparking, there must be a condenser in the circuit leading to the thermostat. A telephone condenser (0.500MF) is satisfactory. Possibly, some of the cheap radio condensers would do.

A bath which, if carefully watched, gives the very best results, can be made by any tinner (Fig. 11). It is simply a triangular piece of brass, 2 or 3 mm. in thickness, with three legs screwed into it. There should be long brass boxes to contain the paraffin. It is neither necessary nor desirable that the boxes have covers. Boxes are easily made from square brass tubing, 1 inch square. Cut the tube into pieces 6 inches long, saw off one side, and solder pieces into the ends with hard solder. The brass plate can be heated with any kind of a flame at the pointed end.

Since the Venetian turpentine method has almost entirely displaced the glycerin method, the turntable is becoming an unfamiliar object in the botanical laboratory; but some objects, like *Nemalion* and moss protonema, are still mounted in glycerin jelly, and so one still finds occasional use for this once necessary apparatus. A serviceable form is shown in Figure 12. The more expensive turntables,

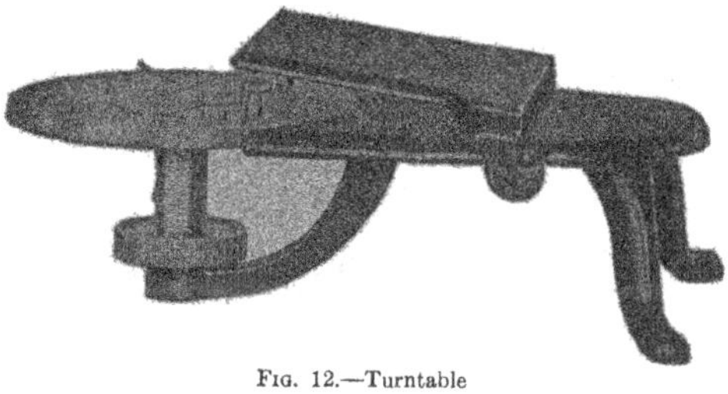

FIG. 12.—Turntable

with devices for automatic centering, present no practical advantages and the centering devices are often in the way.

Scalpels made from thin razor blades have been mentioned already. For trimming paraffin blocks and handling paraffin ribbons a more rigid scalpel is necessary. A scalpel with a straight edge is preferable.

Needles are used so constantly that it is well to have clamping holders. However, if it were not for the trouble of inserting and pulling out needles, nothing is quite equal to a rather large handle whittled out from a piece of light pine.

Scissors are seldom used in the botanical laboratory except for cutting out labels. Rather stout scissors, with blades about $2\frac{1}{4}$ inches in length, are best for general purposes.

It is convenient to have two pairs of forceps, a strong pair for handling slides and a delicate pair for handling covers. If there is to be only one pair, they must be strong enough for the slides but not too clumsy for covers. Curved forceps are not necessary; the cover-glass forceps, used by bacteriologists in staining on the cover, are of no use in botanical work.

Stender dishes are very generally used in staining on the slide. The form shown in Figure 13*A*, about 60×90 mm., is in quite general use. Some prefer the Coplin jar, shown in Figure 13*B*. The latter is more troublesome to clean, but requires less of the reagent. Many other forms are on the market. When large numbers of slides of the same kind are to be stained at one time, small battery jars, holding about a liter, may be used. In this case, it is well to have a rack, holding about thirty slides, so that all may be transferred at the same time from one reagent to another. With this convenience, it is not necessary to handle the slides separately, except at critical stages.

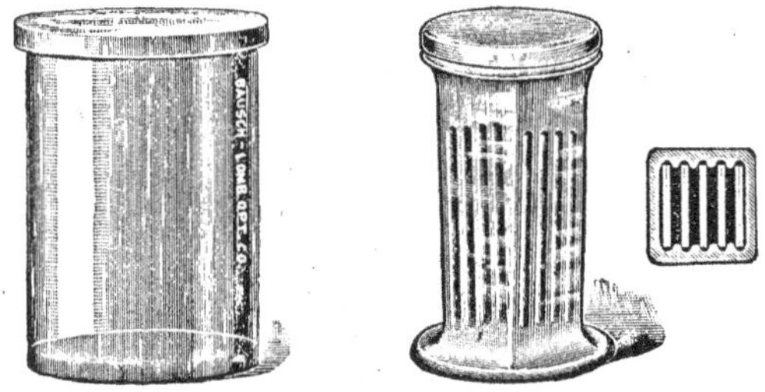

A B

Fig. 13.—Staining-dishes: A, Stender dish; B, Coplin jar

A very cheap and practical device, which can be used in an ordinary Stender dish, is shown in Figure 14. It is simply a coil of brass wire, 0.064 inch in diameter (No. 14, Band's gauge), wound so that the coil is about $\frac{7}{8}$ inch across. Such a coil, carrying 15 slides, will go into an ordinary Stender dish, except that the coil projects enough to prevent the cover from fitting. Taller glasses, from the five and ten cents store, can be used for the absolute alcohol and xylol, which must be kept well covered. We have been using a coil made from wire 0.051 inch in diameter (No. 16 Band's gauge), wound so that the coil is $\frac{13}{16}$ inch across. It holds the slides and the ordinary Stender dish can be covered.

Solid watch glasses, or Minots, as they are often called, are always useful. Each student should have a dozen or more.

Each student should have three bottles of about 1-liter capacity for 90 per cent alcohol, absolute alcohol, and xylol. In addition, a half-dozen bottles, holding about 100 c.c., will be useful. There should be two bottles, holding about 50 c.c., for clove oil. If one is

FIG. 14.—Coil of brass wire holding 15 slides

doing much research work, it will be convenient to have many more bottles for graded series of alcohols and xylols.

There should be a graduate, preferably 50 c.c. or 100 c.c. If the bottles are of uniform sizes, 50 c.c., 100 c.c., 500 c.c., and 1,000 c.c., the student should soon be able to estimate with sufficient accuracy for making up reagents which do not require extreme accuracy.

Three or four pipettes, or medicine-droppers, will be useful. Occasionally, the glass of an ordinary pipette thrust into a small camera bulb will save time in drawing off reagents.

Slides and covers are a constant expense. Many slides now upon the market are imperfect. Beware of slides which are not perfectly flat. Be skeptical in regard to any claim that slides are already clean enough to use. Of course, there should be no bubbles. "White" slides are to be preferred to those which appear greenish in the box. For ordinary class work, slides of medium thickness are more serviceable, but for critical cytological work many investigators prefer very thin slides.

There is never any objection to very thin covers, except that they require care in cleaning. For mounts which are to be used with an immersion lens, it is better to have the cover of the same width as the slide. The advantage is evident, since there is no danger of getting balsam on the cover when wiping off the immersion fluid; besides, one can put sections to the very edge of the slide and still be sure that they will be covered. Since most mounts for research work are mounted under long covers and are intended for examination with immersion lenses, we should recommend covers of 25×50 mm., or even 25×60 mm. Round covers are desirable only when mounts are to be sealed on a turntable. Larger slides and correspondingly larger covers are needed for special purposes.

By consulting a catalogue, which will be furnished by any dealer, the beginner can determine what he needs to buy, and what he can find substitutes for, if it is necessary to be very economical.

CHAPTER II

REAGENTS

No really new reagents have come into general use since the third edition of this book was published in 1915, but there have been valuable modifications of some of the older formulas, and improvements in the use of time-honored combinations. The following account includes those which are used constantly and also a few which are used occasionally. *The Microtomist's Vade-Mecum,* by Lee, is written from the standpoint of the zoölogist, but it contains very complete formulas for stains and other reagents, which are just as useful to the botanist.

A list of reagents, with the quantities used by the average student in a three months' course in methods, is given in chapter xxix. "Stains and Staining" are described in chapter iii.

KILLING AND FIXING AGENTS

Usually the same reagent is used for both killing and fixing. The purpose of a killing agent is to bring the life-processes to a sudden termination, while a fixing agent is used to fix the cells and their contents in as nearly the living condition as possible. The fixing consists in so hardening the material that the various elements may retain their natural condition during all the processes which are to follow. Zoölogists often use chloroform or ether for killing an organism, and then use various fixing agents for various tissues. No promptings of humanity restrain the botanist from the vivisection of plants, but separate reagents for killing and fixing are sometimes used, e.g., material may be killed by placing it for a short time in Flemming's fluid, which is a very rapid killing agent, after which the fixing may be completed in a chromo-acetic solution, without any osmic acid, thus securing the advantages of a very rapid killing without the blackening which results from a prolonged treatment with a solution containing osmic acid.

Probably no process in microtechnic is in more urgent need of improvement than this first step of killing and fixing. Nearly all of our formulas are merely empirical, for very few botanists are expert chemists, and those who have the requisite knowledge of chemistry are interested in physiological problems rather than in microtechnic. The principal ingredients of the usual killing and fixing agents are: alcohol, chloroform, chromic acid, bichromate of potash, potassium iodide, copper acetate, acetic acid, osmic acid, formic acid, picric acid, sulphuric acid, platinum chloride, iridium chloride, corrosive sublimate, and formalin. We shall consider first:

THE ALCOHOLS

a) **Ninety-five Per Cent Alcohol.**—This is in quite general use for material which is needed only for rough work. It is extremely convenient, since it kills, fixes, and preserves at the same time and needs no changing or washing. It really has nothing to recommend it for fine work. It causes protoplasm to shrink, but cell walls usually retain their position, so that 95 per cent alcohol will do for freehand sections of wood and many herbaceous stems, where it is not necessary to preserve cell contents; but even freehand sections of tender stems, like geraniums and begonias, will look better if better reagents are employed. Alcohols weaker than 95 per cent are not to be recommended as fixing agents, although 70 per cent alcohol, or even 50 per cent, will preserve material for habit work. The time required for fixing in 95 per cent alcohol is about the same as for absolute alcohol. The subsequent treatment is the same, except that material to be imbedded in paraffin or celloidin must be dehydrated in absolute alcohol. Material preserved in weaker alcohols and intended only for habit study may be kept in the reagent until needed for use. Unless some glycerin be added, material left in 95 per cent alcohol becomes very brittle. Stems, roots, and similar objects may be kept indefinitely in a mixture of equal parts of 95 per cent alcohol and glycerin.

Methyl alcohol, or wood alcohol as it is commonly called, serves equally well.

b) **Absolute (100 Per Cent) Alcohol.**—This is a fair killing and fixing agent, it causes but little shrinking of the protoplasm, and is a time-saver if material is to be imbedded in paraffin. The time required for fixing in alcohol is very short. For small fungi, like *Eurotium*, 1 minute is long enough. Root-tips of the onion, anthers of the lily, and similar objects require 15 to 30 minutes. Larger objects may require an hour. No washing is necessary, but all plant tissues contain water; consequently, if material is to be imbedded in paraffin, the alcohol used for fixing should be poured off and fresh alcohol added before proceeding with the clearing. If material is to be mounted in Venetian turpentine, as is likely to be the case in small filamentous fungi, the transfer to the stain may be made directly from the absolute alcohol to any stain dissolved in an alcohol not weaker than 85 per cent. Small forms with no vacuoles may be transferred to a weaker alcoholic stain or even to an aqueous stain; but neither the fixing nor the rude transfer would be at all satisfactory with forms like *Zygnema* or *Saprolegnia*.

Acetic acid is used with alcohols to counteract the tendency to shrink. One of the most widely known of the alcohol combinations is

c) **Carnoy's Fluid.**—

Absolute alcohol	6 parts
Chloroform	3 parts
Glacial acetic acid	1 part

The penetration of the reagent is very rapid. An object like an onion root-tip is doubtless killed in less than a minute, and 10 to 15 minutes is long enough for fixing an object of this size. Wash in absolute alcohol, changing frequently, until there is little or no odor of acetic acid. For a root-tip, the entire process of fixing and washing should not require more than an hour. It is better to imbed in paraffin at once, but when this is not convenient the material may be washed in abolute alcohol until the odor of acetic acid disappears, cleared in xylol and, with a block of paraffin about half the bulk of the liquid added, may be left indefinitely. Cyanin and erythrosin, fuchsin and iodine green, and similar combinations give particularly brilliant staining after this reagent.

d) **Acetic Alcohol.**—Farmer and Shove recommend for fixing root-tips of *Tradescantia virginica* a mixture of two parts absolute alcohol and one part of glacial acetic acid. The mixture is allowed to act for 15 to 20 minutes, after which the acid is washed out with absolute alcohol and the material is imbedded as soon as possible.

e) **Formalin Alcohol.**—One of the most satisfactory of the alcohol combinations is formalin alcohol. Various proportions are used by different workers. Professor Lynds Jones, who first brought this combination to my notice, added 2 c.c. of commercial formalin to

100 c.c. of 70 per cent alcohol. We have used a larger proportion of formalin, often as much as 6 c.c. to 100 c.c. of 70 per cent alcohol. Results which seem equally good have been secured by adding 4 to 6 c.c. of formalin to 100 c.c. of 50 per cent alcohol. Material in this fixing agent may be left until needed for use.

f) **Formalin Acetic Alcohol.**—This combination seems even better than the preceding. We use about 5 c.c. of glacial acetic acid and 5 c.c. of commercial formalin to 90 c.c. of 50 per cent or 70 per cent alcohol.

When one is on a long trip, moving frequently from place to place, with little opportunity to make the numerous changes which are necessary when using the chromic formulas, this is the best fixing agent we have found. It will fix and preserve an amount of material equal to its own weight, and the material may be left in the solution for months. The reagent is good for almost any material, except the unicellular and filamentous algae and fungi, which are more satisfactory in media containing no alcohol.

THE CHROMIC-ACID GROUP

Chromic acid, or solutions with chromic acid as a foundation, are the most generally useful killing and fixing agents yet known to the botanist. A 1 per cent solution of chromic acid in water gives good results, but it is better to use the chromic in connection with other ingredients, such as acetic acid, formic acid, osmic acid, etc. Chromic acid does not penetrate well, and this is one reason why it is seldom used alone. Unfortunately it precipitates some liquid albuminoids in the form of filaments and networks, which may be mistaken for structural elements. In botanical work, acetic acid is nearly always mixed with chromic acid. The pickles of the dinner table show that acetic acid is a good preservative, and that it causes little or no shrinking. It penetrates rapidly, and is likely to cause swelling rather than shrinking, thus counteracting the tendency of chromic acid to cause plasmolysis. The swelling is as bad as shrinking. If the proportion of acetic acid is too high, material may even break up; but 2 per cent, or even 3 per cent, may be used to show the topography of an embryo sac of an angiosperm, or the free nuclear stage of the endosperm of a gymnosperm; and for filamentous algae, which are to be mounted whole, 3 per cent is very effective.

It will be found convenient to have in the laboratory the following stock solution of chromo-acetic acid from which various solutions can be made as they are needed:

Chromic-acid crystals	10 g.
Glacial acetic acid	10 c.c.
Water	1,000 c.c.

To make a solution containing 0.5 g. of chromic acid and 2 c.c. of glacial acetic acid to 100 c.c. of water, add 50 c.c. of water to 50 c.c. of the stock solution, and then add to the weakened solution 1.5 c.c. of glacial acetic acid. Any desired proportions can be secured in a similar way. Weighing the crystals for every new proportion is more tedious. The proportions of the various ingredients, for the present at least, must be determined by experiment. With favorable objects like fern prothallia, *Spirogyra*, and other things which can be watched while the fixing is taking place, suitable proportions are rather easily determined, because specimens, after being placed in the reagent, may be examined at frequent intervals, and combinations which cause plasmolysis may be rejected and different proportions tried until satisfactory results are secured. For example, fern prothallia might be placed in the following solution: chromic acid, 2 g.; acetic acid, 1 c.c.; and water, 97 c.c. If plasmolysis takes place, weaken the chromic or strengthen the acetic, since the chromic has a tendency to produce contraction, and the acetic to cause swelling. A 2 per cent solution of chromic acid is rather strong and a ½ per cent is rather weak; for most things, 1 per cent seems to be about right. For the fern prothallia, the stock solution, with the addition of 2 c.c. of glacial acetic to 100 c.c. of the solution, is satisfactory for material to be mounted whole, and also for sections. A combination may be quite satisfactory for fern prothallia and still fail to give good results with *Spirogyra*, and a combination which succeeds very well with *Spirogyra* may not succeed at all with *Vaucheria*. For very critical work the most favorable proportions must be determined for the particular plant under investigation. In observing the effect of the fixing one can determine whether there is any noticeable plasmolysis or distortion, but whether the fixing is thorough can be determined only by noting how the tissues endure the subsequent processes. When the effect of the reagent cannot be observed directly, it is well to make a freehand section and thus determine whether plasmolysis takes place. It is not safe to judge the action of a fixing agent by the appearance of sections cut from material which has been imbedded in paraffin, because shrinking of the cell contents often takes place during the transfer from absolute alcohol to the clearing agent or during infiltration with paraffin, and sometimes even during later processes. When there is doubt as to proportions, we should suggest 2 c.c. chromic acid, 3 c.c. acetic acid, and 300 c.c. water as a good formula for most purposes.

A large quantity of the fixing agent is required and it cannot be used again. The volume of the fixing agent should be at least 25 times that of the material to be fixed. We use about 50 volumes of the fixing agent to one of the material.

The time required for fixing undoubtedly varies with different objects, but even a delicate object, like *Spirogyra*, which is penetrated immediately, should remain in the fixing fluid for 18 to 24 hours. Most botanists leave material like onion root-tips and lily ovaries in the chromo-acetic acid about 24 hours. Some recommend longer periods. Christman, in his work on rusts, left material for three days in Flemming's fluid, a much more vigorous agent than the chromo-acetic acid. We have often imbedded material which had been in chromo-acetic acid for several days, and it seemed to have suffered no injury. It is well known that zoölogists allow fixing agents like Müller's fluid and Erlicki's fluid to act for weeks before the material is passed on to the next stage, and it may well be questioned whether botanists have not made a mistake in allowing the chromic solutions to act for so short a time. More rapid penetration, and consequently more immediate killing, can be secured if the reagent is kept warm (30° to 40° C.). The warming also shortens the time required for fixing, but, for cytological work, it is quite possible that the danger of producing artifacts may be increased by the heat.

After fixing is complete, all reagents containing chromic acid as an ingredient should be washed out with water. Running water is desirable, and where this is not convenient the water must be changed frequently.

About 8 or 10 hours should be long enough for filamentous algae and fungi, which are immediately penetrated by the water. It is a good plan to start the washing in the morning and let the material wash all day. For fern prothallia, onion root-tips, lily anthers, and any material from such a size up to cubes a centimeter square, let the material wash for 24 hours. Even for delicate algae, 24 hours does no damage, and some of the best cytologists prefer the prolonged washing.

Many methods have been devised for insuring thorough washing and for facilitating the process. The most obvious method is to allow a gentle stream of water to flow into the Stender dish or bottle containing the material. There is little danger in this method if the material is heavy enough to remain at the bottom: the only objection is that much of the water never reaches the bottom where it is needed. If material is lighter, tie a piece of cheese-cloth over the mouth of the bottle.

A method devised by Dr. Dudgeon is simple and very efficient, especially for delicate material or objects which have a tendency to float. A glass tube, 1 inch in diameter, is cut into pieces 2½ inches in length. The glass is then heated to round off the sharp edges and, while the glass is still very hot, one end is flared a little, so that a piece of cloth can be tied over or fastened over it with a rubber band. Bolting cloth or bolting silk is best because water passes through it so readily. A number of these tubes can be placed in a dish not quite so deep as the tubes and water can flow into the dish without any danger of making the tubes overflow. With material which does not settle readily, this method is a great time-saver.

An apparatus for washing several collections at one time may be made as follows: Get a piece of ¾-inch lead pipe, bore holes about $\frac{5}{16}$ or ⅜ inch in diameter and about 1½ inches apart, put a short rubber tube in each hole and the glass part of a pipette in the end of each rubber tube. Connect the lead tube with the faucet by a large rubber tube. A still better way is to bore $\frac{3}{16}$-inch holes in the lead tube, screw into these holes short brass tubes, and then fasten the pipettes to the brass tubes with thin rubber tubes.

If there are no facilities for working with metal, take a wooden box about 6 inches wide, 18 inches long, and 4 inches deep; bore ⅜-inch holes in the bottom, and into each hole put a piece of rubber tubing about 4 or 5 inches in length. The pipettes can be fastened in the ends of these rubber tubes. Place the box under the tap. In the botanical laboratory at Woods Hole, Massachusetts, large quantities of material are washed at one time by using an ordinary washtub with the bottom arranged as just described for the box. If one is using such a large box or tub and does not need all the streams of water, the tubes not in use may be closed by means of clamps.

If running water is not available, put the material into a rather large bottle or dish; a 200 c.c. bottle is not too large for half a dozen ¼-inch cubes. Change frequently, especially at first. Nothing is safe with less than 24 hours of this sort of washing. Where running water is not available, Dudgeon's method is particularly valuable, for the tubes can be lifted quickly from one dish into another.

If the washing has not been thorough, the subsequent staining is likely to be unsatisfactory.

Some of the chromic-acid formulas are as follows:

a) **Stock Chromo-Acetic Solution.**—

Chromic acid	1 g.
Glacial acetic acid	1 c.c.
Water	100 c.c.

This solution has been used quite extensively in embryological work upon the higher plants. It fixes thoroughly, but often causes plasmolysis in cells with large vacuoles.

b) **Weak Chromo-Acetic Solution** (Shaffner's formula).—

Chromic acid	0.3 g.
Acetic acid	0.7 g.
Water	99.0 c.c.

This has also been used in embryological work. It causes little or no plasmolysis. Difficult material, like *Aster* heads and ripe *Capsella* pods, cuts more readily after this reagent than after the stronger solutions.

c) **Strong Chromo-Acetic Solution.**—

Chromic acid	1 g.
Glacial acetic acid	3 c.c.
Water	100 c.c.

For fern prothallia, most liverworts, moss capsules before they have begun to get reddish or brownish, and most filamentous algae and fungi, this is a good fixing agent.

d) **Special Chromo-Acetic-Osmic Solution.**—

Chromic acid	1 g.
Glacial acetic acid	3 c.c.
1 per cent osmic acid	1 c.c.

We have been trying these three ingredients, in various proportions, for many years and have found this combination very good for filamentous algae and fungi, even difficult forms like *Hydrodictyon*, *Vaucheria*, and *Saprolegnia* showing practically no plasmolysis. The acetic acid may be too strong for some root-tips and probably other objects; but the reagent is worth a trial. With 2 c.c. of osmic acid, this seems to be as good for mitosis as Flemming's solutions, and it causes less plasmolysis. Besides, in most cases, no bleaching is necessary.

e) **Licent's Formula.**—

1 per cent chromic acid	80 c.c.
Glacial acetic acid	5 c.c.
Formalin	15 c.c.

This formula has been recommended for coenocytic algae and fungi and for embryo sacs.

f) **Flemming's Fluid** (stronger solution).—

A	1 per cent chromic acid	45 c.c.
	Glacial acetic acid	3 c.c.
B.	2 per cent osmic acid	12 c.c.

Keep the mixture A made up, and add B as the reagent is needed for use, since it does not keep well. This fluid is quite expensive on account of the osmic acid. For cytological work it has been very popular, and it is especially recommended for chromosomes, centrosomes, achromatic structures, and mitotic phenomena in general. The fluid should be allowed to act for 24 to 48 hours and the washing in water must be very thorough.

Material should be in very small pieces $\frac{1}{8}$ inch square, or in thin slices $\frac{1}{8}$ inch or less in thickness, for the fluid penetrates poorly. The blackening due to the osmic acid may be removed by peroxide of hydrogen just before the slide is passed from the alcohol into the stain. Harper and Holden, in their work on *Coleosporium*, recommended 4 hours on the slide in a 3 per cent solution of the peroxide of hydrogen. Some prefer a stronger solution of the peroxide of hydrogen, even 20 per cent. The peroxide should be in water, if one is following it by an aqueous stain, but may be in 50 per cent alcohol if it is to be followed by an alcoholic stain. Yamanouchi has used chlorine for bleaching, and the results are fully equal to those obtained with peroxide of hydrogen, and the chlorine is cheaper. Make the bleacher as follows: Place some potassium chlorate crystals—a group about as large as a grain of wheat—in the bottom of a 100 c.c. Stender dish; add one drop of 25 per cent hydrochloric acid in water; immediately fill the Stender full of 30 per cent alcohol and thus dissolve the fumes in alcohol. This will bleach sections in 10 minutes, or even less. Wash in 30 per cent alcohol 2 or 3 hours before staining. Tröndle uses 1 per cent chromic acid in water for bleaching; it is slow, requiring about 8 hours, but he maintains that material stains better than after bleaching with peroxide of hydrogen. According to Miss Merriman, the linin in the nuclei of onion root-tips is not so well preserved in this solution, but the arrangement of the chromatin granules is brought out with greater distinctness. Flemming's safranin, gentian-violet, orange combination gives excellent results after this reagent.

g) **Flemming's Fluid** (weaker solution).—

A	1 per cent chromic acid	25 c.c.
	1 per cent acetic acid	10 c.c.
	Water	55 c.c.
B.	1 per cent osmic acid	10 c.c.

As in case of the stronger solution, mix A and B only as needed for immediate use.

Many prefer the weaker solution, because the blackening is not so extreme and material does not become quite so brittle. Some allow the solution to act for an hour and then transfer the material to solution A for about 24 hours. This secures the rapid killing, which is the principal virtue of the osmic acid, and avoids the disagreeable blackening, so that little or no bleaching may be necessary.

h) **Benda's Fluid.**—

1 per cent chromic acid	16 c.c.
2 per cent osmic acid	4 c.c.
Glacial acetic acid	2 drops

This modification of Flemming's stronger solution has been used in various investigations upon chromatin.

i) **Merkel's Fluid.**—

Equal volumes of a 1.4 per cent solution of chromic acid and a 1.4 per cent solution of platinic chloride. This is also an expensive reagent. It is recommended for mitotic phenomena, but does not seem to equal Flemming's solution.

j) **Hermann's Fluid.**—

1 per cent platinic chloride	15 parts
Glacial acetic acid	1 part
2 per cent osmic acid	4 or 2 parts

This is the most expensive fixing agent yet discovered, and for botanical purposes it does not seem to be any better than the cheaper chromic mixtures. It is mentioned here with chromic mixtures because it originated as a variation of Flemming's fluid, the platinic chloride being substituted for the chromic acid. Recently, it has been resurrected and highly recommended for the structure of the chromosome. Personally, I do not believe it is equal to Flemming's weaker solution; and, even in this weaker solution, the percentage of osmic acid may be too high. The chromo-acetic-osmic solution given under (*d*) fixes chromatin very well and, with 2 c.c. of osmic acid, seems to be equal to any of the modifications of Flemming's solutions.

PICRIC ACID

Use a saturated solution in water or 70 per cent alcohol. One gram of picric acid crystals will saturate about 75 c.c. of water or alcohol. This reagent penetrates well and does not make the material brittle. It is to be recommended when difficulty is anticipated in the cutting. If used cold, the time varies from 1 to 24 hours, depending upon the character of the tissue and size of the specimen. If used hot (85° C.), 5 or 10 minutes will be sufficient. Material should be washed in 70 or 50 per cent alcohol. Water is injurious, and some even go so far as to avoid aqueous stains, unless the material has been thoroughly washed. The washing should be continued until the material appears whitish and the alcohol no longer becomes tinged with yellow. Picro-carmine gives its best results after this reagent. Picric acid can be combined with various other fixing agents, and so we have picro-sulphuric acid, picro-nitric acid, picro-chromic acid, picro-chromic-sulphuric acid, picro-osmic acid, picro-alcohol, and picro-corrosive sublimate. The picric acid in all mixtures should be rather strong.

A picric-acid combination which has gained some popularity for cytological work is

Bouin's Fluid.—

Formalin (commercial)	25 c.c.
Picric acid (saturated solution in water)	75 c.c.
Glacial acetic acid	5 c.c.

Fix about 24 hours. Rinse in water for a few minutes to remove the more superficial picric acid, and then complete the washing in 35 per cent or 50 per cent alcohol. There is likely to be some swelling, but spindles of mitotic figures stain well. The formula has given good results with early stages in the female gametophyte of *Pinus* and would be worth a trial with the embryo sacs of angiosperms.

CORROSIVE SUBLIMATE

Corrosive sublimate, or bichloride of mercury, is soluble in water and in alcohol. About 5 g. will make a saturated solution in 100 c.c. of water. It is somewhat more soluble in alcohol, but for practical purposes 5 g. in 100 c.c. of 50 per cent alcohol may be regarded as a saturated solution. Corrosive sublimate used alone does not give as good results as when mixed with acetic acid, chloroform, or picric acid. Fixing is very rapid, the material being fixed almost as soon as it is penetrated by the fluid. Material which is at all transparent, like some ovules and the endosperm of gymnosperms before the formation of starch, becomes opaque as soon as fixed, and so the time needed for fixing is easily determined. From 10 minutes to one hour should be sufficient for onion root-tips or lily ovaries. Smaller or larger objects require shorter or longer periods. When used hot (85° C.), the fixing is much more rapid. While a few minutes' fixing may be sufficient, we let the reagent reach the boiling-point, then remove the flame and just as soon as the bubbling ceases, put the material in and leave it until the liquid becomes cool. It may be left for hours, without any damage.

Wash out aqueous solutions with water and alcoholic solutions with alcohol. In either case, the washing must be very thorough, since preparations from incompletely washed material are sure to be disfigured by crystals of corrosive sublimate. After material fixed in the aqueous solution has been washed in water for an hour, add a little of the iodine solution used in testing for starch. The liquid will turn brownish or amber-colored, and then clear up; add a little more, until the liquid fails to clear up completely, a very slight amber remaining for an hour or even permanently. After material fixed in the alcoholic solution has been washed in 50 per cent alcohol for an hour or more, add the iodine solution used in testing for starch or even add, drop by drop, tincture of iodine, until the color fails to disappear. With unicellular forms, filamentous forms, and thin things, like fern prothallia, such washing is likely to be sufficient; but with more bulky material, which is to be sectioned, the crystals may appear in the paraffin ribbon. In such cases, the slide should be dipped for a minute in the iodine solution just before staining.

Camphor may be used instead of iodine to hasten the washing, but it does not give any color reaction.

Material should be imbedded as soon as possible, since it gets brittle if allowed to remain in alcohol.

Kinoplasmic structures do not stain well with gentian-violet, but safranin and the haematoxylins stain almost as well as after chromic-acid mixtures, and the carmines give their most brilliant stains, as a result of the formation of mercuric carminate.

The following formulas are merely suggestive:

a) **Corrosive Sublimate and Acetic Acid.**—

Corrosive sublimate	3 g.
Glacial acetic acid	5 c.c.
Alcohol (50 per cent) or water	100 c.c.

b) **Corrosive Sublimate, Acetic Acid, and Formalin.**—

Corrosive sublimate	4 g.
Glacial acetic acid	5 c.c.
Formalin	5 c.c.
Alcohol (50 per cent) or water	100 c.c.

This is our favorite formula. For material which is to be mounted in glycerin, glycerin jelly, or Venetian turpentine, use the aqueous solution; for material which is to be imbedded, use the alcoholic. Cilia are caught and preserved; and even delicate organisms, like *Volvox*, do not collapse.

c) **Corrosive Sublimate, Acetic Acid, and Picric Acid.**—

Corrosive sublimate	5 g.
Glacial acetic acid	5 c.c.
Picric acid, saturated solution in 50 per cent alcohol	100 c.c.

Miss Ethel Thomas recommends this formula for the female gametophyte of *Pinus*.

d) **Corrosive Sublimate and Picric Acid** (Jeffrey's solution).—

Corrosive sublimate, saturated solution in 30 per cent alcohol	3 parts
Picric acid, saturated solution in 30 per cent alcohol	1 part

It would be worth while to try other combinations.

IODINE

Iodine is well known as an antiseptic. It is also a good fixing agent for unicellular, colonial, and filamentous forms. It penetrates rapidly.

To a saturated solution of potassium iodide in distilled water, add iodine to saturation. Filter and dilute with distilled water until the solution has a rich brown color. For fixing, dilute still further to a light-brown color. The solution fixes in 10 to 24 hours, but material may be left in it for several days. Wash thoroughly in tap water which has stood long enough to give off all excess of air. If the staining of the starch does not disappear, a ½ per cent solution of tannic acid in water will remove any excess color.

FORMALIN

Formalin is an excellent preservative. It has been mentioned already as an ingredient in several formulas. Commercial formalin has a strength of 40 per cent. Throughout this book, a 2, 4, or 6 per cent formalin is understood to mean 2, 4, or 6 c.c. of commercial formalin to 98, 96, or 94 c.c. of water, alcohol, or any other ingredient. Commercial formalin is sure to contain some formic acid. For most purposes, it is neither necessary nor desirable to remove the acid. For studying the origin of vacuoles, it is necessary to have neutral formalin, which can be secured from commercial formalin by distillation. Place some sodium bicarbonate in a flask of formalin and distil by heating over a Bunsen flame. It is not worth while to distil more than is needed for immediate use, since the formic acid soon reappears.

For filamentous algae and fungi a 3 to 6 per cent solution of the ordinary commercial formalin in water is very good. Material is left in the solution until needed for use. For marine algae sea-water should be used instead of fresh water. Both marine and fresh-water material should be washed for half an hour in fresh water before staining. A 6 per cent solution will fix one-fourth its volume of material. With material like filamentous algae or leafy liverworts, a 10 per cent solution will fix all one can put into the bottle without crowding.

For class use, material should be washed in water for several minutes, because the fumes are irritating to the eyes and mucous membranes.

For a study of mitochondria or the origin of vacuoles the following combination is recommended:

Bensley's Formula.—

1. Formalin (neutral)	10.0 c.c.
2. Bichromate of potash	2.5 g.
3. Corrosive sublimate	5.0 g.
4. Water	90.0 c.c.

Make the solution 2, 3, 4, and then add the neutral formalin. Fix about 24 hours. Wash in water, but use the iodine—necessary on account of the corrosive sublimate—just before staining sections on the slide.

GENERAL HINTS ON FIXING

It is very desirable that the fixing agent penetrate quickly to all parts of the object. For this reason material should be in small pieces. The best fixing agents do their best work near the surface of the piece. Of course, filamentous algae and fungi, and delicate objects like fern prothallia and root-tips, are simply thrown into the fixing agent. Alcohol, formalin alcohol, or formalin alone, may penetrate $\frac{1}{4}$-inch cubes; but the chromic-acid series, which gives the best results in cytological work, penetrates so poorly that cells more than $\frac{1}{16}$ inch from the surface are not likely to be well fixed. Most objects should be trimmed with a razor so that no part shall be more than $\frac{1}{16}$ inch from the surface. Even then, it must be remembered that a waxy or cutinized or suberized surface presents an almost impassable barrier to the chromic series.

Some objects, although small, cause trouble in various ways. Many buds are hairy and will not sink; if such things are dipped quickly in strong alcohol, they will usually sink. If rather large air bubbles prevent the material from sinking, as in case of perichaetical leaves of some mosses and involucral leaves of liverworts, a little dissection or a careful snip with the scissors will obviate the difficulty. If an air-pump is available some bubbles are easily removed, but air bubbles in cells may resist even the air-pump. Heating followed by rapid cooling is recommended by Pfeiffer and Wellheim for removing air, but, for cytological work, the remedy is worse than the bubbles.

It is often asked whether fixing agents really preserve the actual structure of cell contents. It must be admitted that some things—notably the liquid albuminoids—are much modified in appearance, but the most competent observers are now inclined to believe that such delicate objects as chromosomes, centrosomes, the achromatic figure, and even the structure of protoplasm, can be studied with confidence from material which has been fixed, imbedded, and stained. Extensive investigations upon various objects in the living condition have strengthened this confidence.

It is certain that we have not yet found the ideal fixing agent for cell contents. Such an agent must not be a solvent of any of the cell contents, must penetrate rapidly, must preserve structures perfectly, and must harden so thoroughly that every detail shall remain unchanged during the subsequent processes of dehydrating, clearing, imbedding, sectioning, and staining.

DEHYDRATING AGENTS

Objects which are to be imbedded in paraffin or celloidin, and also all other objects which are to be mounted in balsam or Venetian turpentine must be dehydrated, i.e., they must be freed from water. The slightest trace of water is ruinous. Alcohol is used almost exclusively for dehydrating. The process must be gradual. If material has been fixed in an aqueous solution, it must pass through a series of alcohols of increasing strength, beginning with about 3 per cent alcohol. Ten years ago, most botanists were beginning with 35 per cent alcohol; in the second edition of this book (1905) we recommended 15, 35, 50, 70, 85, 95, and 100 per cent as a safe series, since it causes no obvious plasmolysis of the cell contents. As investigations have become more and more critical, especially investigations upon the structure of chromatin, it has been found that even

15 per cent alcohol is too strong for a beginning. It is maintained that, in addition to the damage done by transferring from water to so strong an alcohol, the final dehydration is not so perfect as it is when the series begins with a weaker alcohol. Yamanouchi, whose work upon delicate algae has been particularly successful, uses the following series: $2\frac{1}{2}$, 5, $7\frac{1}{2}$, 10, 15, 20, 30, 40, 50, 70, 85, 95, and 100 per cent. After such gradual early stages, there seems to be no objection to the less gradual stages which follow. Of course, there is no particular virtue in the fractions: it is convenient to make a 10 per cent alcohol, then dilute it one-half for the 5 per cent, and dilute the 5 per cent one-half for the $2\frac{1}{2}$ per cent. The $7\frac{1}{2}$ per cent is made with sufficient accuracy by adding a little water to the 10 per cent alcohol. For each of the grades up to 50 per cent, one hour is enough for objects like fern prothallia; but an hour and a half should be given to each of the remaining grades. Objects like onion root-tips should have 2 hours in each of the grades up to 50 per cent, and 3 hours in each of the others. Quarter-inch cubes should have 3 or 4 hours in each grade up to 50 per cent, and 4 or 5 hours in the others. The times for larger objects will be longer and will depend upon the size and density. In all cases, the absolute alcohol should be changed 2 or 3 times. The grades below 100 per cent can be used repeatedly. The absolute alcohol should not be used again for this purpose, but may be put back into the 95 per cent bottle. It is always well to filter the alcohols when pouring back into the bottle. Otherwise, there would soon be an accumulation of starch grains, pollen grains, spores, and various other things. Waste alcohol as strong as 85 or 95 per cent will be useful for rinsing one's hands when dealing with Venetian turpentine. If it is necessary to be very economical, the stronger alcohols may be filtered into a single large bottle and the strength of the mixture can then be determined by using an alcoholometer. Knowing the strength of the mixture, one can easily make any weaker grade.

Be very sure that the bottles or Stenders for absolute alcohol are perfectly dry; also, keep the bottles well corked and keep the lids on the Stenders. The importance of excluding moisture cannot be exaggerated. Tightly fitting corks and closely fitting covers are better than absorbents; prevention is better than cure.

The lower grades are made up from 95 per cent alcohol.

Formulas for Alcohols.—The following formulas will enable anyone to make the other grades of alcohol from 95 per cent alcohol and water.

95	95	95	95	95	95	95	95	95
10	15	20	30	40	50	60	70	85
—	—	—	—	—	—	—	—	—
85	80	75	65	55	45	35	25	10

The foregoing are the formulas for various alcohols from 10 to 85 per cent. The first column shows the formula for making 10 per cent alcohol. The percentage of alcohol secured in each case is indicated by the middle number in each column. In the first formula, subtract 10 from 95; the result, 85, is the number of cubic centimeters of water which must be added to 10 c.c. of 95 per cent alcohol in order to obtain 10 per cent alcohol. The mixture contains 95 c.c. of 10 per cent alcohol. If more or less than 95 c.c. of the mixture is needed, take proportional parts of 10 and 85. This simple method is a time-saver, but if the bottles or Stender dishes are to be filled frequently, it will be a still further saving of time to use a long label (Fig. 15) and, after pouring in the 95 per cent alcohol, draw a line showing how high it reaches; and then, after pouring in the water, draw another line. The next time it is necessary to fill the bottles merely pour in 95 per cent alcohol until it reaches the first line, and then pour in water until it reaches the second line. It is not necessary to use distilled water if pure drinking-water is available.

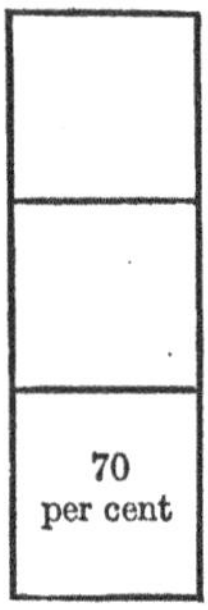

Fig. 15.—Label for staining-dish.

Synthol is used like alcohol, and many believe it to be a good substitute.

Acetone has also been used with more or less success for all grades except absolute alcohol.

Some investigators use more or less complicated diffusion apparatus and make the dehydration process extremely gradual. Judging from the finished preparation, we find no advantage in the method. In the diffusion process, the solution is constantly changing. This may not be an advantage.

Some very minute objects, like bacteria and the smaller Cyanophyceae, may be dehydrated by heating them until all water is drawn off, but, of course, this shows merely the form, with little or nothing of the internal structure.

CLEARING AGENTS

Clearing agents are so named because they render objects transparent. When clearing agents are used to precede infiltration with paraffin, the clearing is merely incidental, the real purpose being to replace the dehydrating agent with a solvent of paraffin. The clearing is useful, even in this case, because it indicates when the replacing has become complete.

When the clearing agent is used to precede infiltration with paraffin, the material should always be most thoroughly dehydrated with absolute alcohol before beginning with the clearing agent. When the clearing agent is used to clear sections or small objects just before mounting in balsam, absolutely perfect dehydration is not necessary with all clearing agents. Bergamot oil, carbolic acid, and Eycleshymer's clearing fluid (equal parts of bergamot oil, carbolic acid, and cedar oil) will clear readily from 95 per cent alcohol. Sections to be cleared in xylol or clove oil should be dehydrated in "absolute" alcohol. If the absolute alcohol is below 99 per cent, xylol will not clear perfectly; but clove oil clears readily from 99 per cent and even from 98 or 97 per cent. If the absolute alcohol is not well above 99 per cent, it is a good practice to go from the alcohol to clove oil; and then, from clove oil to xylol.

Xylol.—In our opinion, xylol is the best clearing agent to precede infiltration with paraffin. After the material has been dehydrated, it should be brought gradually into xylol. Thirty years ago it was customary to bring material directly from absolute alcohol into xylol; twenty years ago, two or three mixtures of absolute alcohol and xylol were used before reaching the pure xylol; at present, those who are doing the most critical work are making this process still more gradual. As cytologists have been studying more and more minute structures, the methods have become more and more critical. As in the case of the alcohol series, the xylol series has its grades closer together at the beginning than at the end. The following series seems to be sufficiently gradual: $\frac{1}{16}$, $\frac{1}{8}$, $\frac{1}{4}$, $\frac{1}{2}$, $\frac{3}{4}$, pure xylol. It is hardly necessary to use a graduate in making up the series. For the $\frac{1}{2}$, use equal parts of xylol and absolute alcohol; for the $\frac{1}{4}$, use equal parts of the $\frac{1}{2}$ and absolute alcohol; for the $\frac{1}{8}$, use equal parts of the $\frac{1}{4}$ and absolute, and for the $\frac{1}{16}$, equal parts of the $\frac{1}{8}$ and absolute. The $\frac{3}{4}$ can be guessed at with sufficient accuracy.

Some prefer a closer series of xylols, using $2\frac{1}{2}$, 5, $7\frac{1}{2}$, 10, 15, 20, 30, 40, 50, 75, and 100 per cent. It is claimed that infiltration with paraffin is more thorough after this series. About one hour in each grade is enough for filamentous algae and fungi, fern prothallia, and similar objects. An hour and a half is enough for onion root-tips. For quarter-inch cubes, try 3 hours. For larger pieces the times should be longer. In all cases, the pure xylol should be changed two or three times. While the pure xylol must not be used again for this purpose, it is still good for dissolving paraffin ribbons when staining on the slide.

Xylol is the best agent for clearing sections just before mounting in balsam. Preparations cleared in xylol harden more rapidly, and this is such a decided advantage that even when sections have been cleared in cedar oil or clove oil it is worth while to give them a minute or two in xylol before mounting. Besides, clove oil is a solvent of many of the most frequently used stains and, consequently, preparations in such stains would fade, if transferred directly from clove oil to balsam.

Xylol evaporates so rapidly that one must take care not to let sections become dry before applying the balsam. Thin sections perfectly dehydrated will clear in a few seconds; but, even with very thin sections, it is better to let the xylol act for at least a minute. Sections 20 μ in thickness should remain in the xylol 2 to 5 minutes before mounting in balsam. If there is much moisture in the air, or if the absolute alcohol is not above suspicion, clear sections in clove oil before transferring to xylol.

Chloroform.—Some botanists use chloroform to precede the infiltration with paraffin. In the later stages of infiltration it is more easily removed than xylol. It seems to possess no other advantages, and for clearing sections just before mounting in balsam it is inferior to xylol or clove oil. Its value in hardening celloidin and as a fixing agent entitles it to a place in the histological laboratory.

Cedar Oil.—It is not always easy to get good cedar oil. If the stuff offered for sale looks like turpentine and smells like it, it is worthless for histological purposes. Good cedar oil has a slightly amber tint, the color resembling a weak clove oil. It should have the pleasant odor of cedar wood. The very expensive cedar oil used with immersion lenses is not needed for clearing or for preceding infiltration with paraffin. It is claimed that material cleared in cedar oil does not become so brittle as that cleared in xylol or chloroform.

Dr. E. J. Krause has used cedar oil extensively in clearing large objects—strawberries and gooseberries either whole or cut in two, sections of apple 2 to 4 mm. thick, and similar objects. This method is proving valuable in vascular anatomy, some material showing the course of bundles very clearly in pieces so large as centimeter cubes.

Xylol can be used in the same way, but is so volatile that specimens often dry up. Dr. Land suggests equal parts of xylol and carbon disulphide for clearing large objects which are to be examined without sectioning.

Clove Oil.—This is an excellent agent for clearing sections and small objects just before mounting in balsam. It clears more readily than xylol. When the absolute alcohol has deteriorated so that xylol no longer clears the sections, clove oil may still clear with ease. While clove oil will clear from 95 per cent alcohol, it is better to use absolute. Since preparations cleared in clove oil harden slowly, it is a good plan to treat them with xylol before mounting in balsam. Gentian-violet is somewhat soluble in clove oil, and this fact makes it possible to secure a beautiful differentiation, because the stain is extracted from some elements more rapidly than from others. The stain may be extracted completely from the chromosomes during the metaphase and still remain bright in the achromatic structures. After the desired differentiation has been attained, the preparation should be placed in xylol to remove the clove oil, since the continued action of the clove oil would cause the preparation to fade. Do not use a Stender dish for clove oil, but keep it in a 50 c.c. bottle. Put on a few drops, and immediately drain them off in such a way as to remove the alcohol as completely as possible. Then flood the slide and pour the clove oil back into the bottle, repeating the process until the proper differentiation has been reached. Replace the clove oil with xylol and mount in balsam. With stains not soluble in clove oil, the xylol is not necessary, except to facilitate the hardening of the preparation.

Clove oil may be used in removing the celloidin matrix from celloidin sections. It is useless as an agent to precede infiltration with paraffin.

Eycleshymer's Clearing Fluid.—This is a mixture of equal parts of bergamot oil, cedar oil, and carbolic acid. It clears readily from 95 per cent alcohol, and consequently is useful in clearing celloidin sections when it is desirable to preserve the celloidin matrix. In sections stained with haematoxylin, or haematoxylin and eosin, the stain may be removed completely from the matrix by the use of acid alcohol, and the matrix may be preserved by clearing from 95 per cent alcohol.

It is not intended that the mixture should be used to precede infiltration with paraffin.

Other Clearing Agents.—Bergamot oil, carbolic acid, turpentine, benzine, gasoline, and other reagents have been tried for clearing, but none seem to be worth more than a warning mention.

MISCELLANEOUS REAGENTS

Canada Balsam is used almost exclusively for mounting Very thick balsam is disagreeable to handle and makes unsatisfactory mounts. Very thin balsam, in drying out, allows bubbles to run under the cover. Xylol is cheaper than balsam, and consequently the balsam on the market is likely to be too thin for immediate use. The stopper may be left out until the balsam acquires the proper consistency. Balsam must not be acid. If there is the slightest acid reaction, most stains will fade.

Paraffin should be of at least two grades, a soft paraffin melting at 40° to 45° C., and a hard paraffin melting at 52° to 54°C. Grübler's paraffin and most imported paraffins melt at the temperature indicated on the wrappers. The melting-point indicated on the wrappers of paraffins sold by some American dealers does not enable one to make even a guess as to the real melting-point. Paraffin marked 70° C. may melt at 60° C., and other grades are likely to melt before the temperature indicated on the labels is reached. The fact that the price rises with the melting-point may explain the discrepancy. Test every grade with a thermometer. If it is desired to get a paraffin melting at 52° C. and your sample melts at 50° C., add a little paraffin with a melting-point above 52° C.; if the sample melts at 55° C., add a little with a melting-point below 52° C.

Paraffin may be used repeatedly. Keeping it in the liquid condition in the bath month after month is an advantage, since it becomes more and more tenacious and homogeneous.

Glycerin, glycerin jelly, Venetian turpentine, and gold size are described in the chapter on "The Glycerin Method" (chap. vii). Celloidin is described in the chapter on "The Celloidin Method" (chap. x), and cellulose acetate in the chapter on "The Cellulose Acetate Method" (chap. xi). The reagents already described are noted further in connection with specific applications. Reagents used in making microchemical tests are described in the chapter on "Temporary Mounts and Microchemical Tests" (chap. v).

A list of reagents with suggestions in regard to quantities and prices will be found in chapter xxix.

CHAPTER III

STAINS AND STAINING

Since the third edition of this book appeared in 1915 no new stains of the first rank have come into favor, but much greater precision has been attained in the use of some which were already popular. For cytological work Haidenhain's iron-haematoxylin holds a firm place at the head of the list, with Flemming's triple stain an easy second. For anatomical work, safranin still holds first place for the lignified elements of the vascular system, but the claim of Delafield's haematoxylin to first place for cellulose tissues is no longer undisputed, for anilin blue is giving excellent results and light green seems to give more accurate views of the phloem than we were securing with any of the other stains. The fact that excellent preparations can be made, almost without trial, by using combinations already perfected doubtless deters investigators from experimenting with other stains. There is still abundant room for experimenting with various stains, and especially in the use of mordants and in the effect of the same stain or combination after various fixing agents. It is to be regretted that botanists who need microtechnic have so little knowledge of chemistry, and that chemists have no interest in developing methods of staining. During the past few years, American stains have been developed until many equal and some even surpass the famous Grübler products; and, besides, the American stains are becoming standardized.

Stains may be classified in various ways: e.g., there are three great groups of stains—the carmines, the haematoxylins, and the anilins. Stains may be classified as basic and acid, or they may be regarded as general and specific. A general stain affects all the elements, while a specific stain affects only certain elements, or stains some elements more deeply than others. Stains which show a vigorous affinity for the nucleus have been called "nuclear stains," and those which affect the cytoplasm more than the nucleus have been termed "plasma stains." Of course, such stains are specific.

We shall consider some of the more important haematoxylins, carmines, and anilins, reserving general directions and theoretical questions for another chapter. The formulas are largely empirical. Some of those given here are taken from *The Microtomist's Vade-Mecum* (Lee), which is easily the most complete compendium of stains and other reagents concerned in microtechnic. It is to be regretted that botanists have no book of this character, but it must be confessed that we have not the material for such an extensive work. Other formulas are from *Botanical Microtechnique* (Zimmermann) and from Stirling's *Histology*, and still others are from current literature and from our own laboratory. The directions for using a stain apply to stains made up according to the formulas which are given here, and may need modification if other formulas are employed. It is hoped, however, that the directions will give the student sufficient insight into the *rationale* of staining to enable him to make any necessary modifications. Since American stains have come into general use, the need for *rationale* is even greater, especially if the American stains are made up according to standard formulas, which are based largely upon the Grübler products. In general, it would seem that the American stains are purer and that they act more rapidly.

The current practice in staining paraffin sections on the slide differs from the practice in staining freehand sections or small objects which are to be mounted whole. In case of paraffin sections, the cell contents are usually as important and often more important than the cell walls; consequently, extreme care must be given to every detail. With freehand sections the cell contents often drop out, but even when they remain, the cell walls are usually the important features; and so the process is considerably shortened.

For staining freehand sections, it is customary to use solid watch glasses, unless the sections are very large. The details of the method are given in chapter vi, on "Freehand Sections."

For staining sections on the slide, nothing is better than the ordinary Stender dish. The arrangement of Stender dishes shown in Figure 16 is very convenient. The advantage is obvious. With two dishes each of xylol, xylol-alchol, and absolute alcohol, one set can be used in passing down to the stain, and the other, which is thus kept free from any paraffin in solution, can be used in passing back to the balsam. Even for paraffin sections, some use only three alcohols, 50, 95 and 100 per cent, and the first two may be simply poured over the slide; in this case, only one Stender dish—for the 100 per cent alcohol—is necessary in the alcohol series, the other two alcohols being kept in bottles. This short method gained great popularity because it was used in Strasburger's laboratory at Bonn. It was the influence of this school and its great master which led to the adoption of the short schedule in the second edition

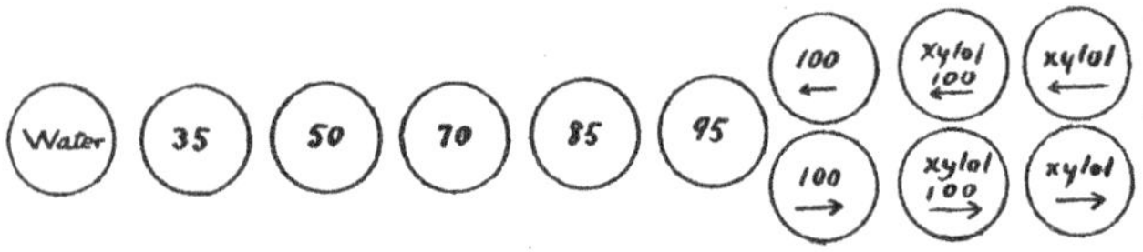

FIG. 16.—A convenient arrangement of staining-dishes

of this book. A few years' trial showed the weakness of the method, and we returned to the longer schedule. The crudeness of the short schedule is doubtless responsible for the tenacity with which the Bonn school has clung to the theory of linin and chromomeres. The young investigator should be warned that during the last twenty years of his life, Strasburger, who had been a leader in technic, cut very few sections and did practically no staining, but used preparations made by assistants.

Let us now consider a few of the most important stains.

THE HAEMATOXYLINS

The most important haematoxylins are Haidenhain's iron-alum haematoxylin, Delafield's haematoxylin, Mayer's haem-alum, and Boehmer's haematoxylin.

All the haematoxylins mentioned contain alum, and, according to Mayer, who has written the most important work on haematoxylin stains,[1] "the active agent in them is a compound of haematin with alumina. This salt is precipitated in the tissues, chiefly in the nuclei, by organic and inorganic salts there present (e.g., by the phosphates), and perhaps also by other organic bodies belonging to the tissues." These salts are fixed in the tissues by the killing and fixing agent, and when the stain is applied a chemical combination results. Haematoxylins stain well after any of the fixing agents described in

[1] "Ueber das Färben mit Hämatoxylin," *Mittheilungen aus der Zoologischen Station zu Neapel,* 10:170–186, 1891, and "Ueber Hämatoxylin, Carmin und verwandte Materien," *Zeitschrift für wissenschaftliche Mikroscopie,* 16:196–220, 1899.

the preceding chapter, but they are most effective when used after members of the chromic-acid series.

Haidenhain's Iron-Alum Haematoxylin.—This stain, introduced by Haidenhain in 1892 immediately gained great popularity and now, after more than 30 years' constant use, still maintains first place in cytological investigations. Two solutions are used, and they are never mixed:

A. 2 to 4 per cent aqueous solution of ammonia sulphate of iron.
B. ½ per cent aqueous solution of haematoxylin.
In making solution A, use the violet ferric crystals, not the ferrous.

The first solution acts as a mordant, i.e., it does not stain, but prepares the tissue for the action of the second solution.

Solution A is at its best as soon as the crystals are completely dissolved and it remains in practically perfect condition for about two months, after which it gradually deteriorates.

The haematoxylin crystals for solution B should be dissolved in water.. This will require about 10 days. The solution should then be allowed to "ripen" for 4 weeks before it is ready for use. Unfortunately, it remains at its best for only a short time, not more than 5 or 6 weeks. This is because the "ripening," which is an oxidation process, continues, and the solution becomes too ripe. During the ripening, a cotton plug should be used instead of a cork, to facilitate the oxidation; but as soon as the stain is ripe, a cork—preferably a perfectly fitting glass stopper—will prolong the maximum efficiency. Some prefer to dissolve the haematoxylin crystals in alcohol—about 10 g. in 100 c.c. of absolute alcohol. This solution should stand until it has a deep wine-red color. This will require 4 or 5 months, and a year is not too long. From this stock solution, make up small quantities as needed. About 4 or 5 c.c. of this stock solution in 100 c.c. of water gives a practically aqueous solution, and it is already ripe.

The American haematoxylin ripens faster and may reach its maximum efficiency in 2 or 3 weeks. One might naturally expect that it would retain its efficiency for a correspondingly shorter time; but we have found that, even for the structure of chromatin, the stain seems to stay at its best for at least two months.

The general method is as follows: treat with A, stain in B, and then return to A to reduce and differentiate the stain. Never transfer directly from A to B, or from B to A; always wash in water before passing from one of the solutions to the other.

While all follow the general method just indicated, no two investigators would prepare exactly the same schedule, even for staining the same object, e.g., root-tips; neither investigator would use the same schedule for a root-tip and an embryo sac; an alga might require different treatment, and all the preceding variations might fail miserably with the pollen tubes of cycads. This stain is so important that every worker must learn it, and the only way to learn it is to become acquainted with the general outline of the process and then adapt every step to the case in hand.

For the sake of illustration, I asked two prominent cytologists, Dr. S. Yamanouchi and Dr. L. W. Sharp, both of whom have been notably successful in staining mitotic figures, to write schedules indicating their methods of using this stain. While both protested that the practice could not be written down, they kindly prepared the following schedules, not for the instruction of their colleagues, but to introduce the method to beginners. Both schedules are for paraffin sections. Throughout the first schedule, I have interpolated comments and suggestions.

Yamanouchi's Schedule.—

1. Xylol, 5 minutes, to dissolve the paraffin.

   Do not heat the slides to melt the paraffin. However, a gentle warming which does not approach the melting-point of the paraffin does no damage and makes the paraffin dissolve more readily. The xylol soon has considerable paraffin in solution, but 100 c.c. of xylol should remove the paraffin from at least 100 slides with ribbons 25 mm.long and 10 μ thick. If the ribbons are only 5 μ thick, 200 slides can be treated.
2. Xylol and absolute alcohol, equal parts, 5 minutes.
3. Absolute alcohol, 5 to 7 minutes.
4. 95, 85, 70, 50, 35 per cent alcohol, 5 minutes each.

   If material has been fixed in a reagent containing osmic acid, it should be bleached. For this purpose, 10 to 15 c.c. of hydrogen peroxide may be added to 100 c.c. of the 50 per cent alcohol.
5. Water, 10 to 20 minutes.

   If any alcohol is left in the sections, the staining will not be brilliant. Change the water several times.
6. Iron-alum.

   Use the 4 per cent solution. For many objects, like the archegonia of gymnosperms and the embryo sacs of angiosperms, 1 hour is usually enough. For chromosomes in root-tips and anthers, 2 hours may be long enough; but for algae, 2 hours is generally a minimum.

7. Wash in water, 5 minutes.

   The water should be changed several times. If the washing is not thorough, the differentiation will not be sharp.

8. Haematoxylin.

   Many objects, like the archegonia of gymnosperms and the embryo sacs of angiosperms, will stain sufficiently in 5 or 6 hours; most algae require at least 20 hours.

9. Wash in water, 5 minutes, changing as often as the water shows any color.

10. Iron-alum, 2 per cent solution.

    No time can be indicated here. The preparation must be watched under the microscope. After some experience, one can form some judgment from the color tone, as the slide stands in the Stender dish of iron-alum, but the finishing must always be done under the microscope. If the stain is coming out rather slowly, as it should, one can handle 6 to 10 slides at one time. Put the slides on a 5×7 glass plate and put the plate on the stage of the microscope. The iron-alum can be added or removed with a pipette. As slide after slide reaches the proper differentiation, it is placed in water.

11. Water, 30 minutes.

    The water should be changed several times. If this washing is not thorough, the preparation will fade, on account of the continued action of the iron-alum. If an aqueous counter-stain is used, apply it at this point.

12. 35, 50, 70, 85, 95, 100 per cent alcohol, 5 minutes in each.

    If an alcoholic counter-stain is used, apply it near the alcohol of the same strength as the stain.

13. Absolute alcohol and xylol, equal parts, 5 minutes.
14. Xylol, 2 to 5 minutes.
15. Balsam.

Sharp's Schedule.—

1. Remove the paraffin with xylol.
2. Rinse in absolute alcohol.
3. 95 per cent alcohol.
4. 50 per cent alcohol.
5. Water.
6. If osmic acid has been used in fixing, place the slides in 10 per cent solution of peroxide of hydrogen in water until bleached.
7. Water.
8. Iron-alum, 2½ per cent, 2 to 3 hours.
9. Wash well.
10. ½ per cent haematoxylin, 24 hours.
11. Wash in water.

12. Extract the stain in 1 per cent iron-alum, watching the process under the microscope.
13. Wash several hours in water.
14. Alcohol series: 10, 30, 50, 70, 80, 95, 100 per cent.
    If a counter-stain is desired, introduce it in one of the alcohols of this series.
15. Absolute alcohol and xylol, equal parts.
16. Xylol.
17. Mount in balsam.

While these two schedules would enable the student to apply the method in case of objects to be mounted whole, like filamentous algae, fern prothallia, etc., a complete schedule is given in chapter viii on "The Venetian Turpentine Method."

The times given above must not be accepted as final. Many prefer to wash in water for several hours after the first immersion in iron-alum. Some think that 4 hours is enough for the entire process. Many put the slide into iron-alum in the morning and finish the process in the afternoon. These short schedules are not likely to prove satisfactory with mitotic figures. A plan which has proved convenient and very successful is to put the slide into the iron-alum in the morning, wash in water for an hour at some convenient time in the afternoon, leave it in the $\frac{1}{2}$ per cent haematoxylin over night and finish the preparation the next morning. It is a long process, requiring care, patience, and judgment, but it is worth the effort.

Chromosomes, centrosomes, and pyrenoids take a brilliant black; or, if the second treatment with iron-alum be more prolonged, a blue black or purple. Achromatic structures stain purple, but the stain can be extracted while it is still bright in the chromosomes. Lignified, suberized, and cutinized structures stain lightly or not at all. Cellulose does not stain so deeply as with Delafield's haematoxylin. Archesporial cells and early stages in sporogenous tissue stain gray. Many details which are not so brilliantly colored often show good definition.

If a counter-stain is desired, anything which gives a serviceable contrast may be used. In any case, the haematoxylin stain must be complete and the washing thorough before the second stain is applied. An aqueous stain should be applied just after the final washing in water; an alcoholic stain should be applied during the process of passing the slides through the alcohols, staining in a solution of safranin in 50 per cent alcohol from the alcohol of a concentration nearest that of the stain; and staining after the final absolute alcohol, if the stain is dissolved in clove oil.

A stain of 3 or 4 minutes in safranin adds an excellent differentiation in case of many algae and does not obscure nuclear details. The exine of pollen grains may take a brilliant red with safranin in 5 or 10 minutes, contrasting sharply with the mouse gray of the intine. Orange G, in clove oil, often gives a pleasing contrast.

Delafield's Haematoxylin.—"To 100 c.c. of a saturated solution of ammonia alum add, drop by drop, a solution of 1 g. of haematoxylin dissolved in 6 c.c. of absolute alcohol. Expose to air and light for one week. Filter. Add 25 c.c. of glycerin and 25 c.c. of methyl alcohol. Allow to stand until the color is sufficiently dark. Filter, and keep in a tightly stoppered bottle" (Stirling and Lee). The addition of the glycerin and methyl alcohol will precipitate some of the ammonia alum in the form of small crystals. The last filtering should take place 4 or 5 hours after the addition of the glycerin and methyl alcohol.

The solution should stand for at least two months before it is ready for using. This "ripening" is brought about by the oxidation of haematoxylin into haematin, a reaction which may be secured in a few minutes by a judicious application of peroxide of hydrogen. However, we prefer to let the haematoxylin ripen naturally. There is no objection to making this stain in considerable quantity, since it does not deteriorate. We have used Delafield's haematoxylin which had been in a cork-stoppered bottle for twenty years, and it still gave the rich characteristic stain.

Transfer to the stain from 50 or 35 per cent alcohol or from water. The length of time required is exceedingly variable. Sometimes sections will stain deeply in 3 minutes, but it is often necessary to stain for 30 minutes or even longer. This stain may be diluted with several times its own volume of water; when this is done, the time required is correspondingly long, but the staining is frequently more precise. The length of time required will be fairly uniform for all material taken from the same bottle. This fact indicates that the washing process, which follows killing and fixing, is an important factor; if the washing has been thorough, the material will stain readily; but if the washing has been insufficient, the material may stain slowly or not at all. The washing is particularly important when the fixing agent contains an acid. Transfer from the stain to tap water. Distilled water is neither necessary nor desirable. Some writers recommend washing for 24 hours, but this is entirely unnecessary; for paraffin sections on the slide, 5 or 10 minutes is long enough, and even for rather thick freehand sections 20 or 30 minutes is sufficient. Use plenty of water and keep changing it as often as it becomes in the least discolored. Precipitates are often formed when slides are transferred directly to alcohol from this stain, and sometimes even after washing in water. A few gentle dips in acid alcohol (2 drops of HCl to 100 c.c. of 70 per cent alcohol) will usually remove the precipitates. This extracts the stain more rapidly from other parts than from the nuclei, and hence gives a good nuclear stain, while at the same time it removes any disfiguring precipitates. Some prefer to stain for a very short time and use no acid alcohol, but, as a rule, it is better to overstain and then differentiate in this way, because sharper contrasts are obtained. Transfer from acid alcohol to 70 per cent alcohol and leave here until a rich purple color replaces the red due to the acid. Since small quantities of the acid alcohol are carried over into the 70 per cent alcohol, it is well to add a *drop* of ammonia now and then to neutralize the effect of the acid. Too much ammonia is to be avoided, for it gives a disagreeable bluish color with poor differentiation, probably on account of the precipitation of alumina. The preparation is now dehydrated in 95 per cent and then in absolute alcohol, cleared in xylol or clove oil, and mounted in balsam.

The following is a general schedule for staining paraffin sections on the slide in Delafield's haematoxylin:

1. Stain (from water or from 35 or 50 per cent alcohol) 10 minutes
2. Rinse in water.............................. 10 minutes
3. 35 and 50 per cent alcohol.................. 3 minutes each
4. Acid alcohol................................ 5 seconds
5. 70 per cent alcohol......................... 3 minutes
6. 85 per cent alcohol......................... 3 minutes
7. 95 per cent and 100 per cent alcohol......... 3 minutes each
8. Xylol and 100 per cent alcohol, equal parts....... 3 minutes
9. Xylol....................................... 3 minutes
10. Mount in balsam.

If, after rinsing in water, the stain is evidently too weak, put the slide or section back into the stain until it appears overstained. Place the slide in acid alcohol. If an acid alcohol with 2 drops of HCl to 100 c.c. of 70 per cent alcohol reduces the stain too much in 4 or 5 seconds, use less acid or stain longer. Transfer to 70 per cent alcohol without any acid. As soon as the color changes from red to purple, examine under the microscope. If it is still overstained, return to the acid alcohol; if the stain is too weak, return to the haematoxylin and try it again. After the haematoxylin is just right, apply a contrast stain, if you wish to double stain. Before transferring to the xylol wipe the alcohol from the back of the slide, or at least rest the corner of the slide upon blotting-paper for two or three seconds, in order that you may not carry over so much alcohol into the xylol. Add a drop of balsam and a cover. Since the xylol is very volatile, this last step must be taken quickly. If blackish spots appear they are usually caused by the drying of sections before the balsam and cover are added; if there are whitish spots or an emulsion-like appearance, the clearing is not thorough; this may be caused by poor xylol (or other clearing agent); by absolute alcohol which is considerably weaker than its name implies (the absolute alcohol must test at least as high as 99 per cent, and ought to test as high as 99.5 per cent, if xylol is to be used for clearing); or by passing too quickly through the absolute alcohol and xylol, or even by moisture on the cover-glass. The last danger is easily avoided by passing the cover quickly through a Bunsen or alcohol flame before laying it on the balsam.

Delafield's haematoxylin is the most generally useful stain in the haematoxylin group. It brings out cellulose walls very sharply, and consequently is a good stain for embryos and the fundamental tissue system in general. With safranin it forms a good combination for the vascular system, the safranin giving the lignified elements a bright red color, while the haematoxylin stains the cellulose a rich purple. It is a good stain for chromatin, and the achromatic structures show up fairly well, but can be brought out much better by special methods. Archesporial cells and sporogenous tissue are very well defined if proper care be taken. Lignified and suberized walls and also starch and chromatophores stain lightly or not at all. Whenever you are in doubt as to the selection of a stain for general purposes, we should advise the use of Delafield's haematoxylin.

Mayer's Haem-Alum.—Haematoxylin, 1 g., dissolved with gentle heat in 50 c.c. of 95 per cent alcohol and added to a solution of 50 g. of alum in a liter of distilled water. Allow the mixture to cool and settle; filter; add a crystal of thymol to preserve from mold (Lee).

It is ready for use as soon as made up. Unless attacked by mold, it keeps indefinitely. Transfer to the stain from water. It is seldom necessary to stain for more than 10 minutes, and 4 or 5 minutes is generally long enough. As a rule, better results are secured by diluting the stain (about 1 c.c. to 10 c.c. of distilled water) and allowing it to act for 10 hours or over night.

This is a good stain for the nuclei of filamentous algae and fungi, since it has little or no effect upon cell walls or plastids. Wash thoroughly in water, transfer to 10 per cent glycerin, and follow the Venetian turpentine method, as described in chapter viii.

Erlich's Haematoxylin.—

Distilled water	50 c.c.
Absolute alcohol	50 c.c.
Glycerin	50 c.c.
Glacial acetic acid	5 c.c.
Haematoxylin	1 g.
Alum in excess.	

Keep it in a dark place until the color becomes a deep red. If well stoppered, it will keep indefinitely. Transfer to the stain from 50 per cent or 35 per cent alcohol. Stain 5 to 30 minutes. Since there is no danger from precipitates and the solution does not overstain, it is not necessary to treat with water or with acid alcohol, but the slide may be transferred from the stain to 70 per cent alcohol. Eosin, erythrosin, or orange G are good contrast stains. Jeffrey uses safranin and Erlich's haematoxylin for woody tissues.

Boehmer's Haematoxylin.—

A	Haematoxylin	1 g.
	Absolute alcohol	12 c.c.
B	Alum	1 g.
	Distilled water	240 c.c.

The solution A must ripen for two months. When wanted for use, add about 10 drops of A to 10 c.c. of B. Stain 10 to 20 minutes. Wash in water and proceed as usual.

Cellulose walls take a deep violet. The closing membrane (torus) of the bordered pits of conifers will usually stain deeply in about 15 minutes. Lignified, suberized, and cutinized structures stain slightly or not at all. When they do stain, the color is not violet, but a light yellow or brown.

THE CARMINES

Botanists have never given the carmines a fair trial, doubtless because the stains were not considered worth it; but the splendid preparations by Professor Powers of various members of the Volvocaceae prove that we should pay more attention to this group. Only a few of the multitudinous formulas will be considered.

The carmine solutions keep for several years, some of them even improving with age, if distilled water has been used in the formulas and the stains have been kept in tightly stoppered bottles. If the solution becomes turbid, it should be filtered.

Greenacher's Borax Carmine.—

Carmine	3 g.
Borax	4 g.
Distilled water	100 c.c.

Dissolve the borax in water and add the carmine, which is quickly dissolved with the aid of gentle heat. Add 100 c.c. of 70 per cent alcohol and filter (Stirling).

The following is a slightly different method for making this stain from the ingredients mentioned above: Dissolve the borax in water, add the carmine, and heat gently for 10 minutes; after the solution cools, add the alcohol and filter; let the solution stand for 2 or 3 weeks, then decant and filter again.

Stain the material in bulk from 50 per cent alcohol 1 to 3 days, then treat with acid alcohol (50 c.c. of 70 per cent alcohol+2 drops of hydrochloric acid) until the color becomes a clear red; this may require only a few hours, but may take 2 or 3 days. The material may then be passed through the rest of the alcohols (6 to 24 hours each), cleared, imbedded, and cut. After the sections are fastened to the slide, the paraffin should be dissolved off with xylol. The balsam and cover may be added immediately, or the xylol may be rinsed off with alcohol and a contrast stain may be added.

Alum Carmine.—A 4 per cent aqueous solution of ammonia alum is boiled 20 minutes with 1 per cent of powdered carmine. Filter after it cools (Lee).

Stain from 12 to 24 hours and wash in water. No acid alcohol is needed, since the solution does not overstain.

Carmalum (Alum Lake).—Use 1 gram of the powdered stain to 100 c.c. of very dilute ammonia water. Filter, if there is any precipitate.

Mayer's Carmalum.—

Carminic acid	1 g.
Alum	10 g.
Distilled water	200 c.c.

Dissolve with heat; decant or filter and add a crystal of thymol to avoid mold.

Alum Cochineal.—

Powdered cochineal	50 g.
Alum	5 g.
Distilled water	500 c.c.

Dissolve the alum in water, add the cochineal, and boil; evaporate down to two-thirds of the original volume, and filter. Add a few drops of carbolic acid to prevent mold (Stirling).

Stain as with alum carmine. It used to be a common practice to stain in bulk in alum cochineal and counter-stain on the slide with Bismarck brown.

Iron Aceto-Carmine 1.—For counting chromosomes in pollen mother-cells mounted whole, Belling and also Sand used a modified aceto-carmine method. The preparations are good for an immediate count, but do not last longer than a few days or a week.

"Ordinary aceto-carmine is prepared by heating a 45 per cent solution of glacial acetic acid to boiling with excess of powdered carmine, cooling, and filtering. The young anthers are teased out with steel blades or needles in a drop of this until it changes slightly toward bluish red. An excess of iron spoils the preparation.

Iron Aceto-Carmine.—"To a quantity of aceto-carmine a trace of solution of ferric hydrate dissolved in 45 per cent acetic acid is added until the liquid becomes bluish red, but no visible precipitate forms. An equal amount of ordinary aceto-carmine is then added.

The anthers are teased out with nickled instruments." A cover-glass is then added and sealed with vaseline. The preparation lasts only a few days, but is much superior to any obtained by the usual *intra vitam* processes.

THE ANILINS

Many of the most brilliant and beautiful stains yet discovered belong to this group. These stains are very numerous, but not so numerous as their names; for different names have been given to the same stain, and the same name has been given to different stains. Fortunately, the Committee on Standardization of Biological Stains is doing a good work in standardizing the nomenclature as well as the stains themselves. A valuable list of synonyms, with the preferred designations, was published in *Science*, **57**:743–746, 1923, and other references to the work of the commission are given in the Bibliography on page 341.

General Formula.—Make a 10 per cent solution of anilin oil in 95 per cent alcohol; when the anilin oil is dissolved, add enough water to make the whole mixture about 20 per cent alcohol; add 1 g. of cyanin, erythrosin, safranin, gentian-violet, etc., to each 100 c.c. of this solution. Solutions containing anilin oil do not keep as well as aqueous or alcoholic solutions.

The anilins keep well in balsam, but not so well in glycerine. Xylol is a good clearing agent for all of them; but clearing in clove oil improves stains like gentian-violet, which are more or less soluble in clove oil. Even in such cases, xylol should follow the clove oil, or the preparation will fade.

While the anilins are not as permanent as the haematoxylins, most of them keep fairly well if the staining has been carefully done. Preparations fade if exposed long to bright sunlight. Keep the slides in the box when not in use, and even when in use, do not leave them on the laboratory table, exposed to the sun. We have preparations, made more than 25 years ago, in which the safranin and gentian-violet are still bright; and others made more than 10 years ago, in which Magdala red and aniline blue have not faded.

Some of the anilins are acid, some basic, and some are neutral.

The rapidity with which sections must be transferred from one fluid to another makes many of them more difficult to manage than the haematoxylins or the carmines, but the stains are so valuable that even the beginner should spend most of his time with the anilins.

Many anilins stain quite deeply in 1 to 20 minutes, but if the stain washes out during the dehydrating process, stain longer, even 10 to 24 hours if necessary. Often the brilliancy of the stain can be increased by leaving the slide for 5 minutes in a 1 per cent solution of permanganate of potassium before staining. The permanganate acts as a mordant.

The following are the more important anilins now in use by botanists. The directions apply to solutions made up according to the formulas given with the different stains.

Safranin.—Two safranins are sold by dealers, one soluble in water and the other soluble in alcohol. The alcoholic is somewhat soluble in water and the aqueous is somewhat soluble in alcohol, but both make better solutions when used with their intended solvents.

The best aqueous solution is simply a 1 per cent solution in distilled water.

The alcoholic solution is made by dissolving 1 g. of the alcoholic safranin in 100 c.c. of 95 per cent or absolute alcohol and, after the safranin is completely dissolved, adding 50 c.c. of distilled water.

According to Flemming, dissolve 0.5 g. of alcoholic safranin in 50 c.c. of absolute alcohol, and after 4 days add 10 c.c. of distilled water.

A method which we have used for many years with good results is to make a 1 per cent solution of the aqueous safranin in distilled water; then make a 1 per cent solution of the alcoholic safranin in 95 per cent alcohol; then mix equal volumes of the two solutions. This makes a strong solution of safranin in about 50 per cent alcohol.

The first American safranins were very unsatisfactory but there have been great improvements and one company, the National Anilin and Chemical Company, has produced a safranin with 90 per cent dye content, much stronger than any of the European stains. This is the safranin which has been certified by the Committee on Standardization of Stains. Other companies are also improving. Coleman and Bell's Safranin Y, for bacilli, is good for staining xylem.

All safranins keep indefinitely, solutions 20 years old staining as well, or better, than when fresh.

An anilin safranin may be made according to the general formula.

The transfer to the stain depends upon the formula. If the stain is aqueous, transfer to the stain from water; if made up according to the general anilin oil formula, transfer to the stain from water or, if coming down from higher alcohols, from 35 per cent alcohol; if the mixture of aqueous and alcoholic safranins is used, transfer from 35 per cent alcohol, if going up in the series, and from 70 per cent alcohol, if coming down from stronger alcohols. For freehand sections of woody tissues we always use the mixture. If sections are cut from living material, leave them in 95 per cent alcohol for half an hour and then transfer to the stain. Sections cut from alcoholic material may be transferred directly to the stain. If cut from formalin-alcohol material, leave the sections in 50 per cent or 70 per cent alcohol for ten minutes before transferring. If cut from formalin material, leave them in water for 10 minutes, then in 35 per cent alcohol for 10 minutes before staining.

The time required for staining varies with the tissue, the fixing agent, and the quality of the stain. In general, it may be said that 2 hours is a minimum and 24 hours a maximum. If the staining be too prolonged, delicate structures, like starch grains, crystals, and various cell constituents, may wash out. The mere fact that the whole section does not wash off does not mean that everything is fastened to the slide. On the other hand, with a short period, it is difficult to get a sharp differentiation. In staining a vascular bundle, one should be able to wash the safranin from the cellulose walls and still leave a brilliant red in lignified structures. For paraffin sections, 3 to 6 hours will usually be sufficient. It is a good practice to put the slides into the stain in the morning and finish the mounts any time in the afternoon. For freehand sections of woody tissues, 24 hours is not too long.

From the stain transfer to 50 per cent alcohol. If the sections are deeply stained, and sufficient differentiation is not secured within 5 or 10 minutes, a drop of hydrochloric acid added to 50 c.c. of the alcohol will hasten the extraction of the stain. If staining vascular tissue, draw the stain from the cellulose walls, but stop before the lignified walls begin to fade. If a contrast stain is to be added, like light green, which weakens the safranin; or aniline blue or Delafield's haematoxylin, which need to be followed by an acid; the safanin should be strong enough to allow the necessary reduction. If staining mitotic figures, draw the stain from the spindle, but stop before the chromosomes begin to weaken. When the desired differentiation has been reached, wash out the acid in 50 per cent alcohol, if acid has been used. About 5 minutes should be sufficient.

If safranin is to be used alone, pass through 50, 70, 85, 95, and 100 per cent alcohol, through the xylol-alcohol, then through xylol to balsam. If clove oil is used, omit the xylol-alcohol, but follow the clove oil with xylol to hasten the hardening of the preparation.

If a second stain is to be added, transfer from the 50 per cent alcohol to any alcoholic stain. If the second stain is an aqueous stain, rinse the slide or sections for a minute in water before applying the stain.

Safranin is the most generally useful of all the red stains, and, fortunately, it is quite durable. Lignified, suberized, cutinized, and chitinized structures stain red, as do also the chromosomes, nucleoli, and centrosomes.

Acid Fuchsin.—Use a 1 per cent solution in water or in 70 per cent alcohol. The solution in alcohol is preferable if sections are to be mounted in balsam. This stain often acts with great rapidity, 2 or 3 minutes being sufficient. The method for using acid fuchsin with woody tissues is given in the chapter on "Freehand Sections" (chap. vi). In staining embryo sacs, pollen grains, and such structures, longer periods are better. Stain 1 or 2 hours, and then differentiate in a saturated solution of picric acid in 70 per cent alcohol. This may require 30 seconds, or even several minutes. Rinse in 70 per cent alcohol until a bright red replaces the yellowish color due to the acid, and then proceed as usual.

Congo Red.—This is an acid stain resembling acid fuchsin. For cytological work use a ½ per cent aqueous solution; for anatomical work use a saturated solution. It is a good stain to use after malachite green or anilin blue. Transfer to the Congo red from water, stain 15 minutes, wash in water, transfer—for wood sections—to 85 per cent alcohol, and wash until the green or blue color of the previous stain begins to show through the red. Then treat quickly with absolute alcohol, clear xylol, and mount in balsam.

Eosin.—This has long been a favorite stain, but for most purposes it has been replaced by similar stains giving better differentiation. The dry stain is made in two forms, one for aqueous and the other for alcoholic solution. Each should be used with its intended solvent. Make a 1 per cent solution in alcohol or water.

For material to be mounted whole in glycerin, glycerin jelly, or Venetian turpentine, stain several hours, or over night; pour off the stain, which may be used repeatedly; treat, without washing in water, with a 2 per cent aqueous solution of acetic acid for 5 to 10 minutes, changing the acid several times; transfer to 10 per cent glycerin, leaving about 1 c.c. of the acid to keep the whole solution slightly acid. When the glycerin becomes thick, mount in glycerin jelly. If the Venetian turpentine method is to be used, wash the glycerin out in alcohol slightly acidulated with acetic acid (a couple of drops of acetic acid to 50 c.c. of alcohol), and do not drain off the last alcohol too completely before transferring to the 10 per cent Venetian turpentine. According to Lee, the glycerin should be slightly alkaline. The alkalinity can be brought about by adding half a gram of common salt to 100 c.c. of the 10 per cent glycerin. We have found that eosin keeps better when the media are slightly acid.

For staining paraffin sections, the alcoholic solution is better and the time may not be more than a few minutes, especially if the eosin is being used as a contrast stain.

We have found the Eosin Y, of Coleman and Bell, very satisfactory, especially for fungi to be mounted whole. With the rapid improvement in the manufacture of stains, it is very probable that other dealers will have equally good products. Investigators will save time and money by keeping track of the findings of the commission on Standardization of Biological Stains.

Haematoxylin and eosin and methyl blue and eosin are good combinations. The eosin should follow the other stain.

Erythrosin.—This is really an eosin, but there is some difference in the method of manufacturing. It is more precise and a more transparent stain than eosin and is to be preferred for nearly all staining of paraffin sections. Make a 1 per cent solution in distilled water or in 70 per cent alcohol. It gives good results when made up according to the general formula.

Erythrosin stains rapidly, 30 seconds to 3 minutes being sufficient. When used in combination with other stains, erythrosin should come last.

Magdala Red.—At least two Magdala reds are sold by dealers, one the *echt* (genuine) Magdala red, and the other simply Magdala red. The latter is much cheaper and, in our experience, much superior to the *echt* stain. The directions apply to the cheaper stain.

For staining algae which are to be mounted in Venetian turpentine, use a 1 per cent solution in 85 or 95 per cent alcohol. Stain for 6 to 8, or even 24, hours. Rinse in 95 and 100 per cent alcohol for a few minutes. Transfer to 10 per cent Venetian turpentine and allow the turpentine to concentrate as described in chapter viii.

In staining sections to be mounted in balsam, the same stain may be used, but it is better to dilute it one-half with water. Stain for 6 to 24 hours, dehydrate in 95 per cent and absolute alcohol, clear in clove oil, and mount in balsam.

Magdala red stains lignified, suberized, and cutinized structures, and also chromosomes, centrosomes, nucleoli, and pyrenoids. It is likely to overstain, but the differentiation is easily secured by placing the finished mounts upon a white background in the direct sunlight. When the desired differentiation has been reached, it is better to avoid direct sunlight, although the mounts do not seem to fade in the ordinary light of a room.

Except for special purposes, it is better to use this stain in combination with blue, green, or violet.

Magdala red (not *echt*) has always been erratic in its behavior, and recent analyses of samples show that it is variable in composition. A stain called *phloxine,* made by the National Anilin and Chemical Co., behaves like the best Magdala red and seems to give uniform results.

Gentian-Violet.—This is one of the most important stains in the botanical laboratory. It may be made according to the general formula for anilin stains, but that solution does not keep well. A 1 per cent solution in distilled water keeps indefinitely and seems to be as good as, if not better than, the anilin solution. Gentian-violet dissolves readily in clove oil and some prefer to use it this way, rather than in water or alcohol.

With the aqueous or anilin-oil solutions, the following directions will enable the student to become acquainted with the behavior of the stain. Transfer to the stain from water and allow the stain to act for 1 to 30 minutes. The time depends upon the fixing and upon the structures to be stained. The brilliancy of the stain in achromatic structures may often be increased by leaving the slide from 2 to 5 minutes in a 1 per cent aqueous solution of permanganate of potassium before applying the stain. The greatest objection to the aqueous and anilin-oil solutions of gentian-violet is that the stain washes out so rapidly in alcohols that it is impossible to run the slide up through the series. The usual practice is to dip the slide in water to remove most of the stain and thus avoid carrying it into the alcohol: then transfer directly from water to 95 per cent alcohol, allowing the alcohol to act for only 2 or 3 seconds, then allow the absolute alcohol to act for 5 or 6 seconds, and then, while the stain is still coming out in streams, begin the treatment with clove oil. Holding the slide in one hand, pour on a few drops of clove oil, and immediately drain off in such a way as to carry off the alcohol. This clove oil should not be used again. Then flood the slide repeatedly with clove oil, pouring the clove oil back into the bottle. A 50-c.c. bottle of clove oil is large enough. About 100 mounts can be cleared with 50 c.c. of this oil. The clove oil is a solvent of gentian-violet, but it dissolves the stain from some structures more rapidly than from others; e.g., the stain may be completely removed from the chromosomes while it is still bright in the achromatic structures. As soon as the stain is just right, drain off the clove oil and leave the slide in xylol for a minute or two before mounting in balsam. This is a necessary step, because the continued action of clove oil would cause the preparation to fade. As may be inferred from what has preceded, alcohol would soon extract the stain, without any application of clove oil. The clove oil is used, not only because it extracts the stain more slowly, but because it dissolves the stain from some structures more rapidly than from others; e.g., the stain may be completely removed from the chromosomes while it is still bright in the achromatic structures, so that with safranin and gentian-violet one can get red chromosomes on a violet spindle.

Some still use cedar oil to follow the clove oil. This stops the action of the clove oil, but the preparations harden slowly.

Gentian-violet is an excellent stain for achromatic structures in all stages of development. Chromatin, in many of its stages, is also stained. In metaphase and anaphase one should be able to get red chromosomes and violet spindles with safranin and gentian-violet. If the chromosomes also persist in retaining the violet, shorten the stain in gentian-violet. Cilia stain well; starch grains stain deeply, chromatophores less deeply, and lignified walls may not stain at all. One should be able to get red lignified walls and violet cellulose walls with safranin and gentian-violet.

Cyanin.—This stain is also called Quinoleïn blue and Chinolin blue. Dissolve 1 g. of cyanin in 100 c.c. of 95 per cent alcohol and add 100 c.c. of water. The cyanin would not dissolve in 50 per cent alcohol. We have not found Grübler's cyanin at all satisfactory with the foregoing formula. With the general formula the Grübler's cyanin will not dissolve. We use a cyanin prepared by H. A. Metz & Co., 122 Hudson Street, New York. This cyanin dissolves completely when made up according to the general formula. It stains rapidly, 5 to 10 minutes usually being sufficient. Chromosomes take a deep blue, but the spindle is only slightly affected. Lignified structures stain blue, while cellulose walls are scarcely affected and the stain is easily washed out.

Iodine Green.—Use a 1 per cent solution in 70 per cent alcohol. Stain for an hour, rinse in 70 per cent alcohol, dehydrate in 95 per cent alcohol and absolute alcohol, clear in xylol or clove oil, and mount in balsam. If the stain washes out too rapidly and does not give sufficient differentiation, stain longer, over night or even 24 hours.

Lignified structures stain green, but, after proper washing, cellulose is scarcely affected. A bright green may be left in the chromosomes after all the stain has been washed out from the spindle.

Acid fuchsin, erythrosin, and eosin are good contrast stains for mitotic figures. Acid fuchsin or Delafield's haematoxylin are good for cellulose walls.

Light Green (*Licht Grün*).—Light green is an acid stain, soluble in water, alcohol, or clove oil. It stains quickly and forms a sharp contrast with safranin or Magdala red.

Stain in safranin and then, with little or no washing out, stain in a weak alcoholic solution of acid green (about 0.2 g. in 100 c.c. of 95 per cent alcohol). From 20 seconds to about 1 minute may be sufficient. The green rapidly reduces the safranin, and consequently the staining must not be too prolonged. A successful preparation should show red chromosomes and green spindle. Lignified walls should be red and cellulose walls green.

Malachite Green.—A 1 to 3 per cent aqueous solution is good for cellulose walls. The stain contrasts well with Congo red.

Methyl Green.—A 1 per cent solution in water is good for staining lignified structures. Lee recommends that the solution be acidulated with acetic acid. This is not necessary for staining lignified membranes nor for staining chromosomes. Methyl green has long been a favorite stain for living tissues. It is more easily controlled than iodine green, especially in double staining to differentiate lignified and cellulose walls.

Acid Green.—Make a solution according to the general formula, or simply make a 1 per cent solution in water. This stains cellulose walls and achromatic structures, but scarcely affects lignified walls or chromosomes.

Anilin Blue.—Strong alcoholic solutions are best for botanical work. Even though the dry stain may be intended for aqueous solution, make a 1 per cent solution in 85 or 95 per cent alcohol.

This stain can be recommended for cellulose walls, achromatic structures of mitotic figures, for cilia, and it is particularly valuable for algae. Directions for using it with algae are given in chapter viii.

Orange G.—Make a 1 per cent solution in water, in 95 per cent alcohol, or in clove oil. We prefer the solution in clove oil.

The orange dissolves very slowly if put directly into clove oil. It is better to dissolve 3 or 4 grams of orange in 50 c.c. of absolute alcohol. In a well-corked bottle in the paraffin oven at 52° C., this much orange may go into solution. Then remove the cork and allow about half of the alcohol to evaporate. Pour on 200 c.c. or more of clove oil and let it stand for several hours. If any of the stain has not gone into solution, pour off the clear fluid, which is now ready for use, and pour some more clove oil on the residue, allowing the residue to go slowly into solution. In staining, we use a small bottle of the orange, pouring it on the slide and draining it back into the bottle. The absolute alcohol, carried into the clove oil in this way, does no damage, except that it dilutes the stain a little.

Transfer to the aqueous stain from water; to the alcoholic stain from 85 per cent alcohol, since the stain is always applied as a second or third stain; use the solution in clove oil after the dehydration in absolute alcohol. Times are always short and are to be reckoned in seconds rather than in minutes. If the solution in clove oil has been used, the slide should be transferred to xylol before mounting in balsam.

This is a plasma stain. It is distinctly a general rather than a selective stain, but is valuable as a background for other structures which have been stained violet or blue or green. It first came into prominence as the third member of the triple stain, safranin, gentian-violet, orange.

Gold Orange.—This stain, which many incorrectly suppose to be the same as orange G, is much more readily soluble in clove oil and stains with much greater rapidity.

Bismarck Brown.—Use a 2 per cent solution in 70 per cent alcohol.

This is a good stain for cellulose walls, although it is not so precise as haematoxylin. Embryo sacs stained in one of the carmines are improved by 1 or 2 minutes' staining in Bismarck brown. Material fixed in alcohol stains better than that which has been fixed in reagents containing chromic acid. A faint background of Bismarck brown is quite effective in staining sections containing bacteria.

Nigrosin.—Make a 1 or 2 per cent solution in water. A few drops of this solution to a watch glass full of water stains filamentous algae or fungi in 1 to 3 hours. It keeps well in glycerin or Venetian turpentine. It also keeps well in balsam, but it is of little value in staining microtome sections.

COMBINATION STAINS

Sometimes preparations are stained with a single stain, selected to emphasize some particular feature, but in the great majority of cases two or more stains are used. In staining a vascular bundle, one stain may be selected which stains the xylem, but not the phloem, while another of a different color stains the phloem, but not the xylem, thus affording a sharp contrast. In staining mitotic figures, one stain may stain the chromosomes, while another of a different color may be used to stain the spindle.

Success in double staining can be obtained only by noting the effect of each stain upon the various plant structures.

Flemming's Safranin, Gentian-Violet, Orange.—Safranin has long been a famous stain for mitosis. This triple combination was published in 1891, but its value in plant cytology was not thoroughly appreciated until five or six years later, when its application was developed to a high degree of perfection by various investigators of the Bonn (Germany) school. Three methods, which may be designated as A, B, and C, will be described.

A. According to Flemming,[1] stain 2 or 3 days in safranin (dissolve 0.5 g. safranin in 50 c.c. absolute alcohol, and after 4 days add 10 c.c. distilled water); rinse quickly in water; stain 1 to 3 hours in a 2 per cent aqueous solution of gentian-violet; wash quickly in water, and then stain 1 to 3 minutes in a 1 per cent aqueous solution of orange G. Transfer from the stain to absolute alcohol, clear in clove oil, and mount in balsam.

[1] Flemming's original method is included only as interesting history.

B. The following formulas and method seem to be better for mitotic phenomena in plants: Make a 1 per cent solution of alcoholic safranin in absolute or 95 per cent alcohol, and after the safranin is completely dissolved, add an equal volume of a 1 per cent solution of aqueous safranin in water, thus making a 1 per cent solution of safranin in 50 per cent alcohol. Use a 1 per cent aqueous solution of gentian-violet, and a 1 per cent aqueous solution of orange G.

Transfer paraffin sections to the stain from 95 per cent alcohol after the xylol or turpentine used in dissolving away the paraffin has been rinsed off. Stain 3 to 24 hours. If the period be too short, the washing out is so rapid that it is difficult to stop the differentiation at the proper point and, besides, the red is likely to be less brilliant. Rinse in 50 per cent alcohol until the stain is properly differentiated. Leave the slide in the 50 per cent alcohol until the stain is washed out from the spindle and cytoplasm, but stop the washing out before the chromosomes begin to lose their bright red color. If the washing out takes place too slowly, treat with slightly acidulated alcohol (one drop of HCl to 50 c.c. of 50 per cent alcohol) for a few seconds. The acid must be removed by washing for 15 to 30 seconds in alcohol which has not been acidulated.

Then dip the slide 5 or 6 times into water and stain in gentian-violet. The time required is so variable that definite instructions are impossible. The gentian-violet should stain the spindle, but not the chromosomes. If the stain be too prolonged, it may be impossible to get it out from the chromosomes and still leave it bright in the spindle. If the period be too short, the stain will wash out from the spindle. For mitotic figures in the germinating spores of the liverwort, *Pellia,* 30 minutes is not too long. In this case, the stain washes out easily from the chromosomes without the use of acid, and the spindle takes a rich violet which is not easily washed out. In embryo sacs of *Lilium* try 10 minutes. In pollen mother-cells try 5 to 10 minutes. For root-tips try 2 to 10 minutes. Chromatin in the early prophases and in telophases will stain with the violet, and the violet will not wash out, but in phases in which fully formed chromosomes are visible the violet can be washed out if the period has not been too long.

Remove the slide from the gentian-violet and dip it 5 or 6 times in water and then stain 30 seconds to 1 minute in orange G. The orange stains cytoplasm and at the same time washes out gentian-violet.

Transfer from the orange G to 95 per cent alcohol, dipping the slide a few times in this merely to save the absolute alcohol. Dehydrate in absolute alcohol 3 to 30 seconds.

Clear in clove oil, as already described in the paragraph on gentian-violet. Transfer to xylol and mount in balsam.

Safranin and gentian-violet are often used without the orange. In this case, transfer from the gentian-violet directly to 95 per cent alcohol, and proceed as before.

An objection to both these methods is that the gradual series of alcohols cannot be used, because the gentian-violet washes out so rapidly. If one should try a filament of *Spirogyra* with either of these methods, it would hardly be recognizable when it reached the balsam; but with thin sections, especially when well fastened to the slide, conditions are different and there does not seem to be any serious damage.

Just now, we are using a third method, which seems better than either of the two just described.

C. Use the safranin solution described in B, an aqueous solution of gentian-violet, and a solution of orange G or gold orange in clove oil.

Transfer to the safranin from 50 per cent alcohol and stain as directed under B. Rinse in 50 per cent alcohol for a minute, then in water for a minute, and stain in gentian-violet 3 to 30 minutes. Rinse quickly in water, dehydrate for about 5 seconds in 95 per cent alcohol and for about 10 seconds in absolute alcohol; then pour on the slide the orange dissolved in clove oil and allow it to act for 10 to 30 seconds. Drain off the clove-oil orange, which can be used again, pour on pure clove oil and watch until the gentian is satisfactory, then transfer to xylol and balsam.

With freehand sections, which are likely to be much thicker, the process is the same but the times will be longer.

Safranin and Light Green.—Stain in aqueous safranin; wash in water; pass through 50, 95, and 100 per cent alcohol, about 2 minutes in each grade; stain in light green dissolved in clove oil; rinse in xylol and mount in balsam.

If the stain does not rinse off readily, transfer to absolute alcohol, then to equal parts xylol and absolute alcohol, clear in xylol and mount in balsam.

Cyanin and Erythrosin.—Both solutions may be made according to the general formula for anilins, or 1 per cent aqueous solutions may be used. Miss Thomas recommends 1 gram dissolved in 100 c.c. of 95 per cent alcohol. After the solution is complete, add 100 c.c. of distilled water. Since the combination is used only for paraffin sections or for small organisms dried down on the slide, her formula is preferable. Transfer to the alcoholic cyanin from 50 per cent alcohol, stain 5 to 10 minutes or longer; rinse quickly in 50 per cent alcohol, transfer to erythrosin and stain 30 seconds to one minute. Rinse quickly in 50 per cent alcohol, then in 95 per cent and absolute alcohol. Clear in xylol and mount in balsam.

If aqueous stains are used, transfer to the cyanin from water, rinse in water, stain in erythrosin, rinse in water, and transfer directly to 95 per cent alcohol. If the cyanin washes out, stain for 1 hour, and if it still washes out, omit the rinsing and transfer directly from the cyanin to the erythrosin.

The erythrosin may be used first; in this case stain for 5 minutes in erythrosin, transfer directly to cyanin, and stain for about 10 seconds. Dehydrate in 95 per cent and in absolute alcohol, clear in xylol or in clove oil, and mount in balsam.

The stains wash out so rapidly that the series of alcohols cannot be used.

Chromosomes and nucleoli stain blue and achromatic structures red. Lignified structures stain blue and cellulose walls red. The various cell constituents are often sharply differentiated. It was this combination which suggested the now obsolete terms, "cyanophilous" and "erythrophilous."

Magdala Red and Anilin Blue.—Make both solutions as directed in chapter viii on "The Venetian Turpentine Method."

For paraffin sections, stain 3 to 24 hours in Magdala red, dip in 95 per cent alcohol to rinse off the stain, and then stain 2 to 10 minutes in the anilin blue. Dip in 95 per cent alcohol to rinse off the stain, and treat for a few seconds with alcohol slightly acidulated with hydrochloric acid (one drop to 50 c.c. of 95 per cent alcohol). In the acid alcohol the blue will become more intense, but the red would soon be extracted. Wash in 95 per cent alcohol to remove the acid. If the acid has weakened the Magdala red, put a pinch of sodium carbonate into the 95 per cent alcohol. The red may brighten. If the red is too weak, return to the Magdala red and try again. From the 95 per cent alcohol, transfer to absolute alcohol, to xylol, and then mount in balsam.

Acid Fuchsin and Iodine Green Mixtures.—Two solutions are kept separate, since they do not retain their efficiency long after they are mixed:

A	Fuchsin acid	0.1 g.
	Distilled water	50.0 c.c.
B	Iodine green	0.1 g.
	Distilled water	50.0 c.c.
C	Absolute alcohol	100.0 c.c.
	Glacial acetic acid	1.0 c.c.
	Iodine	0.1 g.

Mix equal parts of A and B. Transfer to the stain from water. The proper time must be determined by experiment. For a trial, 24 hours might be recommended. Transfer from the stain directly to solution C and from C to xylol.

A. Acid fuchsin	0.5 g.
B. Iodine green	0.5 g.

Mix a pipette full of A with a pipette full of B; stain 2 to 8 minutes; transfer to 85 per cent or 95 per cent alcohol, dehydrate rapidly, clear in xylol, and mount in balsam. Both these formulas are good for mitosis.

Acid Fuchsin and Methyl Green.—Both may be used in 1 per cent aqueous solutions.

For mitotic figures, stain in green for about an hour, wash in water or alcohol until the green is extracted from the spindle, and then stain for about 1 minute in the fuchsin. Dehydrate in 95 and 100 per cent alcohol, clear in xylol or clove oil, and mount in balsam. If the green washes out, stain longer; if it is not readily extracted from the spindle, shorten the period. If the fuchsin stains the chromosomes, shorten the period, and lengthen it if the fuchsin washes out from the spindle. The chromosomes should take a brilliant green and the spindle a bright red.

Delafield's Haematoxylin and Erythrosin.—Stain first in the haematoxylin, and after that stain is satisfactory, stain for 30 seconds or 1 minute in erythrosin. This is a good combination, and, for most plant structures, gives a far better differentiation than the traditional haematoxylin and eosin, since the erythrosin has all the advantages of the eosin and is more transparent. Orange G is also a good stain to use with Delafield's haematoxylin.

Directions for staining in safranin and Delafield's haematoxylin are given in the chapter on "Freehand Sections" (chap. vi).

Haidenhain's Iron-Haematoxylin and Orange G.—This haematoxylin is very satisfactory when used alone. A light staining in orange G, however, sometimes improves the mount. After the last washing in water, stain for about 30 seconds in orange G, or, if the orange is in clove oil, stain after dehydrating in absolute alcohol. Eosin, erythrosin, and nearly all plasma stains fail to increase the effect of a good stain in iron-haematoxylin.

Combinations might be described almost without limit. Several more will be suggested in connection with the various groups of plants in Part II.

We have not attempted to make the list of stains complete. It is better to master a few stains than to use many stains indifferently. A successful photographer once advised an amateur to stick to one brand of plate and one formula for developer. His hint might well have a wider application. If one really masters two or three good combinations, he is well prepared to develop methods for meeting special needs.

CHAPTER IV

GENERAL REMARKS ON STAINING

Many things may be examined alive without killing, fixing, staining, or any of those processes. A filament of *Spirogyra* shows the chromatophore nicely if merely mounted in a drop of water; the nucleus may be visible and the pyrenoids can usually be located. Of course, such a study is necessary if one is to understand anything about the plant, and in an elementary class this might be sufficient; but a drop of iodine solution applied to the edge of the cover would emphasize certain details, e.g., the starch would appear blue, the nucleus a light brown, and the cytoplasm a lighter brown. This illustrates at least one advantage to be gained by staining; it enables us to see structures which would otherwise be invisible, or almost invisible. Much of the recent progress in morphology and cytology has been due to the development of critical methods of staining. Some of the combinations and methods recommended by various workers are good in themselves, while others, not so good, have yielded results because they have been so skilfully used.

SELECTION OF A STAIN

With so many stains at our disposal, it at once becomes a problem just which stain or combination to use in each particular case. Beautiful and instructive preparations occasionally result from some happy chance, but uniform success demands skill and judgment in manipulation, and also a knowledge of the structures which are to be differentiated. Let us take a vascular bundle for illustration. Safranin stains the xylem a bright red, but, with judicious washing, is entirely removed from the cambium and cellulose elements of the phloem. A careful staining with Delafield's haematoxylin now gives a rich purple color to the cellulose elements which were left unstained by the safranin, thus contrasting sharply with the lignified elements. If cyanin and erythrosin be used, the xylem takes the blue while the cambium and phloem take the red.

The mere selection of two colors which contrast well is not sufficient. Green and red contrast well, but safranin and iodine green would be a poor combination, for both would stain chromosomes and neither would stain the spindle; both would stain lignified structures and neither would give satisfactory results with cellulose walls. Both stains are basic. Acid green would have given a contrast in both these cases, because it stains achromatic structures and cellulose walls. In general, an acid stain should be combined with a basic one, but there are so many exceptions that it is hardly worth while to learn a list of basic and acid stains. Stains which stain chromosomes are likely to be basic, and those which do not stain chromosomes are likely to be acid or neutral. If it were true that acid stains affect only basic structures, and basic stains affect only acid structures, a classification of stains would be of great value. Safranin and gentian-violet are both basic, but with proper washing out the chromosomes are red and the spindle is violet, the safranin being washed out from the spindle, while the gentian-violet is washed out from the chromosomes. The only way to insure success is to become familiar with the action of each stain upon the various structures.

THEORIES OF STAINING

In 1890 Auerbach, a zoölogist, published the results of his studies upon spermatozoa and ova. He found that, if preparations containing both spermatozoa and ova were stained with cyanin and erythrosin, the nuclei of the spermatozoa took the cyanin, while the nuclei of ova preferred the erythrosin; hence he proposed the terms "cyanophilous" and "erythrophilous." Auerbach regarded these differences as an indication of sexual differences in the cells.

Rosen (1892) supported this theory, and even went so far as to regard the tube nucleus of the pollen grain as female, on account of its erythrophilous staining. In connection with this theory it was suggested that the ordinary vegetative nuclei are hermaphrodite, and that in the formation of a female germ nucleus the male elements are extruded, leaving only the erythrophilous female elements; and, similarly, in the formation of a male nucleus the female elements are extruded, leaving only the cyanophilous male elements.

As long ago as 1884 Strasburger discovered that with a mixture of fuchsin and iodine green the generative nucleus of a pollen grain stains green, while the tube nucleus stains red. In 1892, in his *Verhalten des Pollens,* he discussed quite thoroughly the staining reactions of the nuclei. The nuclei of the small prothallial cells of gymnosperm microspores are cyanophilous like the male generative nuclei. The nuclei of a nucellus surrounding an embryo sac are also cyanophilous, while the nuclei of structures within the sac are erythrophilous. His conclusion is that the cyanophilous condition in both cases is due to poor nutrition, while the erythrophilous condition is due to abundant nutrition. A further fact in support of the theory is that the nuclei of the adventitious embryos which come from the nucellus of *Funkia ovata* are decidedly erythrophilous, while the nuclei of the nucellus to which they owe their food-supply are cyanophilous.

In division stages nuclei are cyanophilous, but from anaphase to resting stage the cyanophilous condition becomes less and less pronounced, and may even gradually change to the erythrophilous.

An additional fact in favor of this theory is that in *Ephedra* the tube nucleus, which has very little cytoplasm about it, is cyanophilous. Strasburger claimed that there is no essential difference between male and female generative nuclei, and subsequent observation soon showed that within the oöspore the sex nuclei rapidly become alike in their reaction to stains.

Malfatti (1891) and Lilienfeld (1892–1893) claim that these reactions are dependent upon the amount of nucleic acid present in the structures. During mitosis the chromosomes consist of nearly pure nucleic acid and are intensely cyanophilous, but the protoplasm, which has little or no nucleic acid, is erythrophilous. There is a gradual transition from the cyanophilous condition to the erythrophilous, and vice versa, the acid structures taking basic stains and basic structures the acid stains.

The terms "erythrophilous" and "cyanophilous" soon became obsolete, and many claimed the affinity is for basic and acid dyes, rather than for blue or red colors. That the terms were misnomers became evident when a combination like safranin (basic) and acid green (acid) was used, for the cyanophilous structures stained red, and the erythrophilous green.

According to Fischer (1897 and 1900), stains indicate physical but not chemical composition. Fischer experimented with substances of known chemical composition. Egg albumin was shaken until small granules were secured. These were fixed with the usual fixing agents, and then stained with Delafield's haematoxylin. The extremely small granules stained red, while the larger ones became purple. Since the granules are all alike in chemical composition, Fischer concluded that the difference in staining must be due to physical differences. With safranin, followed by gentian-violet, the larger granules stain red and the smaller violet; if, however, the gentian-violet be used first, then treated with acid alcohol and followed by safranin, the larger granules take the gentian-violet and the smaller the safranin. In root-tips similar results were obtained. Safranin followed by gentian-violet stained chromosomes red and spindle fibers violet, while gentian-violet followed by safranin stained the chromosomes violet and the spindle red. One often reads that chromosomes owe their strong staining capacity to nuclein, and especially to the phosphorous, but, according to Fischer, this is shown to be unfounded, since albumin gives similar results, yet contains no phosphorous, and is not chemically allied to nuclein.

Probably the most important reason which led Fischer to undertake this series of experiments was the claim that certain granules of the Cyanophyceae should be identified as chromatin because they behaved like chromatin when stained with haematoxylin. Fischer's experiments not only proved that chromatin cannot be identified in this way but raised the question whether staining reactions ever indicate chemical composition. At present, it would seem that, in most cases, the staining indicates only physical differences. However, in some cases there is a chemical reaction, e.g., when material fixed in bichloride of mercury is stained in carmine, mercuric carminate is formed.

It would be very convenient if we knew just how much dependence should be placed upon staining reactions as a means of analysis. If two structures stain alike with Delafield's haematoxylin, does this mean that they have the same chemical composition; or if, on the other hand, they stain differently, must they necessarily be different in their chemical composition? Delafield's haematoxylin, when carefully used, gives a rich purple color, but a careful examination will often show that in the same preparation some structures stain purple, while others stain red. Does this mean that the purple and red structures must have a different chemical composition? Many people believe that structures which stain differently with a given stain must be chemically different, but they readily agree that structures which stain alike are not necessarily similar in chemical composition. Chromosomes of dividing nuclei and lignified cell walls stain alike with safranin; chromosomes and cellulose cell walls stain much alike with Delafield's haematoxylin; but everyone recognizes that the chromosome is very different in its chemical composition from either the cellulose or the lignified wall.

However, in an indirect and somewhat uncertain way, one can infer the nature of certain structures from the staining. For instance, if sections of various objects have been stained with safranin, we may draw the following inferences with more or less confidence: if cells in the xylem region of a vascular bundle stain red, their walls are lignified; if cortical cells, which may appear quite similar in transverse section, stain red, they are likely to be suberized; if the outer walls of epidermal cells stain red, they are cutinized; but if the outer boundary of the embryo sac of a gymnosperm stains red, it is chitinized. Of course, these inferences can be made only because the various structures have been tested by more accurate methods.

Whatever doubt or uncertainty there may be in regard to theories of staining or in regard to the value of stains as a means of analysis, there is no doubt that stains are of the highest importance in differentiating structures, and in bringing out details which would otherwise be invisible.

PRACTICAL HINTS ON STAINING

The number of stains in the catalogues is becoming so great that it is impossible to become proficient in the use of all of them. As we have already intimated, it is better to master a few of the most valuable stains than to do indifferent work with many. An experienced technician knows that it is impossible to judge from a few trials whether a given stain or combination is really valuable or not. As a matter of fact, some of the most valuable combinations, like Haidenhain's iron-alum haematoxylin and Flemming's safranin, gentian-violet, orange, require patient study and long practice before they yield the magnificent preparations of the trained cytologist. The beginner, especially if somewhat unacquainted with the details of plant structure, may believe that he has an excellent preparation when it is really a bad, or at most an indifferent, one. To illustrate, let us suppose that sections of the pollen grain of a lily have been stained in safranin and gentian-violet. If the preparation merely shows a couple of dense nuclei and a mass of uniform cell contents surrounded by a heavy wall, the mount is poor. If the two nuclei are quite different and starch grains are well differentiated in the tube cells and the wall shows a violet intine contrasting sharply with a red exine, the mount is good. Anything intermediate is indifferent. If mitotic figures have been stained with cyanin and erythrosin, a first-class preparation should show blue chromosomes and red spindles; if stained with safranin and gentian-violet, the chromosomes should be red and the spindles violet.

In staining growing points, apical cells, young embryos, antheridia, archegonia, and many such things, the cell walls are the principal things to be differentiated, if the preparations are for morphological study. As a rule, it is better in such cases not to use double staining, but to select a stain which stains the cell walls deeply without obscuring them by staining starch, chlorophyll, and other cell contents. For example, try the growing point of *Equisetum.* The protoplasm of such growing points is very dense. If Delafield's haematoxylin and erythrosin be used, the haematoxylin will stain the walls and nuclei, and will slightly affect the other cell contents, but the erythrosin will give the cytoplasm such a dense stain that the cell walls will be seriously obscured. It would be better to use haematoxylin alone. For counting chromosomes, it is better to stain in iron-alum haematoxylin alone, or in safranin alone. The same suggestion may well be observed in tracing the development of antheridia, archegonia, embryos, and similar structures.

In using combinations, it must be remembered that the second stain often affects the first, e.g., if safranin is to be followed by Delafield's haematoxylin in staining a vascular bundle, it will not do to make the safranin just right and then apply the haematoxylin, for the acid which must be used to differentiate the haematoxylin and to avoid precipitates will also reduce the safranin, and the red will be too weak. You must overstain in safranin so that the reduction will finally leave it just right. The same hint will apply if safranin is to be followed by anilin blue, since here, also, acid must be used; if light green is to follow the safranin, the stain itself is so acid that the safranin must be rather strong before the light green is applied. Orange, whether in water or in clove oil, reduces many stains and, consequently, such stains must be strong enough to allow the weakening. These hints are only samples: the student must observe the behavior of the various stains when used singly and when used in various combinations.

The best advice that can be given, not only to the beginner but to the investigator is this: Master a few stains. For nuclear phenomena, strive for perfection in the use of Haidenhain's iron-alum haematoxylin; then, if you can afford the time, practice the safranin, gentian-violet, orange combination. For vascular anatomy, learn to stain xylem with safranin; and, for a contrast, stain cellulose walls with Delafield's haematoxylin, gentian-violet, light green or anilin blue. For filamentous algae and fungi, master the iron-alum haematoxylin method; and then try the Magdala-red and anilin-blue combination. Do not pass judgment against a standard method or even a new method just because you fail to get results at the first trial. After you have become proficient with the iron-alum haematoxylin for mitotic figures, you are sure to fail if you try the same procedure with *Rhizopus;* but, nevertheless, the stain is just as good for *Rhizopus* as for figures in pollen mother-cells or root-tips.

Permanent preparations are an absolute necessity for the greater part of most advanced work, but let us not imagine that we cannot examine anything until we have made a permanent mount. It would be impossible to make a permanent mount of the rotation of protoplasm. It is better for many purposes to look at motile spores while they are moving. Use *Spirogyra* while it is fresh and green, and use permanent preparations only to bring out nuclei and other details which are not so easily seen in living material. Examples might be multiplied.

CHAPTER V

TEMPORARY MOUNTS AND MICROCHEMICAL TESTS

Skill in making freehand sections, without any microtome, and in teasing with needles and in making delicate dissections under the simple microscope are absolutely necessary in any investigation dealing with the structure and development of plants. Preliminary study with the aid of such methods not only gives a broader view of structures in all dimensions and helps the interpretation of stained microtome preparations, but is necessary in determining whether material is worth all the labor of making permanent mounts. That particular class of temporary mounts intended only for chemical tests is considered separately in the second part of this chapter.

TEMPORARY MOUNTS

A preliminary examination of almost any botanical material may be made without any fixing, imbedding, or staining. If a little starch be scraped from a potato, and a small drop of water and a cover-glass be added, a very good view will be obtained, and if a small drop of iodine solution be allowed to run under the cover, the preparation, while it lasts, is better than some permanent mounts. The unicellular and filamentous algae can be studied quite satisfactorily from such mounts. The protonema of mosses and the prothallia of ferns should be studied in this way, even if a later study from sections is intended. The addition of a little iodine identifies the starch and makes the nucleus more plainly visible. If the top of a moss capsule be cut off at the level of the annulus, a beautiful view of the peristome may be obtained by simply mounting in a drop of water, or, in a case like this where no collapse is to be anticipated, the object may be mounted in a small drop of glycerin—just enough to come to the edge of the cover without oozing out beyond—and the preparation may be made permanent by sealing with balsam, gold size, or any good cement. The antheridia and archegonia of mosses may be examined if the surrounding leaves are carefully teased away with needles. Freehand sectioning with a sharp razor and judicious teasing with a pair of needles will give a fair insight into the anatomy of the higher plants without demanding any further knowledge of technic. This rough work is a very desirable antecedent to the study of microtome sections, because most students see in a series of microtome sections *only* a series of sections when, in the mind's eye, they ought to see the object building itself up in length, breadth, and thickness as they pass from one section to another.

The movements of protoplasm can, of course, be studied only in the living material. Every laboratory should keep *Chara* growing at every season of the year. Mount a small portion and note the movements in the internodal cells. Avoid any pressure and any lowering of the temperature. A gentle raising of the temperature will accelerate the movements. A leaf of *Elodea* shows the movements very clearly, especially in the midrib region. The stamen hairs of *Tradescantia* have long been used, their color, resembling a faint haematoxylin stain, making them particularly favorable. Stinging hairs show a brisk movement if they are mounted quickly and without injury. Fortunately, the common onion always furnishes favorable material for demonstrating the movements of protoplasm. Strip the epidermis from one of the inner scales of the bulb and mount in water. The granules may appear to better advantage in yellow light, like that of an ordinary kerosene lamp.

The discharge of spores and gametes should be observed in the living material: the difference in the behavior of spores and gametes is very striking and can be appreciated only while they are alive. Most aspects of growth and movement can be studied best in the living condition. In short, it is well to make a preliminary study of everything. The germination of spores and the growth of pollen tubes can be studied in the hanging drop. For facilitating such cultures there are many devices, such as hollow-ground slides, glass rings, rubber rings, etc. (Fig. 17). A device which is better for most purposes, and which is easily made by any student, is shown in Figure 18.

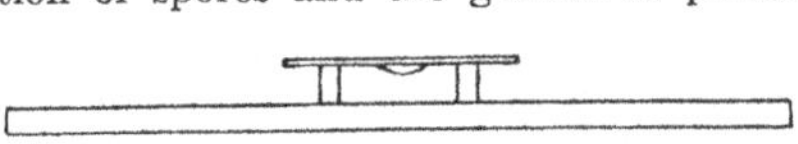

FIG. 17.—The hanging-drop culture

A square or round hole $\frac{5}{8}$ inch in diameter is cut in a piece of pasteboard $\frac{1}{8}$ inch thick, 1 inch wide, and $1\frac{1}{2}$ inches long. The pasteboard is then boiled to sterilize it and to make it fit more closely to the slide. While the pasteboard is still wet, press it to the slide, make the culture in a drop of water or culture solution on the cover, and invert the cover over the hole. A little water added at the edge of the pasteboard from time to time will keep it from warping and will at the same time provide a constant moist chamber.

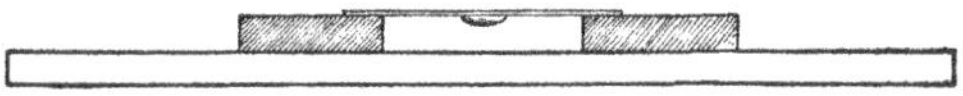

FIG. 18.—Another hanging-drop culture

In collecting material for mitotic figures in anthers it is necessary to examine fresh anthers, if one wishes to avoid a tedious and uncertain search after the anthers have been imbedded. By teasing out a few cells from the apex and a few from the base of the anther the stage of development is readily determined, and anthers which do not show the desired stages can be rejected. By allowing a drop of eosin or methyl green to run under the cover the figures are more easily detected. The actual progress of mitosis has been observed in living stamen hairs of *Tradescantia.*

MICROCHEMICAL TESTS

Botanical microchemistry has developed to such an extent that it has become an independent subject, like bacteriology. We shall consider only the commonest tests which are needed constantly by students of morphology. For a thorough presentation of the chemistry of the cell, we are looking forward with great anticipation to a forthcoming book by Dr. Sophia Eckerson, whose critical tests and analyses we have observed for many years. In the meantime, *Pflanzenmikrochemie,* by Dr. O. Tunmann (Gebrüder Bornträeger, Berlin), is recommended to those who read German. Zimmerman's *Botanical Microtechnique* (Henry Holt & Co., New York) is still recommended to those who must rely upon English. We shall give only a few tests, but in considering the various stains we shall indicate the effect of each stain upon the various plant structures.

Starch.—Mount the starch or starch-containing structures in water, and allow a drop of iodine solution to run under the cover. Starch assumes a characteristic blue color. The solution may be prepared by dissolving 1 g. of potassium iodide in 100 c.c. of water and adding 0.3 g. of sublimed iodine. A strong solution of iodine in alcohol (about 1 g. in 50 c.c. of absolute alcohol) keeps well. A drop of this solution added to 1 c.c. of water is good for testing. With too strong a solution, the starch first turns blue but rapidly becomes black.

Grape-Sugar.—In cells containing grape-sugar, bright-red granules of cuprous oxide are precipitated by Fehling's solution. It is better to keep the three ingredients in separate bottles, because the solution does not keep long after they are mixed. The solutions may be labeled A, B, and C.

A	Cupric sulphate	3 g.
	Water	100 c.c.
B	Sodium potassium tartrate (Rochelle salt)	16 g.
	Water	100 c.c.
C	Caustic soda	12 g.
	Water	100 c.c.

When needed for use, add to 10 c.c. of water 5 c.c. from each of the three solutions. The sections, which should be two or three cells in thickness, are warmed in the solution until little bubbles are formed. Too much heat must be avoided. Mount and examine in a few drops of the solution. The twig or organ may be treated with the solution, and the sections may be cut afterward. Other substances precipitate copper, and may be mistaken for grape-sugar by the beginner.

Cane-Sugar.—Cuprous oxide is not precipitated from Fehling's solution by cane-sugar, but after continued boiling in this solution the cane-sugar is changed to invert-sugar and the copper is precipitated. The solution becomes blue.

Proteids.—The proteids turn yellow or brown with the iodine solution. It is better to use a stronger solution than when testing for starch. It must be remembered that many other substances also turn brown when treated with iodine.

When proteids are warmed gently in concentrated nitric acid, the acid becomes yellow. The color may be deepened by the addition of a little ammonia or caustic potash.

When proteids are heated with Millon's reagent, the solution becomes brick-red or rose-red. This reaction takes place slowly even in the cold. The following is one formula for this reagent:

Mercury	1 c.c.
Concentrated nitric acid	9 c.c.
Water	10 c.c.

Dissolve the mercury in the nitric acid and add the water.

Fats and Oils.—The fatty oils are not soluble in water and are only slightly soluble in ordinary alcohol. They dissolve readily in chloroform, ether, carbon disulphide, or methyl alcohol.

Alcannin colors oils and fats deep red. The test is not decisive, because ethereal oils and resins take the same red color. Dissolve commercial alcannin in absolute alcohol, add an equal volume of water, and filter. The fats and oils in sections left in this solution for 24 hours should be bright red. The reaction is hastened by gentle heating.

Osmic acid, as used in fixing agents, colors fats and oils brown or black. The dark color is removed by bleaching in a 3 to 10 per cent solution of hydrogen peroxide.

In case of fats and oils, solubility and color reactions are useful, but must be regarded as corroborative evidence, not as decisive proof. For more critical and detailed methods, consult the book by Tunmann, which will also give the literature of the subject.

The Middle Lamella.—Even the origin and development of the middle lamella are none too well known; its microchemistry has progressed but little beyond the color-reaction stage. The middle lamella consists largely of pectin or pectic compounds. The easy isolation of cells, when treated with Schultze's maceration, depends upon the ready solubility of pectins in this reagent. Many intercellular spaces arise through the natural solution or gelatinization of the lamella.

In polarized light, with crossed Nichols, the middle lamella is resolved into three lamellae, the middle one appearing dark, and the two outer lamellae, light.

Ruthenium red is a good stain, since it gives as good results as any and has the advantage of keeping well in balsam or glycerin jelly. Make a very weak solution—1 g. to 5,000 c.c. of water, or even weaker—and keep it in the dark. It stains many other things besides the lamella, but is, nevertheless, a good stain.

Pectin is not at all confined to the middle lamella, but is found in other membranes, particularly in spore coats.

Cellulose.—In concentrated sulphuric acid cellulose swells and finally dissolves. It is also soluble in cuprammonia. The cuprammonia can be prepared by pouring 15 per cent ammonia water upon copper turnings or filings. Let the solution stand in an open bottle. It does not keep well, but its efficiency is readily tested. Cotton dissolves almost immediately as long as the solution is fit for use.

With iodine and sulphuric acid cellulose turns blue. Treat first with the undiluted iodine-potassium-iodide solution described in the test for starch, then add a mixture of two parts of concentrated sulphuric acid and one part of water.

With chloroïodide of zinc cellulose turns violet. Dissolve commercial chloroïodide of zinc in about its own weight of water and add enough metallic iodine to give the solution a deep-brown color.

The cell walls of fungi consist of *fungus cellulose*. When young, they give a typical cellulose reaction; when older, they become insoluble in cuprammonia and, with iodine and sulphuric acid, show only a yellow or brown, instead of the typical blue. With chloroïodide of zinc, the wall stains yellow or brown, instead of violet.

Reserve cellulose, which is common in thick-walled endosperm of seeds, shows the same microchemical reactions as ordinary cellulose.

Callose.—The thickening on the sieve plate differs from cellulose in its staining reactions, and in its solubility. It is insoluble in cuprammonia, but will dissolve in a 1 per cent solution of caustic soda.

Stain in a 4 per cent aqueous solution of soda (Na_2CO_3) for 10 minutes, and transfer to glycerin. The callus should take a bright red. If stained very deeply and then transferred to a 4 per cent soda (without the corallin), the stain is extracted from the cellulose but remains in the callus. Unfortunately, the preparations are not permanent.

If stained for about an hour in a dilute aqueous solution of anilin blue, the stain may be extracted with glycerin until it remains only in the callus. After the blue is satisfactory, a few minutes in aqueous eosin will afford a good contrast. The preparation may be mounted in balsam and is fairly permanent.

Lignin.—Lignified walls are insoluble in cuprammonia. The iodine and sulphuric acid or the chloroïodide of zinc, used as in testing for cellulose, give the lignified walls a yellow or brown color. After a treatment with Schultze's maceration fluid, lignified membranes react like cellulose.

Phloroglucin in a 5 per cent aqueous or alcoholic solution applied simultaneously with hydrochloric acid gives lignified walls a reddish-violet color. The preparations do not keep.

Cutinized and Suberized Walls.—These are insoluble in cuprammonia or concentrated sulphuric acid. They are colored yellow or brown by chloroïodide of zinc, or by iodine and sulphuric acid, when applied as in testing for cellulose or lignin. With alcannin, they take a red color, but the red is not as deep as in case of fats and oils. After soaking in an aqueous solution of caustic potash, suberized membranes take a red-violet color when treated with chloroïodide of zinc.

If a strong, fresh alcoholic solution of chlorophyll be allowed to act upon suberized membranes for 15 to 30 minutes in the dark, they stain green, while lignified and cellulose walls do not take the stain. The preparations are not permanent.

A solution of alcannin in 50 per cent alcohol stains suberized and cutinized walls red, but the color may not be very sharp.

Cyanin can be recommended. First, treat with *Eau de Javelle* (potassium hypochlorite), which can be obtained ready for use at any drug-store. This destroys tannins, and the lignified walls lose their staining capacity. Make a 1 per cent solution of cyanin (Grübler's) in 50 per cent alcohol and add an equal volume of glycerin. This should show blue suberized walls, while the lignified walls remain unstained.

Gum, Mucilage, and Gelatinized Membranes.—These are all soluble in water and are further characterized by their strong power of swelling. They are insoluble in alcohol. A series of forms with various color reactions is included under this heading.

Crystals.—Nearly all crystals which are found in plants consist of calcium oxalate. Crystals of calcium carbonate, calcium tartrate, and calcium sulphate also occur. Calcium oxalate is soluble in hydrochloric acid or nitric acid. It is better to use the concentrated acids. The crystals are insoluble in water and acetic acid. Sulphuric acid changes calcium oxalate into calcium sulphate. When treated with barium chloride, crystals of calcium sulphate become covered with a granular layer of barium sulphate, while crystals of calcium oxalate are not affected.

Calcium carbonate, when treated with hydrochloric acid or acetic acid, dissolves with effervescence. The acetic acid should be rather dilute.

CHAPTER VI

FREEHAND SECTIONS

Sections which may be cut without imbedding, whether they are really cut freehand or with the aid of a microtome, will be considered here. The chapter will also deal with other small or thin objects which may be treated like freehand sections.

The beginner is advised to start with the freehand section, because the processes are rapid, and it is comparatively easy to find the causes of imperfections and failures. In the paraffin method, where the processes are more complicated, it is often difficult, or even impossible, to determine the exact cause of a failure.

As a matter of fact, real freehand sections, cut by holding the object in one hand and the knife in the other, are becoming less and less frequent in well-equipped laboratories. However, the laboratory is no place for one who is awkward with the hands; a certain amount of manual dexterity must be acquired if there is to be any success in morphological studies which demand critical preparations. Although we know the student will turn at once to the microtome, we venture a few remarks in regard to real freehand sections.

A sharp razor is a necessity. For cutting sections of twigs, roots, rhizomes, and similar objects, a razor like the one shown in Figure 8*A*, should be used; while for sections of soft tissues, like young asparagus stems, young ovaries of plants, most leaves, and such things, the type of razor shown in Figure 8*B*, is much better. In cutting, brace the forearms against the sides, hold the object firmly in the left hand and cut with a long, oblique stroke from left to right. The edge of the razor and the direction of the stroke should be toward the body, not away from it as in whittling. If the material is fresh, the object and the razor should be kept wet with water, the razor being dipped in water for every stroke. For hard objects, like twigs of oak or maple, the razor will need sharpening after cutting a dozen sections. It is a waste of time to put off sharpening until the razor has become noticeably dull, for all sections except those cut when the razor is perfectly sharp are sure to be inferior. With softer material the razor may hold its edge for hundreds of sections. Those sections which seem to be worth further treatment should be placed at once in water or in a fixing agent and, of course, the choice of a fixing agent should be determined before the sections are cut.

With the advent of a cheap, efficient sliding microtome, the hand microtome began to fall into disuse and, today, it has almost disappeared.

The sliding microtome (Fig. 2) reduces to a minimum the necessity for manual dexterity, but it is a more complicated machine. Study the various parts before you begin to cut sections. How is the knife adjusted? How is the object clamp raised and lowered? How is the thickness of the section determined? In case of a simple microtome like the one shown in Figure 2, the student should soon answer such questions without any help from the instructor. In case of more complicated microtomes, a demonstration by the instructor will save both time and machine.

In cutting sections of wood or herbaceous stems, the knife should be set obliquely so as to use as much as possible of the cutting edge. In most cases it is neither necessary nor desirable to cut very thin sections by this method; 10 μ is very thin, and 20, 30, or even 40 μ is usually thin enough.

Cut with a firm, even stroke, wetting both knife and object after every section. Use water, if the material is fresh; if preserved, use the preservative. Some use a brush in removing sections from the knife, but nothing is quite equal to one's finger; anyone who is in danger of a cut while performing this act is in need of this little practice in manual dexterity.

WOODY AND HERBACEOUS SECTIONS

Safranin and Delafield's Haematoxylin.—In order to make the directions as explicit as possible, let us follow the processes from collecting the material to labeling the slide. The rhizome of *Pteris aquilina* is a good object to begin with. Dig down carefully until the rhizome is exposed; then with a sharp knife cut off pieces a few inches in length, taking the greatest care not to strain the tissues. If the rhizome has been cut carelessly or pulled up, as is usually the case, the finished mount will show ruptures between the bundles and bundle sheaths, disfiguring what might have been a beautiful preparation.

While the material is still fresh and moist, cut the sections and place them at once in 95 per cent alcohol, where they should remain 20 to 30 minutes. It is not necessary to use a large quantity of alcohol; 10 c.c. is enough for 100 thin sections of the rhizome.

Pour off the alcohol and pour on an alcoholic solution of safranin (a 1 per cent solution of safranin in 50 per cent alcohol. See chapter xxix on "Formulas for Reagents"). It is better to let the safranin act over night, or even for 24 hours.

Pour off the safranin (which may be used repeatedly) and pour on 50 per cent alcohol. The alcohol will gradually wash out the safranin, but this stain is washed out more rapidly from cellulose walls than from those which are lignified. The sections should remain in the alcohol until the stain is nearly—but not quite—washed out from the cellulose walls, while still showing a brilliant red in the large lignified tracheids. If 5 or 10 minutes in the alcohol draws the safranin from the lignified walls as well as the cellulose, stain longer; if the differentiation is not secured in 5 or 10 minutes, a small drop of hydrochloric acid added to the alcohol will hasten the process. Some recommend staining for only 1 or 2 hours, but the washing-out process is likely to be rapid and uncertain.

Pour off the alcohol and wash the sections thoroughly in ordinary drinking-water. The washing should be particularly thorough if acid has been used to hasten the previous process, for the preparations will fade if any acid remains.

Stain in Delafield's haematoxylin 3 to 30 minutes. Usually 5 minutes will be about right. Delafield's haematoxylin will stain the cellulose walls, but will have little or no effect upon lignified structures.

Transfer to drinking-water, not distilled water. The red color of the whole section, as it appears to the naked eye, will be rapidly replaced by a rich purple. Continue to wash in water for 2 or 3 minutes after the purple color appears. If the cellulose walls show only a faint purplish color, put the sections back into the stain and try a longer period. If the color is a deep purple or nearly black, add a little hydrochloric acid (1 drop to 50 c.c. is enough) to the water. It is better to put the drop into a bottle of water and shake thoroughly before letting the acidified water act upon the sections. As soon as the sections begin to appear reddish, which may be within 4 or 5 seconds, pour off the acidified water and wash in drinking-water, changing the water 3 or 4 times a minute, until the reddish color caused by the acid has been replaced by the rich purple color so characteristic of haematoxylin. The acid not only secures differentiation by dissolving out the stain from lignified structures more rapidly than from cellulose walls, but it also removes the disfiguring precipitates which almost invariably accompany staining with Delafield's haematoxylin. The acid also washes out the safranin; it is for this reason that the washing after safranin should be stopped while there is still some red color in the cellulose walls. The acid should not only reduce the density of the haematoxylin and remove precipitates, but should also remove the little safranin which may remain in the cellulose walls. After the purple color has appeared, the sections should be left in water for 20 or 30 minutes. They might be left for several hours.

Now place the sections in 50 per cent alcohol for 1 minute, then in 95 per cent alcohol for 1 minute, 100 per cent alcohol for 5 minutes, and then transfer to xylol. As soon as the sections become clear—in about 1 to 5 minutes—they are ready for mounting in balsam. If the sections do not clear readily, as may be the case if the air is damp, or if the alcohol or xylol is not quite pure, transfer from the absolute alcohol to clove oil, which will clear, even if the absolute alcohol is rather poor. Then transfer from clove oil to xylol; the objection to mounting directly from clove oil is that preparations harden more slowly than when mounted from xylol. With a section-lifter, or scalpel, or brush, transfer 3 or 4 sections to a clean, dry slide, put on 1 or 2 drops of balsam, and add a cover, first heating it gently to remove moisture. If xylol has been used for clearing, it is necessary to work rapidly; for the sections must never be allowed to dry. Use square or oblong covers for such mounts, reserving round covers for glycerin mounts. If material is abundant, use as many sections as you can cover conveniently. If you have used several stains with the same material, select for each mount sections from the different stains. In ordinary wood sections each mount should show the three most important views, transverse, longitudinal radial, and longitudinal tangential sections. It is wasteful to use three slides and three covers to show these three views, or to make a mount containing only a single section of the rhizome of *Pteris*.

Put the label at the left. Write first the genus and species; then indicate what part of the plant has been mounted. The date on which the material was fixed is often valuable. After a year or so, the date of making the mount may be of interest in indicating the relative durability of stains. The beginner is likely to write also the stains used, and other details, which he will find quite unnecessary after a little experience. Figure 19 illustrates a good style of labeling and mounting.

FIG. 19.—The label

The following is a convenient summary of the foregoing processes, beginning with the sections in 95 per cent alcohol:

1. Sections in 95 per cent alcohol.
2. Safranin, 12 to 24 hours.
3. 50 per cent alcohol, with or without acid, until color is right, generally about 2 to 10 minutes.
4. Water, 5 minutes, changing frequently.
5. Delafield's haematoxylin, 3 to 30 minutes.
6. Water, 5 to 10 minutes, changing frequently.
7. Water slightly acidulated, 5 to 10 seconds.
8. Water, to wash out acid, 20 to 30 minutes.
9. 50 per cent alcohol, 1 minute.
10. 95 per cent alcohol, 1 minute.
11. 100 per cent alcohol, 5 minutes.
12. Xylol, 1 to 5 minutes.
13. Balsam.
14. Cover and label.

If clove oil seems necessary, finish as follows:

12. Clove oil, 2 to 5 minutes.
13. Xylol, 1 to 5 minutes.
14. Balsam.
15. Cover and label.

Another method has given excellent results, especially with old stems, pieces of dry boards, etc. Cut pieces 5 to 10 mm. square and 2 to 3 cm. long, boil in water, and, when cool, transfer to equal parts of 95 per cent alcohol and glycerin. The material may be left in this mixture indefinitely.

After sections have been cut, wash them in tap water, then in distilled water, and stain half an hour or more in weak Delafield's haematoxylin—about 5 parts of the solution, as given in the formula, to 100 parts of water—and then wash in distilled water. Stain in weak safranin—about 2 parts of the stock solution to 100 parts of water—over night or even for several days. Wash in tap water, then wash in 95 per cent alcohol for 30 seconds or longer, according to the appearance of the stain. Dehydrate in absolute alcohol, clear in clove oil, transfer to xylol, and mount in balsam. This method is very good for gymnosperm woods.

Since it usually happens that processes are commenced, but cannot be completed at a single laboratory period, it is necessary to know where sections may be left for several hours or until the next day without suffering injury. At 1, 2, or the pure water of 8 in the schedule given above, sections may be left until the next day. If it is not desirable to mount all of the sections which have been prepared, they may be kept indefinitely in clove oil or xylol. If the sections are to remain for a year or more in the clearing agent, xylol is to be preferred. Shells with good corks are best for keeping such material.

For the study of vascular anatomy, this is the most permanent stain which has come into general use.

More recently, safranin combined with anilin blue or with light green has been coming into favor. Both these methods will be described.

Safranin and Anilin Blue.—Use the alcoholic safranin already described, and a 1 per cent solution of anilin blue in 90 per cent alcohol.

With this combination we should recommend a long stain in safranin, not less than 24 hours. Wash in 50 per cent alcohol, but do not extract all the safranin from the cellulose walls. Stain 2 to 10 minutes in anilin blue. Rinse a few seconds in 95 per cent alcohol, then treat for about 5 seconds with 95 per cent alcohol slightly acidulated with hydrochloric acid. The weak blue should at once change to a bright blue and, at the same time, the acid will remove some of the safranin. It is for this reason that we proceed while the sections are still somewhat overstained in safranin. Wash for

1 or 2 minutes in 95 per cent alcohol to remove the acid. A trace of sodium carbonate, just enough to make the alcohol alkaline, may be added to the 95 per cent alcohol. If any acid remains, the safranin will fade. Dehydrate in absolute alcohol 1 to 5 minutes, clear in xylol, or first in clove oil and then in xylol, and mount in balsam.

For convenient reference, the process may be summarized, but it must be remembered that all the schedules are intended merely to introduce the method to the beginner.

1. Sections in 95 per cent alcohol.
2. Stain in safranin, 24 hours.
3. 50 per cent alcohol until the stain becomes weak in cellulose walls, but not until it is removed entirely.
4. Anilin blue, 2 to 10 minutes.
5. 95 per cent alcohol, 2 to 5 seconds.
6. 95 per cent alcohol, slightly acidulated with hydrochloric acid, 5 seconds.
7. 95 per cent alcohol, with or without a trace of sodium carbonate, 1 or 2 minutes.
8. Absolute alcohol, 1 to 5 minutes.
9. Xylol, 1 to 5 minutes. The xylol may be preceded by clove oil.
10. Mount in balsam.

Lignified and suberized walls should stain bright red and cellulose walls bright blue. To make this beautiful combination a success, it is necessary to be very careful. If too much safranin is extracted at stage 3, the acid at stage 6 will still further weaken the red stain and the contrast will not be sharp.

Safranin and Light Green (Land's Schedule).—This is another beautiful combination and the student should be successful from the first, since the light green is simpler to apply than either Delafield's haematoxylin or anilin blue.

Land uses either aqueous, anilin, or alcoholic safranin, and uses the light green in clove oil, or in a mixture of clove oil and absolute alcohol. Make a saturated solution of light green in clove oil. Since the solution takes place slowly, the mixture should stand several days before using. If a small quantity of absolute alcohol be added to the clove oil, the stain dissolves more readily. For some structures the stain is more brilliant than with the simple clove-oil solution.

Sections from fresh material are fixed in 95 per cent alcohol; sections from preserved material are rinsed in alcohol or water before staining. The following schedule will summarize the method:

1. Safranin, 2 to 24 hours.
2. 50 per cent alcohol, until differentiated.
3. Dehydrate in 95 and 100 per cent alcohol.
4. Light green (in clove oil), 3 to 30 minutes.
5. Xylol: 2 or 3 c.c. of absolute alcohol may be added to each 100 c.c. of xylol, if the free light green shows a tendency to precipitate.
6. Mount in balsam.

This stain is particularly good for phloem. Since the light green is not likely to overstain and does not extract the safranin, the combination is a rather easy one. Even the beginner can hardly fail to get a good preparation.

Malachite Green and Congo Red.—I am indebted to Dr. Sharp for this method, which has been popular in Professor Grégoire's laboratory at Louvain.

Sections of fresh material should be treated with 95 per cent alcohol and then transferred to water.

1. 3 per cent aqueous solution of malachite green or methylin blue, 6 hours or more.
2. Wash in water.
3. Congo red, 1 per cent aqueous solution, 15 minutes.
4. Wash in water.
5. Rinse in 80 per cent alcohol. As soon as the malachite green or anilin blue color appears through the red, transfer quickly to
6. Absolute alcohol.
7. Xylol.
8. Balsam.

Iodine Green and Acid Fuchsin is another good combination for such sections. The stain will be particularly brilliant if sections from fresh material are fixed in 1 per cent chromo-acetic acid for 10 to 24 hours; and then washed for an hour in water. Beginning with the sections in water, the procedure is as follows:

Stain is aqueous iodine green for 12 to 24 hours. Then wash in water until the stain is nearly all washed out from the cellulose walls, but is still brilliant in the lignified walls. If the stain acts for too short a time, the washing-out process necessary to remove the stain from the cellulose walls will leave only a pale-green color in the lignified walls. Stain in aqueous acid fuchsin for 2 to 10 minutes. This should stain the cellulose walls sharply, but should not act long enough to affect the lignified tissues. Pour off the stain (which may be used repeatedly), and pour on 95 per cent alcohol, and immediately pour it off and add absolute alcohol. The 95 per cent alcohol should not act for more than 5 or 10 seconds, its only function being to save the more expensive absolute alcohol. From 10 to 30 seconds will usually be long enough for the absolute alcohol. Too long a period in the alcohols will weaken the stain. Clear in xylol or clove oil, and mount in balsam.

If a 50 or 70 per cent alcoholic solution of iodine green has been used, the stain should be washed out in 50 per cent alcohol; otherwise the treatment is the same.

Methyl Green (aqueous solution) **and Acid Fuchsin** is a good combination, and the student may find it easier to get a good differentiation than with iodine green. Follow the directions for the aqueous iodine green and acid fuchsin. It may be necessary to wash more rapidly, since the methyl green is easily extracted.

Safranin and Gentian-Violet.—This is a good combination for vascular anatomy. Stain over night in safranin, rinse in 50 per cent alcohol until the stain is reduced to a light-pink color in the cellulose walls; then rinse in water and stain 5 to 10 minutes in aqueous gentian-violet. Rinse in water, dehydrate in 95 and 100 per cent alcohol, clear in clove oil, transfer to xylol and mount in balsam.

Orange may be added to this combination, making a triple stain. In this case, do not reduce the safranin at all, but rinse quickly in 50 per cent alcohol, then in water, stain in gentian-violet, rinse in 95 per cent alcohol and stain for 1 or 2 minutes in orange dissolved in clove oil. This will not only reduce and differentiate the gentian-violet, but will reduce the safranin. Transfer to xylol and mount in balsam. If the safranin is drawn out too rapidly, stain for 15 to 30 seconds in the orange, transfer to clove oil without any orange until the gentian-violet is satisfactory; then transfer to xylol and mount in balsam.

The bordered pits of conifers, the Bars of Sanio, and the middle lamella are beautifully stained by this method.

Other Combinations might be suggested, e.g., iodine green or methyl green with Bismarck brown, methyl green with Delafield's haematoxylin; orange G might be added after the safranin and

Delafield's haematoxylin, and various other stains might be tried. In double staining it is usually best to combine a basic with an acid stain. Green and red make a good contrast, but a section stained with iodine green and safranin would be a failure, because both stains would stain the xylem and neither would stain the cellulose. Both stains are basic. Red lignin and green cellulose could be secured by using safranin and acid green. Green lignin and red cellulose, as already indicated, can be got with iodine and acid fuchsin.

The Time Required for the different processes varies greatly, and the time required for a subsequent process is often more or less dependent upon the time given to processes which preceded it. Good mounts of sections of the petiole of *Nuphar advena* have been secured from material which had been cut, fixed, stained in safranin and Delafield's haematoxylin, and mounted in balsam, the entire time being less than 30 minutes. This is an extreme case, and nothing is gained, except time, and the saving of time is apparent rather than real, for the histologist always has something to do while the sections are in the stain.

Preserved, Fresh, and Dry Material.—If sections are to be cut from material preserved in formalin, the piece should be washed in water, since the odor is annoying and the fumes are injurious to the eyes.

The sections are placed in the stain from water. Sections from alcoholic material are transferred directly to the stain. If the material is in a mixture of alcohol and glycerin, the sections should be washed in water or 50 per cent alcohol until the glycerin has been removed before transferring to the stain.

Some material cuts well when fresh, but cuts with difficulty when preserved. On the other hand, some material cuts well when preserved, but hardly at all when fresh. Some material which is too soft to cut when fresh can be cut with ease after it has been in formalin alcohol for a week or more.

Very hard material, like oak, hickory, maple, etc., should be boiled in water and treated with hydrofluoric acid before any sectioning is attempted. Cut the material into blocks suitable for sections and boil in water for several minutes; then transfer to cold water and, after several minutes, repeat the boiling. The alternate boiling and cooling, which should be repeated several times, drives out the air. Transfer to equal parts of commercial hydrofluoric acid and water.

From 1 to 3 weeks will be enough for most woods. Some oaks, ebony, apple, etc., may require a longer time and the acid may be used pure. One week in 25 per cent acid may be enough for corn stems. Wash thoroughly in water for a day or two. Then leave in equal parts of 30 per cent alcohol and glycerin for several days before cutting. Material may be left indefinitely in the mixture of glycerin and alcohol.

An article by Dr. La Dema M. Langdon, dealing with the preparation and sectioning of hard, woody tissues, appeared in the *Botanical Gazette* of July, 1920. By her method, hard, woody tissues are softened so that they cut readily and can even be imbedded in paraffin successfully.

OBJECTS MOUNTED WITHOUT SECTIONING

Fern Prothallia, mounted without sectioning, make very useful preparations. Select desirable stages and fix in chromo-acetic acid for 10 to 24 hours; wash in water for 3 or 4 hours, changing the water frequently; stain in Delafield's haematoxylin for 5 to 30 minutes; wash in slightly acidulated water for a few seconds, and then wash thoroughly in pure water. The prothallia must now be brought through a graded series of alcohols, 15, 35, 50, 70, 85, 95, and 100 per cent being sufficiently close to prevent plasmolysis. Then use mixtures of alcohol and xylol, 3 parts absolute alcohol and 1 part xylol, 2 parts alcohol and 2 parts xylol, 1 part alcohol and 3 parts xylol, and then pure xylol. Then bring the sections into a mixture of xylol and balsam, using at least 10 parts of xylol to 1 of balsam. If left in a shell, without corking, the xylol will soon evaporate, so that in a few days the prothallia may be mounted. Use the balsam in which the material has been standing, because any other balsam may have a different concentration. At every step in the process the prothallia should be examined under a microscope, so that any plasmolysis may be detected. If each succeeding step is tested with a single prothallium, a general disaster may be avoided. If plasmolysis takes place, weaken the reagent and try another prothallium. When a safe strength is found, bring on the bulk of the material, and use the same method with succeeding steps. The dangerous places are likely to be the transfer from alcohol to xylol and the transfer from xylol to balsam. The process is tedious, but the mounts are very firm and durable. The Venetian turpentine method is less tedious and gives better results than have been secured by the method just described. Fix in the chromo-acetic-osmic mixture (*d*), described on page 26. Stain some prothallia in iron-alum haematoxylin and some in Magdala red and anilin blue. When both have reached the thick turpentine, fine preparations can be made by mounting prothallia from both lots under the same cover.

Sori of Ferns.—Instructive mounts of sori or of individual sporangia may be made without sectioning. It is better to choose ferns with thin leaves, since leaves thicker than those of *Asplenium thelypteroides* are likely to be unsatisfactory. If this fern is at hand, cut off several of the small lobes which bear three to six pairs of sori. Fix in chromo-acetic acid; wash in water; stain in Delafield's haematoxylin, or omit staining altogether; pass through a series of alcohols, allowing each grade to act for at least 10 minutes; clear in clove oil, and mount in balsam. If the sori have begun to turn brown, better views of the annulus will be obtained without staining.

Mosses and Liverworts.—Nearly all mounts are more successful by other methods, for which the student should consult the chapters on Bryophytes (chaps. xix and xx). Excellent mounts of the peristome of the moss can be made as follows: From fresh or preserved capsules cut off the peristome just below the annulus. Treat with 95 per cent alcohol 1 minute, absolute alcohol 2 to 5 minutes, clear in clove oil or xylol, and mount in balsam. It is a good plan to put at least three peristomes on a slide, one with the outside up, one with the inside up, and another dissected to show details of the teeth.

Fairly good unstained mounts of the archegonia and antheridia of small mosses can be obtained by following the directions for mounting the sori of ferns.

Beautiful and instructive mounts of the more delicate foliose Jungermanniaceae can be made by staining lightly in Delafield's haematoxylin whole plants, or pieces as long as can be covered conveniently. The method is that just given for fern prothallia. The mount should show both dorsal and ventral views.

The Epidermis shows its best surface views without sectioning. Select some form with large stomata, like *Lilium* or *Tulipa*, strip pieces of epidermis from both sides of the leaf, and place them immediately in absolute alcohol for 1 or 2 minutes. Stain in Delafield's haematoxylin; after this stain is satisfactory and all acid has been washed out, stain for 1 or 2 minutes in aqueous eosin, erythrosin, or acid fuchsin; place directly into 95 per cent alcohol for a few seconds (merely to save the absolute alcohol), then into absolute alcohol for about 30 seconds, and then into clove oil. Mount in balsam. The epidermis is likely to curl and, unfortunately, patience seems to be the only remedy. In mounting, be careful to get pieces from both sides of the leaf, and be sure that the pieces are outside up. The inside of the epidermis is usually more or less rough, on account of the mesophyll torn off with it. *Sedum purpurascens* will show various stages in the development of stomata, even in epidermis stripped from mature leaves. The epidermis of the *Sedums* strips off very easily. If the large *Sedum maximum* is available, it is not difficult to strip off pieces 2 or 3 centimeters wide and several centimeters long. There is not much tendency to curl. The pieces may be spread out flat in a Petri dish, fixed in the chromo-acetic-osmic solution, just recommended for fern prothallia, or in this solution without the osmic acid. Wash in water, stain in Delafield's haematoxylin and eosin, or in safranin and gentian-violet. Then wash in water and run up through a series of alcohols—10, 20, 35, 50, 70, 85, and 95 per cent, about 2 minutes in each grade. If the gentian-violet is lost, stain for a minute in an alcoholic gentian-violet or in light green, for about 30 seconds; then transfer to clove oil, xylol, and balsam. This is a more tedious method, but it is worth the trouble. We have had the best results with the haematoxylin and eosin combination.

Other Objects.—The cases just given will suggest other objects which might be mounted by such methods. Nearly all objects which used to be mounted in balsam without sectioning are now handled more successfully by the Venetian turpentine method.

CHAPTER VII

THE GLYCERIN METHOD

It used to be an almost universal custom to mount unicellular and filamentous forms in glycerin or in glycerin jelly. The method is simple and easily mastered, but mounts must be sealed, and even when well sealed they do not long survive the ordinary use and abuse of the laboratory. However, some things are still mounted in glycerin or glycerin jelly. The greens and browns of moss protonema still keep much of their natural color in preparations which have been in use for many years.

We have abandoned, completely, any mounting in glycerin; but the glycerin method, except the final mounting, precedes mounting in glycerin jelly, and it also constitutes the first part of the Venetian turpentine method. Consequently, it is very necessary to learn the capabilities and limitations of glycerin.

The method, from fixing to mounting, as used in connection with staining and without staining, will now be described.

Stained Preparations.—The familiar *Spirogyra* is a good form to begin with. Fix in the chromo-acetic-osmic solution described under (*d*) p. 26, or in this solution without the osmic acid, for 24 to 48 hours. Wash in running water over night or 24 hours. If running water is not available, wash for 24 hours, changing the water frequently.

The most satisfactory stain is *Haidenhain's iron-alum haematoxylin.*

Treat for 2 to 4 hours in a 2 per cent iron-alum; wash 10 to 20 minutes in water, and then stain 4 hours, or over night, in ½ per cent haematoxylin. Wash in water 10 to 20 minutes and transfer to iron-alum again. This time the material must be watched under the microscope until the stain is just right. If the stain becomes weak in 2 or 3 minutes, the time in the stain was too short; if it requires an hour or more to reduce the stain to the proper tint, the time was too long. Different species of *Spirogyra* and even different collections, fixed in the same reagent, will differ in their reaction to stains; and different unicellular and filamentous forms, in different fixing fluids, will present so much difference in times that only general suggestions can be given. When the stain is satisfactory, wash in running water for an hour. If this second iron-alum is not washed out thoroughly, its continued action will cause the preparation to fade.

Put the material into 10 per cent glycerin (1 part glycerin and 9 parts water), and then allow the water to evaporate gradually in a place as free from dust as possible. Nothing is better than a Petri dish for this purpose, because it presents a large surface for the evaporation of the water in the mixture. If there is much dust, cut a piece of filter paper just the size of the dish and let it float on the 10 per cent glycerin. The liquid will soak through the paper and evaporate without exposing the material itself to the dust. The process may be hastened, safely, by warming up to 35° C. The temperature of a paraffin bath—45° to 52° C.—causes such rapid evaporation that the material is likely to shrink.

When the glycerin has become about as thick as pure glycerin, the material is ready for mounting. Place a small drop of glycerin, with the material, in the center of the slide, taking care not to put on so much that there will be a confusing tangle. Use scissors constantly so as not to injure filaments by trying to tease them out. Put on a round cover. There should be *just enough* glycerin to come to the edge of the cover-glass, but *not any more*, for it is impossible to seal a mount if glycerin has oozed out beyond the cover.

The mount should now be sealed. Canada balsam, various asphalts, cements, flat varnish, gold size, and other things have been used. Canada balsam is always at hand and seems to be as good as any. Preparations which had been sealed with gold size more than fifty years before have been exhibited in perfect condition, but they must have been hidden away in some museum, for a glycerin mount would never survive fifty years of laboratory use. The gold size, as painters use it, is likely to be too thin for sealing mounts. Put some of it in a 1-ounce bottle with a wide neck and leave the cork out until the gold size thickens a little. Should it become too thick, thin it with turpentine.

Nothing but practice will enable one to spin a good ring, but a good camel's-hair brush, a good turntable, and a balsam neither too thick nor too thin will facilitate matters. Give the turntable a spin, and with the brush touch first the slide about as far out from the cover as you wish the ring to extend, then gradually approach the cover. Dip the brush in the balsam again, and gradually extend the ring until it is about $\frac{1}{16}$ inch wide on the cover. The touch must be extremely gentle or the cover will be moved. Do not try to put on a thick ring the first time, but let a thin ring harden for an hour (months would do no damage), and then a thicker ring can be added without any danger. Thin rings are too likely to be broken, and thick rings are in the way if the preparation is to be examined with high powers. A medium ring is best, and it should consist of two coats, for a crack would seldom appear at the same place in both coats. A good shape and thickness for a ring are shown in Figure 20.

FIG. 20.—Slide, natural size, showing size and form of the ring

The following is a summary of the foregoing processes:

1. Fix in chromo-acetic-osmic acid, 24 to 48 hours.
2. Wash in water, 24 hours.
3. Iron solution, 2 hours.
4. Wash in water, 10 minutes.
5. $\frac{1}{2}$ per cent haematoxylin, 3 to 24 hours.
6. Wash in water, 10 minutes.
7. Iron solution until stain is right.
8. Wash in water, 1 hour.
9. 10 per cent glycerin.
10. Mount and seal.

If the material has been fixed in formalin, it should be washed in water for 5 to 10 minutes before staining. Material preserved in 70 per cent alcohol should be placed successively in 50 per cent, 35 per cent, 15 per cent alcohol, and then in water, allowing each to act for 15 to 30 minutes before being placed in the stain.

Mayer's haem-alum is also a good stain for filamentous algae and fungi which are to be mounted in glycerin. The process, after fixing and washing in water, is as follows:

1. Transfer to the stain from water.

   It is seldom necessary to stain longer than 10 minutes. As a rule, it is better to dilute the stain (about 1 c.c. to 10 c.c. of distilled water) and allow it to act for 10 hours or over night.
2. Wash in water, 20 minutes.
3. 10 per cent glycerin until sufficiently concentrated.
4. Mount and seal.

Eosin is a good stain for many algae and fungi, when sharp outlines rather than cell contents are to be brought out. After the material has been fixed and washed in water, stain in an aqueous solution of eosin for 12 to 24 hours. Pour off the eosin, which can be used repeatedly, and pour on a 1 or 2 per cent solution of acetic acid in water. Pour this off and pour on some more of the acid, until very little stain washes out. The process may require 2 to 5 minutes. Then place in 10 per cent glycerin containing about ½ per cent acetic acid, and allow the glycerin to concentrate. The acetic acid is to prevent the stain from washing out. When the glycerin has reached the proper concentration, mount and seal as before.

The following is a rapid method for forms like *Eurotium* and *Penicillium:* Fix in 100 per cent alcohol about 2 minutes; stain in aqueous eosin 5 minutes; wash in water about 1 minute; fix in 1 per cent acetic acid 1 minute; then mount directly in 50 per cent glycerin to which about 1 per cent acetic acid has been added. It is hardly worth while to try this method with forms which have large cells; they are almost sure to collapse. If a form like *Eurotium* passes through the earlier processes without danger, but collapses when put into the 50 per cent glycerin, put it into the 10 per cent glycerin and allow the glycerin to concentrate.

Mounting without Fixing or Staining.—It is sometimes desirable to retain the natural color of an object. The chlorophyll green can usually be preserved by mounting directly in glycerin without any previous fixing. Other colors also are often preserved in this way. Moss protonema makes beautiful preparations by this method. If possible, select protonema showing the very young moss plants. The brown protonema and brown bulbils preserve their color perfectly. Wash the dirt away from the protonema, which is then placed in 50 per cent glycerin. The brown or black spores of fungi are readily mounted in this way.

The method is very useful when one finds a single specimen of *Pediastrum,* or any small form which would be lost in the more complicated processes. Place a large drop of 10 per cent glycerin on a slide; with a pipette, transfer the object to the drop, and allow the glycerin to concentrate. Then add a cover and seal the mount.

GLYCERIN JELLY

It is almost never necessary to mount anything in glycerin, because material can be transferred directly from glycerin to glycerin jelly.

If the glycerin jelly is well made, it is quite firm and mounts will last for a year or two, without sealing; but it is better to seal them with balsam. A very good formula is known as *Kaiser's gelatin.* It is made as follows: One part by weight of the finest French gelatin is left for about 2 hours in 6 parts by weight of water; 7 parts of glycerin are added, and for every 100 grams of the mixture, 1 gram of concentrated carbolic acid. The whole is warmed for 15 minutes, stirring all the while until all the flakes produced by the carbolic acid have disappeared. Filter while warm through a fine-mesh cheese-cloth.

To make a mount, take a small piece of the glycerin jelly, not more than half as large as a grain of wheat—the exact size will depend upon the material—warm it until it melts; and then transfer to it the material which has already been brought into thick glycerin. It is a good plan to touch the material to filter paper in order to remove as much glycerin as possible; for the less glycerin the firmer the mount will be. The mount may be sealed as soon as it is cool; but some prefer to let it stand for a week or two before sealing. In any case, it is a fairly firm mount, so that there is no danger of moving the cover.

Everything which can be brought safely into pure glycerin can be mounted in glycerin jelly and the preparation is much more stable than a glycerin mount.

CHAPTER VIII

THE VENETIAN TURPENTINE METHOD

Just as glycerin jelly is superior to glycerin as a mounting medium, so Venetian turpentine is superior to glycerin jelly. While the method was described by Pfeiffer and Wellheim[1] thirty years ago, it received no recognition in the United States or even in Europe. I made a casual trial of the method when preparing the first edition of this book, but the preparations were such miserable failures that the process did not seem worth mentioning. The method was next brought to my attention during a demonstration in Strasburger's laboratory at Bonn. He was using preparations of *Zygnema* and *Spirogyra*, the staining of which surpassed anything I had ever seen. He remarked that it was not worth while to consult the lengthy article, because his preparations had been made by the authors and no one else had made a success of the method. However, when I returned, I made a careful study of the process, and finally learned to use it successfully. The details as given in this paper were too indefinite for practical use, but, after one has learned the method the article can be read with profit.

The great practical advantages of the method are that preparations are as hard and durable as balsam mounts, and that a much greater variety of staining is possible than in case of glycerin mounts.

The principal features of the method are that material is brought from pure glycerin into alcohol, and from alcohol into Venetian turpentine, without passing through xylol or other clearing agent.

After fixing and washing in water and staining in an aqueous stain, e.g., iron-alum haematoxylin, the process is as follows:

1. 10 per cent glycerin until concentrated.
2. Wash the glycerin out thoroughly in 95 per cent alcohol.
3. Complete the dehydration in 100 per cent alcohol.
4. 10 per cent Venetian turpentine in an exsiccator until the turpentine becomes thick enough for mounting.
5. Mount in the Venetian turpentine.

[1] Ferdinand Pfeiffer and R.v. Wellheim, "Zur Preparation der Süsswasseralgen," *Jahrbüchern für wissenschaftliche Botanik*, **26**: 674–732, 1894.

Even after the method became established, there occurred a period of several years during which it was practically impossible to get a Venetian turpentine suitable for histological use. Consequently, it was necessary to resort to glycerin jelly or to try various schemes for bringing material into Canada balsam; but good Venetian turpentines have appeared again and are even more satisfactory than those which established the method in popular favor. We have tested two brands which are giving uniform and excellent results: these are the "Venice Turpentine (True)," sold by the Fuller-Morrison Company, of Chicago; and the "Turpentine Venetian" (No. 2605), sold by the National Anilin and Chemical Company, of New York. There are probably other good turpentines and still others are likely to appear. Venetian turpentine is made from the resin of *Larix europea.* Since this resin is soluble in absolute alcohol, material can be transferred directly from absolute alcohol to a thin solution of the mounting medium, without the intervention of xylol or any other oil, thus eliminating the most dangerous stage in the preparation of filamentous algae and fungi and any forms which are to be mounted whole.

The general outline, just given, is not sufficiently definite for a working introduction. The following concrete examples, describing the use of Venetian turpentine with an aqueous stain, with an alcoholic stain, and with a combination of aqueous and alcoholic stains, will be more practical than general directions. The steps from fixing to mounting, as used with an aqueous stain, will be described first, since this will introduce the method in its least complicated form.

Haidenhain's Iron-Haematoxylin.—Using *Spirogyra* as a type, proceed as follows:

1. Fix 24 hours in chromo-acetic acid.

   1 per cent chromic acid........................ 100 c.c.
   Glacial acetic acid........................... 3 c.c.

   The volume of the fixing agent should be at least 50 times that of the material to be fixed.
2. Wash in running water 10 hours; if running water is not available, 24 hours, changing often.
3. 2 per cent aqueous solution of iron-alum (ammonia sulphate of iron), 2 hours.
4. Wash in running water, 20 minutes. If running water is not available, wash in a large quantity of water and change frequently.

5. Stain over night, or 24 hours, in ½ per cent aqueous solution haematoxylin.
6. Wash in water, 20 minutes.
7. 2 per cent aqueous solution of iron-alum, until the stain is satisfactory. This can be determined only by examining frequently under the microscope.
8. Wash in water, 2 hours. If this washing is not thorough, the continued action of the iron-alum will cause the preparations to fade.
9. Transfer to 10 per cent glycerin, and allow the glycerin to concentrate until it has the consistency of pure glycerin. It is not necessary to use an exsiccator. Merely put the glycerin into shallow dishes, and leave it exposed to the air, but protected from dust. If the material is in Petri dishes or other dishes with a large surface, 3 or 4 days will be sufficient. This process may be hastened by warming, if the temperature does not go above 35° C. If the reduction from 10 per cent glycerin to pure glycerin is accomplished in less than 24 hours, the change in the concentration is so rapid that material is likely to suffer.
10. Wash out the glycerin with 95 per cent alcohol. It will be necessary to change the alcohol several times. From 10 to 20 minutes will be sufficient if the alcohol is changed frequently. This alcohol cannot be used again for the same purpose, but it will be useful in cleaning one's hands and in cleaning dishes which have contained Venetian turpentine.
11. Complete the dehydration in 100 per cent alcohol: 10 minutes should be sufficient.
12. *Most failures are now ready to occur.*

From the absolute alcohol the material is transferred to a 10 per cent solution of Venetian turpentine in absolute alcohol. The turpentine thickens as the alcohol evaporates, and when it reaches the consistency of pure glycerin the material is ready for mounting. *The 10 per cent Venetian turpentine is very sensitive to moisture,* and most failures are due to this characteristic; consequently the concentration cannot be allowed to take place with the turpentine exposed to the air of the room. Use an exsiccator. This will not only absorb the moisture from the air, but will soon remove the alcohol from the turpentine mixture. Make an exsiccator as follows: Place a saucer full of soda lime (sodium hydroxide with lime) on a plate of glass, and cover with a bell jar. This is a simple and effective exsiccator. Instead, you may simply scatter soda lime in the bottom of any low museum jar with tight-fitting cover. The tin cans, with tight covers, in which you get your pound of "Improved Vacuum Coffee" make good exsiccators for small amounts of material. You may improvise other forms; the essential thing is to provide a small, air-tight place in which the soda lime may work.

Instead of soda lime you may use fused calcium chloride or the white sticks of sodium hydroxide.

We are now ready for the transfer from absolute alcohol to the 10 per cent Venetian turpentine. *Make the transfer quickly.* Pour off the absolute alcohol and place the dish, with the material, in the exsiccator; then pour on the 10 per cent turpentine, *and immediately put on the cover.* This is better than to pour on the turpentine and then try to get the dish well placed in the exsiccator.

The greater the surface of soda lime exposed, the more rapid will be the concentration of the Venetian turpentine. The concentration must not be *too* rapid. Not less than 2 days should be allowed for the concentration of 30 c.c. of the turpentine in an ordinary Minot watch glass.

Great care must be taken not to let any of the soda lime, or other drier, get into the turpentine.

When the lime has become saturated, it may be heated until dry, and then used again. If material is put into an exsiccator with nearly saturated lime, the turpentine becomes milky. If the material is very valuable, wash in absolute alcohol until entirely free from any milky appearance, and start again in 10 per cent turpentine.

As soon as the turpentine has attained the consistency of pure glycerin, it may be exposed to the air without any danger from moisture; but the turpentine would soon become too thick for mounting. If the turpentine has become too thick, thin it with a few drops of absolute alcohol or with 10 per cent or any thin solution of Venetian turpentine.

Mount the material in a few drops of the Venetian turpentine and add a cover. Tapping on the cover with the handle of a needle or scalpel will often separate the filaments so that they are more convenient for examination. Square covers may be used since it is entirely unnecessary to seal the mounts, which are as hard and durable as those mounted in balsam.

Material in the thickened Venetian turpentine, when not needed for immediate mounting, may be put into small bottles. The corks should be of the best quality; otherwise the turpentine will become too thick. While it can be thinned by adding thin turpentine, it is better, for easy mounting, not to let the turpentine become too thick. If the turpentine is only a little too thick, warming it gently will thin it enough for making mounts; but if any material is to be put away, a few drops of absolute alcohol or of a thin Venetian turpentine should be added. Material in Venetian turpentine, well corked and kept in the dark, does not fade or deteriorate in any way.

Magdala Red and Anilin Blue.—Fix in chromo-acetic acid and wash in water, as described in the previous schedule. Transfer from water to 10 per cent glycerin and allow the glycerin to concentrate. It is not necessary to use an exsiccator since there is no danger from moisture in the air. When the glycerin attains the consistency of pure glycerin, wash the glycerin out with 95 per cent alcohol and then proceed with the staining.

1. Stain in Magdala red. At least two Magdala reds are sold by dealers. The one marked "*echt*" is more expensive, but, in our experience, is inferior to the one marked simply "Magdala red." Make a 1 per cent solution in 90 per cent alcohol. We use the stain much stronger than recommended by Pfeiffer and Wellheim. The pure stain, allowed to act for 24 hours, does not stain too deeply.

2. Rinse the material for a minute in 90 per cent alcohol.

3. Stain in anilin blue, using a 1 per cent solution in 90 per cent alcohol. We prefer to make a fresh solution every time we have anything to stain. It is not necessary to measure it. A little of the powder—about half the bulk of a grain of wheat—in 30 c.c. of 90 per cent alcohol, will give an efficient solution. The time required for successful staining will vary from 3 to 30 minutes. Do not put all the material into the anilin blue at once, but, by trying a few filaments at a time, find out what the probable periods may be.

4. Rinse off the stain in 90 per cent alcohol, and then treat for a few seconds in acid alcohol (1 very small drop of HCl to 30 c.c. of 90 per cent alcohol). The acid alcohol fixes and brightens the anilin blue, but extracts the Magdala red. If the anilin blue or the acid alcohol acts for too short a time, the blue will be weak; if they act too long, the red is lost entirely. If the blue overstains too much, wash it out in 95 per cent alcohol. If the red overstains, wait until the mount is finished, and then reduce the red by exposing the slide to direct sunlight.

5. Absolute alcohol, 5 or 6 seconds.

6. Transfer *quickly* to 10 per cent Venetian turpentine and proceed as in the previous schedule.

The surprising beauty of successful preparations will compensate for whatever failures may occur. Nuclei and pyrenoids should show a brilliant red, while the chromatophores and cytoplasm should be dark blue. The cell walls should show a faint bluish color.

Haidenhain's Iron-Alum Haematoxylin and Eosin.—Follow the schedule for iron-haematoxylin until the glycerin has been washed out in 95 per cent alcohol. Then stain for a minute in a solution of eoşin in 95 per cent alcohol. Wash for a minute in 95 per cent alcohol, then a minute in absolute alcohol, and then transfer to the 10 per cent Venetian turpentine.

Haidenhain's Iron-Alum Haematoxylin and Safranin.—Follow the schedule for iron-haematoxylin until the glycerin has been washed out in alcohol, and then add to the 95 per cent alcohol several drops of a solution of safranin in 95 or 100 per cent alcohol and allow the stain to act for 30 minutes or an hour. Then dehydrate in absolute alcohol and transfer to 10 per cent Venetian turpentine.

Other Stains may be used. Aqueous stains should be used before starting with 10 per cent glycerin. Alcoholic stains should be in strong alcohol—about 90 per cent—and should be applied just after washing out the glycerin.

This method is equally good for filamentous fungi and also for the prothallia of *Equisetum* and ferns, for delicate liverworts and mosses, and similar objects.

If you have a good turpentine, good stains, *and avoid moisture*, the Venetian turpentine method should not be difficult, and the results with filamentous and unicellular forms and other small objects surpass anything yet secured by other processes.

CHAPTER IX

THE PARAFFIN METHOD

The paraffin method is still the most important of all histological methods now in use. The results obtained by this method would have been regarded as almost miraculous by the histologists of one hundred years ago. At that time it was customary to observe things dry, and no cover-glasses were used. Section-cutting with sharp knives or razors did not become general until about 1830. The need for an instrument which would cut sections without demanding an extreme degree of manual dexterity was soon felt, but a successful microtome did not appear until much later. The latest microtomes, while rather complicated, give wonderful results. The Spencer microtome, shown in Figure 21, with the cooling attachment devised by Dr. Land, will cut even ribbons, 1 μ in thickness, from such material as the antheridial receptacles of *Marchantia*. This means that a series of sections can be cut from pollen grains or spores too small to be seen by the naked eye. Many of the principles involved in this method are general in their application, and some of the processes are common to other methods. Before attempting the free-hand sectioning, the beginner should read the following paragraphs on killing and fixing, washing, hardening and dehydrating, and on clearing.

KILLING AND FIXING

As stated in the chapter on "Reagents" (chap. ii), the purpose of a killing agent is to bring the life-processes to a sudden termination, while a fixing agent is used to fix the cells and their contents in as nearly the living condition as possible. The fixing consists in so hardening the material that the various elements may retain their natural condition during all the processes which are to follow. Usually the same reagent is used for both killing and fixing. Zoölogists, from humane motives, may use chloroform for killing, while other reagents are used for fixing. In fixing root-tips, anthers, and other material for a study of mitotic figures, it is necessary that killing be very prompt. In a weak solution of chromo-acetic acid, nuclei which have begun to divide may complete the division, although the reagent might hinder nuclei from entering upon division. By treating for 20 minutes to 1 hour with Flemming's weaker solution, or with a chromo-acetic solution containing a much smaller proportion of osmic acid, the killing will be greatly accelerated and the proportion of nuclei in division will be correspondingly greater. If filamentous

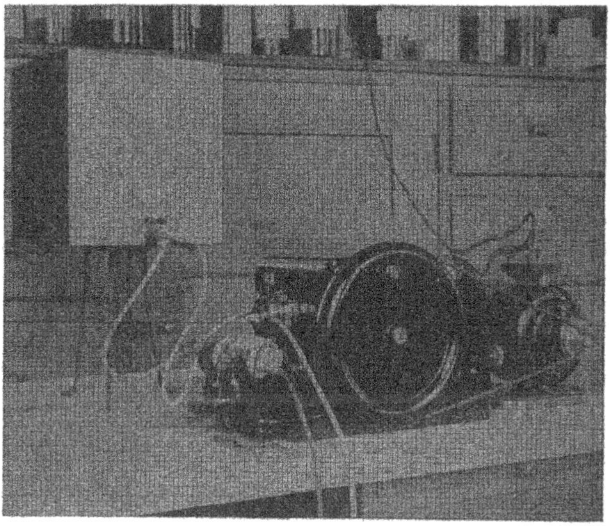

Fig. 21.—Spencer rotary microtome with electric motor and Land's apparatus for temperature control.

algae are placed for 10 or 20 minutes in a chromo-acetic solution containing a little osmic acid, all the advantages of immediate killing will be secured. Material is then transferred to chromo-acetic acid containing no osmic acid. The short treatment with an osmic solution is not likely to cause any serious blackening.

Take the killing and fixing fluids into the field. If one waits until the material is brought to the laboratory there may be some fixing, but it will, in many cases, be too late to do much killing. Material which has begun to wilt is not worth fixing. Material like

Spirogyra, however, may be wrapped in several thicknesses of newspaper, placed in a botany can and brought into the laboratory. Before fixing, it should be placed in water for half an hour. Such forms suffer more from lack of air when placed in a bottle or a can than from lack of water when wrapped in wet newspaper. Branches with developing buds may be brought in and kept in water. Cones of the cycad, *Ceratozamia*, sent from Jalapa, Mexico, have arrived in Chicago with cell division still going on at a rapid rate. But such cases are extremes; as a rule, take the killing and fixing fluids into the field.

Always have the material in very small pieces, in order that the reagents may act quickly on all parts of the specimens. Pieces larger than cubes of 1 cm. should be avoided whenever possible. While one sometimes needs sections 2 or even 3 cm. long, it is not likely to be necessary to fix pieces more than 4 or 5 mm. in thickness. For very fine work no part of the specimens should require the reagent to penetrate more than 1 or 2 mm.

For fixing agents of the chromic-acid series, the volume of the reagent should be about 50 times that of the material.

Fixing agents with alcohol as an ingredient will fix a larger proportion of material. It must be remembered that the water, which is always present in living tissues, weakens the fixing agent.

The time required for fixing varies with the reagent, the character of the tissue, and the size of the piece. About 24 hours is a commonly recommended period for chromic-acid solutions, but 2 or even 3 days will do no harm.

Directions for making and using the various fixing agents are given in the chapters on "Reagents" (chaps. ii, xxix).

WASHING

Nearly all fixing agents, except the alcohols, must be washed out from the material as completely as possible before any further steps are taken, because some reagents leave annoying precipitates which must be removed, and others interfere with subsequent processes. Aqueous fixing agents with chromic acid as their principal ingredient are washed out with water; aqueous solutions of corrosive sublimate are also washed out with water; but alcoholic solutions should be washed out with alcohol of about the same strength as the fixing agent; picric acid, or fixing agents with picric acid as an ingredient, must not be washed out with water, but with alcohol, whether the picric acid be in aqueous or alcoholic solution. When washing with water, running water is best, and where this is not convenient the water should at least be changed frequently. The washing-out process usually requires about 24 hours.

HARDENING AND DEHYDRATING

After the material has been washed, it is necessary to continue the hardening and also to remove the water. Alcohol is used almost entirely for these purposes. It completes the hardening and at the same time dehydrates, that is, it replaces the water in the material, an extremely important consideration, for the least trace of moisture interferes seriously with the infiltration of the paraffin.

The process of hardening and dehydrating must be gradual; if the material should be transferred directly from water to absolute alcohol, the hardening and dehydrating would be brought about in a very short time, but the violent osmosis would cause a ruinous contraction of the more delicate parts. In recent years, cytologists have been making the dehydration process more and more gradual. Twenty years ago most workers began the dehydration process with 35 per cent alcohol and used the series 35, 50, 70, 85, 95, and 100 per cent alcohol. Some placed an intermediate grade between water and 35 per cent alcohol. If plasmoylsis—the tearing away of the protoplast from the cell wall—was avoided, the series was thought to be sufficiently gradual; but a series which may avoid plasmolysis may not be adequate if one is to study the finer details of cell structure. The following series is recommended: 2½, 5, 7½, 10, 15, 20, 30, 40, 50, 70, 85, 95, and 100 per cent. There is no particular virtue in the fractions: it is convenient to make 10 per cent alcohol, dilute with an equal volume of water for the 5 per cent, and dilute the 5 per cent with an equal volume for the 2½ per cent. It will be noted that the series begins with very close grades and that the intervals are gradually increased. The claim is that by beginning with very weak alcohols in close grades, more perfect dehydration can be secured at the end of the series. Various devices, like constant drip and osmotic apparatus, have been proposed to secure a more gradual transfer, but it is very doubtful whether these possess any real advantages. It is not necessary to use a large amount of alcohol: 2 or 3 times the volume of the material is sufficient.

The grades of alcohol may be used several times, but it must be remembered that pollen grains, fungus spores, starch grains, and various granules are likely to be left in the alcohol, so that it is wise to pour back through a filter each time, thus keeping the alcohols clean.

As the alcohols absorb water from the material, they become weaker and weaker. If the various alcohols be poured in a large "waste alcohol" bottle, when a couple of liters has been accumulated, the strength may be determined by testing with an alcoholometer. Then any grade of less strength can be made from this stock.

The time necessary for each of the stages has not been determined with any certainty. About 2 hours seems to be long enough for each of the grades from 2½ to 70 per cent; for 70, 85, and 95, about 4 hours each; for absolute alcohol, 4 to 12 hours, changing 2 or 3 times. If material is to be kept in alcohol, leave it in 85 per cent, but where labor is no object, it is better to go on and imbed it in paraffin.

CLEARING

Let us suppose that the material has been thoroughly dehydrated, so that not the slightest trace of water remains. If the supposition chances to be contrary to fact, all the work which has preceded, as well as all which is to follow, is only an idle waste of time. The purpose of a clearing agent is to make the tissues transparent, but clearing agents also replace the alcohol. At this stage the latter process is the essential one, the clearing which accompanies it being incidental. The clearing, however, is very convenient, since it shows that the alcohol has been replaced and that the material is ready for the next step.

Various clearing agents are in use. Xylol is the most generally employed, and for most purposes it seems to be the best. Bergamot oil, cedar oil, clove oil, turpentine, and chloroform are used for the same purpose. Cedar oil and chloroform may, in some cases, be as good as xylol.

Only a small quantity of the clearing agent is necessary, enough to cover the material being sufficient.

The transfer from absolute alcohol to the clearing agent should be *gradual*, like the hardening and dehydrating processes. The most successful workers have been making this transfer more and more gradual. Thirty years ago it was customary to transfer from absolute alcohol directly to xylol; then a mixture of equal parts of absolute alcohol and xylol was interpolated; in the second edition of this book three grades were placed between the absolute alcohol and xylol. It is undoubtedly better to make the transfer still more gradual. The following series seem to be safe, $2\frac{1}{2}$, 5, 10, 15, 25, 50, 75, and 100 per cent xylol. These mixtures of absolute alcohol and xylol can be made with sufficient accuracy without measuring in a graduate. The 50 per cent grade is made by mixing equal parts of absolute alcohol and xylol; the 25 per cent, by adding to the 50 per cent an equal volume of absolute alcohol; make the 10 per cent grade from the 25 per cent by adding a little more than an equal volume of absolute alcohol; in the same way, make the 5 per cent from the 10 per cent, and the $2\frac{1}{2}$ per cent from the 5 per cent. The different grades may be kept in bottles and may be used repeatedly. A couple of drops of safranin dissolved in absolute alcohol, added to the 50 or 75 per cent xylol, will color the material a little and will often be helpful in orienting after the imbedding in paraffin.

About 2 or 3 hours is enough for each grade. The pure xylol should be changed once or twice. Throughout the dehydrating and clearing it is a good plan to keep the material in Number 4 shells, which are made from glass tubing about 25 mm. in diameter.

Other clearing agents may be used, but the process must be just as gradual.

THE TRANSFER FROM CLEARING AGENT TO PARAFFIN

This should also be a *gradual* process. The most convenient method is to place a small block of paraffin in the pure clearing agent with the material, but the block of paraffin should not rest directly upon the objects. Dr. Land uses coarse wire gauze, cut into strips about 15 mm. wide and tapered at both ends. The strip is then bent so that the pointed ends rest upon the bottom of the Number 4 shell, while the middle portion forms a flat table upon which the paraffin may rest. Dip the wire gauze table into xylol and then slip it carefully into the Number 4 shell. The table portion should be 10 or 15 mm. above the material, and there should be enough xylol to extend a few millimeters above the table. Place on the table a block of paraffin about equal to the volume of the xylol in the shell. The table not only prevents the paraffin from injuring the material by mechanical pressure, but insures considerable diffusion before the mixture of paraffin and xylol reaches the specimens. After 24 hours (or several days, if time permits) at room temperature, place the shell on a pasteboard box—slide boxes are good—on top of the paraffin bath. Do not place the shell directly upon the metal of the bath, since it is better to minimize heat. As soon as the paraffin is dissolved, add some more, this time leaving the cork out, in order that the xylol may evaporate. About 24 hours on the top of the bath should be sufficient.

THE PARAFFIN BATH

This step is usually called infiltration, but when the transfer from the clearing fluid to paraffin is made gradually, as has just been indicated, the process of infiltration is already begun. It is now necessary to get rid of the xylol or other clearing agent. This is accomplished, to a considerable extent, by pouring off the mixture of xylol and paraffin and replacing it with pure melted paraffin. Pour off the pure paraffin *immediately*. This is important. You will notice that often, when the pure paraffin is poured on, a froth or scum will appear on the surface. Much of the xylol will be in this scum, and, if allowed to remain, it would diffuse into the mass and greatly prolong the time needed for infiltration. So, pour it off and add more pure paraffin, for some xylol remains in the tissues and must be removed. Do not put the shell into the bath, but use a flat dish of some sort. The main object is to have a fairly large surface exposed, so that the remaining xylol may evaporate as rapidly as possible. Change the paraffin 2 or 3 times. Soft paraffin (about 45° C.) may be used at first, but the second should be the paraffin of the grade in which the material is to be imbedded. If there are two baths, one should be kept at 46° C. and the other at 52° C., if the material is to be imbedded in 52° C. paraffin. While using the soft paraffin, keep the material in the 46° C. bath; for the harder paraffin, use the 52° C. bath. If there is only one bath, there is no object in using the 45° C. paraffin.

Do not throw away the paraffin which you pour off, but put it in a waste jar or beaker, or, still better, in a small tin lard pail, in which you have made a lip to facilitate pouring. This can be placed in the bath, or, in winter, on the radiator, and the xylol will gradually evaporate. After long heating, the paraffin not only becomes as good as new, but even better, since it becomes more homogeneous and tenacious. If it contains dust or débris of any kind, it may be filtered with a hot filter.

The time required varies with the character of the material and the thoroughness of the dehydrating and clearing. If this schedule has been followed up to this point, the time will be much shorter than most investigators now deem necessary. Fern prothallia infiltrate perfectly in 15 to 20 minutes; onion root-tips in 20 to 30 minutes; ovaries of *Lilium* at the fertilization stage, 30 minutes to 1 hour; 5 or 6 mm. cubes of endosperm of cycads, containing archegonia, 2 to 2½ hours; median longitudinal sections, 4 or 5 mm. thick, through ovulate cones of *Pinus Banksiana* may require 6 or 8 hours; if serial sections through the entire cone are wanted, Miss Aase found that the time must be prolonged to 24 or even 48 hours. When one is dealing with many lots of the same kind of material, as in research work, the time required for infiltration is easily determined. As a rule, *minimize heat.* It is, probably, never necessary to use paraffin with a melting-point higher than 52° C. With Land's cooling device sections 1 μ in thickness can be cut from 52° C. paraffin.

IMBEDDING

Material may be imbedded in paper trays, watch crystals, or any apparatus made for the purpose. The most satisfactory imbedding-dish we have used is a thin rectangular porcelain dish glazed inside. This dish, called a *Verbrennungsschale,* is made by the Königliche Porzellan-Manufactur, Berlin, Germany. The most convenient sizes are 40×50×10 mm., 68×45×10 mm., and 91×58×15 mm. As listed, these dishes are not glazed; care should be taken to indicate that the dishes must be glazed inside (*innen glasiert*). Where these dishes are not available, any dishes of convenient size and shape can be used. The paper tray, if well made, is as good as anything. Thick ledger linen or thin, smooth cardboard make good trays.

Smear the dish or tray with glycerin or soapy water to prevent sticking. Another way to prevent sticking is to put a piece of tissue paper in the dish, pour on water and make the tissue paper fit the inside of the dish, and then pour on the paraffin with the material to be imbedded. The paraffin will not stick to the paper. If several objects are to be imbedded in one dish, it is best to have the dish as near the temperature of melted paraffin as possible; otherwise the objects may stick to the bottom, and it will be impossible to arrange them properly. Hot needles are good for arranging material. Great care should be taken not to have the dish too hot, since too high a temperature not only injures the material, but also prevents a thorough imbedding. Pour the paraffin with the objects into the imbedding-dish and arrange them so as to facilitate the future cutting-out from the paraffin cake. Look at Figures 22 and 23, representing the arrangement of root-tips in a paraffin cake. From a cake like that in Figure 22 it is easy to cut out tips for sectioning. The arrangement, or rather the lack of it, shown in Figure 23 should be remembered only as an exasperating example.

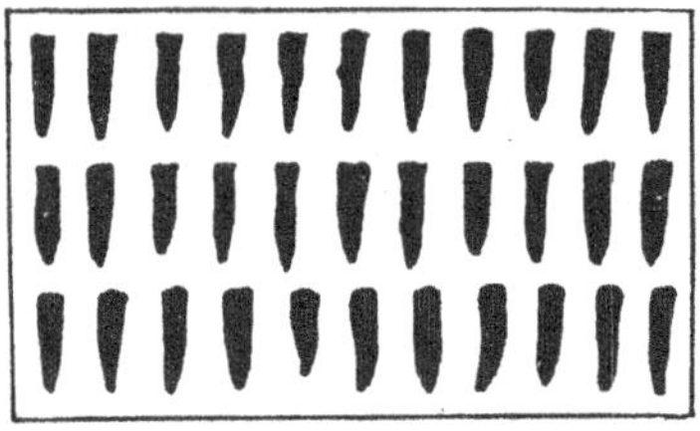

FIG. 22

FIG. 23

FIGS. 22, 23.—Paraffin cakes of root-tips, the upper (Fig. 22) showing a good arrangement, the lower (Fig. 23) showing fewer tips and most of these not in position to be blocked without injury.

After the objects have been arranged, cool the cake rapidly by allowing the bottom of the dish to rest upon cold water. As soon as a sufficiently firm film forms on the surface of the cake, let water flow gently over the top. After the cake has been under water for a few minutes, the paraffin will either come out and float on the water or, at least, it will be easily removed from the dish. If paraffin cools slowly it crystallizes and does not cut well. The layer of paraffin should be just thick enough to cover the objects, not only as a matter of economy, but because a thick layer retards the cooling. Very small objects, like the megaspores of *Marsilea*, ovules of *Silphium*, etc., may simply be poured out upon a *cool* piece of glass, which has been smeared with glycerin or soapy water. In this way, thin cakes are made which harden very rapidly.

If one is doing much imbedding, it is worth while to have the paraffin cakes uniform in size and to have a convenient method of filing. In our own collection, there are more than 6,000 paraffin cakes. They are filed in pasteboard boxes 28 cm. long, 10 cm. wide, and 2 cm. deep.

With the generic name written on the box, it is easy to find anything in the large collection.

CUTTING

As soon as the paraffin is thoroughly cooled, it is ready for cutting. Trim the paraffin containing the object into a convenient shape, and fasten it upon a block of wood. Blocks of pine $\frac{3}{4}$ inch long and $\frac{3}{8}$ inch square are good for general purposes. Put paraffin on the end of the block so as to form a firm cap about $\frac{1}{8}$ inch thick. Warm the cap and the bottom of the piece containing the object, and press them lightly together; then touch the joint with a hot needle, put the whole thing into cold water for a minute, and it is ready for cutting. Cutting can be learned only by experience, but a few hints may not come amiss:

a) Keep the knife *sharp*. There should be two hones, one for use when the knife is rather dull and the other for finishing. For the first hone, nothing equals a fine carborundum hone. About 5.5×22.5 cm. is a good size. A hard Belgian hone, of the same size, may be a little better for finishing. Flood the stone with water, and rub it with the small slip which accompanies all high-grade hones; this not only makes a lather which facilitates the sharpening, but it also keeps the surface of the hone flat. As soon as the edge of the knife appears smooth and even under a magnification of 30 or 40 diameters, the sharpening is completed with a good strop. It is better to sharpen the knife every time you use it. A first-class microtome knife, in perfect condition, is unsurpassed for cutting paraffin sections, but it requires both time and skill to keep the edge perfect. More than 20 years ago we began to experiment with the Gillette safety-razor blade and devised several holders for it, some of them more or less successful. Mr. Strickler finally perfected a holder which has already been mentioned. In using this holder the blade should not project more than 1 mm. The Gillette blade is harder than the ordinary microtome knife, and is so sharp that the edge appears smooth, even under a high-power dry lens. The bevel is about the same as that of a microtome knife which has been "backed" for sharpening. With this blade in the holder as made by Mr. Strickler or Mr. Larsen, we have cut smooth ribbons of *Selaginella* strobili, sections through the sporangium region of the whole plant of *Isoetes*, sections of stems of *Cucurbita*, in fact, we have not used an ordinary microtome knife for cutting paraffin ribbons for more than 15 years. Many fail at the first attempt and go back to the continual drudgery of sharpening microtome knives. If the holder is placed in the microtome at the angle used for a microtome knife, failure is certain, because the blade will scrape rather than cut. The angle should be considerably nearer vertical than in the case of a microtome knife. A study of Figure 6 should enable anyone to secure the proper angle.

b) Keep the microtome well oiled and *clean*.

c) Trim the block so that each section shall be a *perfect rectangle*.

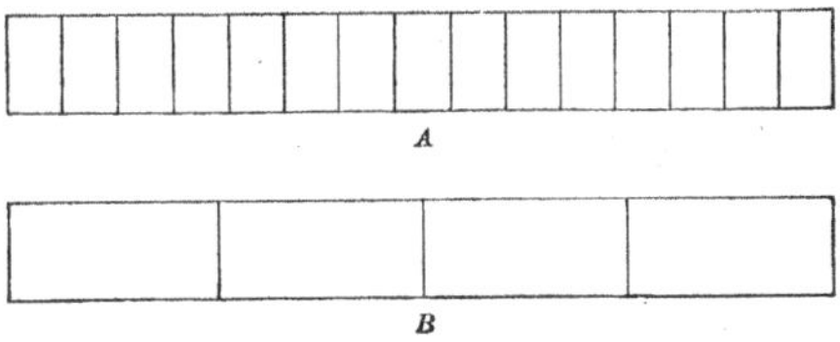

FIG. 24.—The ribbon

A ribbon of sections like that shown in Figure 24*A* is much better than one like *B* of the same figure, because sections will usually come off in neater ribbons if the knife strikes the longer edge of the rectangle, so that the sections are united by the longer sides rather than by the shorter. Crooked ribbons are caused by wedge-shaped sections, and are always to be avoided, because they make it difficult to economize space, and also because they present such a disorderly appearance. The knife, which should be placed at a right angle to the block and not obliquely, should strike the *whole edge* of the block at once, and should leave in the same manner.

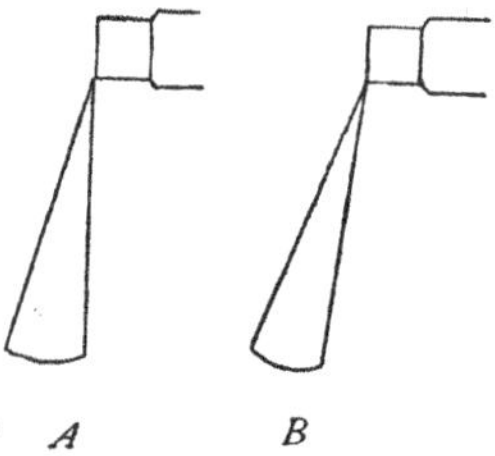

FIG. 25.—Position of knife

If sections stick to the knife, it may be that the knife is too nearly parallel with the surface of the block, as in Figure 25*A*. By inclining the knife as in Figure 25*B*, this difficulty is often obviated. A split or scratch in the paraffin ribbon may be caused by a nick in the knife. Use some more favorable position of the edge, or sharpen the whole knife. A split or a scratch in the ribbon is often caused by some hard granule which becomes fastened to the inner side of the edge of the knife. This is the most common cause of the difficulty. Simply wipe the knife by an upward stroke of the finger, slightly moistened with xylol. Do not use a cloth.

Sometimes good sections can be cut with a rather slow stroke when a rapid stroke fails. When paraffin is rather hard, sections may sometimes cut nicely at 5 μ, when, at 10 μ, ribbons cannot be secured. If very thin sections are desired and the paraffin seems too soft, cool the paraffin and the edge of the knife with ice, by Land's cooling device, or by pressing a piece of ice against the paraffin block and the knife. Sometimes hard paraffin does not ribbon well. This difficulty may be remedied by dipping a hot needle in soft paraffin and applying it to the opposite edges of the block to be cut. Often the mere warming of the opposite edges of the block with a hot needle is sufficient.

Another method, suggested by Dr. Land to facilitate the cutting of difficult material, has been tested in this laboratory and has been found to be very effective. Paraffin absorbs a small amount of water, or water penetrates between the crystals of paraffin. At any rate, water reaches cell walls and, perhaps, other structures which have not been completely infiltrated, and thus softens them. The paraffin cakes may be left for weeks in water. Cakes of class material may be put in water in a fruit can and kept until ready for use. After such treatment, smooth ribbons may be cut from material which would hardly cut at all without it.

A ribbon carrier is very convenient. A good carrier can be made by mounting a couple of spools 15 or 20 inches apart, with a strong piece of cloth for a band. More elaborate carriers may be made if one has tools.

FIXING SECTIONS TO THE SLIDE

Mayer's Fixative.—Sections must be firmly fixed to the slide, or they will be washed off during the processes involved in staining. Mayer's albumen fixative is excellent for this purpose. Formula:

White of egg (active principle)	50 c.c.
Glycerin (to keep it from drying up)	50 c.c.
Salicylate of soda (antiseptic, to keep out bacteria, etc.)	1 g.

Shake well and filter through cheese-cloth. It will keep from 2 to 6 months, but, to say the least, it is never better than when first made up. Of course, white of egg may be used alone, since the other two ingredients are merely incidental. Put a small drop of fixative on the slide, smear it evenly over the surface, and then wipe it off with a clean finger until only a scarcely perceptible film remains; then add several drops of distilled water and float the sections or ribbons on the water. Warm gently until the paraffin becomes smooth and free from wrinkles. Wrinkled or curved ribbons may be straightened by touching with a needle at each end and pulling gently, just as the ribbon begins to smooth out in the warming. Be careful not to melt the paraffin, for the albumen of the fixative coagulates with less heat than is required to melt the paraffin. If the paraffin should melt, run some cold water under it, and transfer the ribbon to another slide, prepared with fixative and water. It is a very good plan to put the slide on a metal bath or warming plate, like that shown in Figure 11. After the sections have become smooth, remove the surplus water and leave them on the bath with a couple of thicknesses of blotting paper under them for 3 or 4 hours, or, better, over night. If the fixative is used alone, as is often the case when sections are very thick, none of the delay is necessary, since the sections are merely laid upon the fixative and pressed down gently with the finger.

Land's Fixative.—Mayer's fixative is so easily prepared and it keeps so well that it is in universal use; but, in many cases, it will not hold the section to the slide. Moss archegonia and moss capsules are likely to wash off, especially if cut rather thick. Large sections of cones of conifers are almost sure to float off as soon as the slide comes into the xylol or alcohol. Sections of ovules of cycads, as soon as they attain a length of 1.5 to 2 cm., are likely to wash off. For handling these more difficult cases, Dr. Land devised a fixative which has proved satisfactory, even in such extreme cases as sections of ovulate cones of *Pinus Banksiana* 2 cm. long. Formula:

Gum arabic	1.0 g.
Bichromate of potash	0.2 g.
Water	100.0 c.c.

The mixture will not keep; the formula is given merely to indicate its composition. Make a 1 per cent solution of gum arabic in water, which will keep as well as Mayer's fixative; but make the bichromate solution immediately before using. Do not make the solution stronger than 1 per cent; usually 0.2 per cent is strong enough. Dr. Land does not measure, but simply adds enough bichromate crystals to make the water pale yellow.

Smear a few drops of the 1 per cent solution of gum arabic on the slide; flood with the bichromate solution; warm to straighten the ribbons; drain off the excess water and let the preparation dry in the light. The exposure to light renders the gum insoluble in water. LePage's glue or Mayer's albumen fixative may be used instead of gum arabic.

The foregoing directions are taken from Dr. Land's notes.

With the ordinary Mayer's albumen fixative the bichromate of potash, without the gum arabic, may be used in floating out ribbons, and makes a stronger fixative than the Mayer's formula.

Szombathy's Fixative.—

Gelatin....................................	1 g.
Distilled water.............................	100 c.c.
Salicylate of soda (a 2 per cent solution)........	1 c.c.
Pure glycerin...............................	15 c.c.

Dissolve the gelatin in water at 30° C., add the salicylate of soda, shake well, cool, and filter through cheese-cloth; then add the 15 c.c. of glycerin. The solution should be perfectly clear.

A couple of drops of the fixative, with a couple of drops of 2 per cent formalin, is rubbed on the slide. The sections are then added, and straightened out. The formalin makes the gelatin insoluble. The fixative is much like Land's and is used for difficult material which is not held by Mayer's fixative.

REMOVAL OF THE PARAFFIN

To remove the paraffin, place the slide in a Stender dish of xylol. About 5 minutes will be sufficient for sections 10 μ thick. The time may be shortened a little by gently warming the slide. *Never heat the slide enough to melt the paraffin. Never attempt to warm the paraffin over a lamp.* Overheating is ruinous.

Many prefer to remove the paraffin by pouring on xylol or turpentine. Hold the slide at an angle of 45°, and pour on a little xylol or turpentine. If the slide has been slightly warmed this should carry off the paraffin immediately. The reagent used in this first pouring cannot be used again. Now flood the slide several times with the turpentine or xylol, pouring the reagent back into the bottle.

REMOVAL OF XYLOL OR TURPENTINE

To remove the xylol, place the slide in equal parts of xylol and absolute alcohol in a Stender dish. After 5 minutes, transfer to absolute alcohol, which should also be allowed to act for 5 minutes.

If the pouring process is preferred and turpentine has been used to remove the paraffin, remove the turpentine by flooding the slide with 95 per cent alcohol. About 100 c.c. of turpentine and 200 c.c. of 95 per cent alcohol should be sufficient for 50 slides, even if the sections are to be mounted under the longest covers. By keeping both reagents in bottles and pouring the liquid on the slide, the reagents are always fresh. A given quantity of the reagent will prepare as many slides by one method as by the other.

TRANSFER TO THE STAIN

After the paraffin has been removed with xylol or turpentine, and the xylol or turpentine has been rinsed off with alcohol, the next step is the staining. If the stain is a strong alcoholic one (85 to 100 per cent alcohol), transfer directly to the stain. If the stain is in 70 per cent alcohol, pass through 95 and 85 per cent alcohol, 2 minutes in each, before staining. If an aqueous stain is to be used, pass down the whole series—95, 85, 70, 50, 35, and water—2 minutes in each, before placing the slide in the stain.

This is rather tedious, but, for cytological work, it seems to be necessary. For general morphological work, the slide may be transferred directly from the absolute or 95 per cent alcohol to any stain.

DEHYDRATING

After the sections have been stained, they must be dehydrated. If they have been stained in a strong alcoholic solution, transfer to 95 and then to 100 per cent alcohol, 2 minutes in each, if the stain does not wash out too rapidly. If stained in an aqueous solution, pass through the series, water, 35, 50, 70, 85, 95, and 100 per cent alcohol, about 2 minutes in each.

With stains which wash out rapidly, the times must be shortened and some of the alcohols must be omitted. With aqueous gentian-violet, all must be omitted except the 95 and 100 per cent, and even in these the time must be shortened to a few seconds.

CLEARING

After the sections have been dehydrated, they must be cleared, or made transparent by some clearing agent. The clearing agent must be a solvent of balsam, but it is not at all necessary that the balsam shall be dissolved in the particular clearing agent which has been used. Xylol balsam is used not only when preparations have been cleared in xylol, but also when they have been cleared in clove oil, cedar oil, bergamot oil, or other clearing agents.

Xylol is the most generally useful clearing agent. Place the slide in equal parts of xylol and absolute alcohol and then in pure xylol, allowing each to act for about 2 minutes.

Clove oil is also an excellent clearing agent. The clove oil should follow the absolute alcohol, without any mixtures. Pour on a few drops of clove oil, and drain them off at once in such a way as to carry with them whatever alcohol may still remain. Then flood the slide repeatedly with clove oil, draining the clove oil back into the bottle. If judiciously used, 50 c.c. of clove oil is enough to clear 100 preparations. Sections are usually cleared in a few seconds. The only objection to clove oil is that mounts harden slowly. To overcome this difficulty, the slide should be dipped in xylol for a minute before mounting in balsam.

Synthetic oil of wintergreen is much less expensive and some claim that it is just as good as clove oil.

For clearing sections on the slide, other clearing agents are hardly worth mentioning.

MOUNTING IN BALSAM

After the sections are cleared, wipe the slide on the side which does not bear the sections. Put on a drop of Canada balsam and add a clean,[1] thin cover. Before the cover is put on, pass it through the flame of an alcohol lamp to remove moisture, for it would be a pity indeed to injure a preparation at this stage of the process. Add a label, and the mount is complete.

[1] Slides and covers should be treated with hydrochloric acid, or equal parts of hydrochloric acid and water, for several hours. They should then be thoroughly rinsed in water and wiped with a cloth perfectly free from lint. After rinsing in water, they may be kept in 95 per cent alcohol. When a cover is needed for use, it is Dr. Land's practice to rest the corner of the cover on a piece of filter paper to remove the drop of alcohol; then pass the cover through the flame of a Bunsen or alcohol lamp. The film of alcohol will burn and the cover may warp, but it will usually straighten, and it will be clean and dry.

The mixture of sulphuric acid and bichromate of potash, used for cleaning laboratory glassware, is equally good for slides and covers.

A TENTATIVE SCHEDULE FOR PARAFFIN SECTIONS

It will be useful to give several tentative schedules for the use of beginners. It cannot be too strenuously insisted that *these schedules are only tentative*, their sole object being to give the beginner a start. The following is a tentative schedule for the ovary of a lily at any period before fertilization. The pieces should not be more than 12 mm. in length.

1. Chromo-acetic acid, 1 day.
2. Wash in water, 1 day.
3. $2\frac{1}{2}$, 5, 10, 15, 20, 30, and 50 per cent alcohol, 3 hours each; 70, 85, and 95 per cent alcohol, 4 hours each; absolute alcohol, 6 to 10 hours, changing 2 or 3 times.
4. Mixtures of absolute alcohol and xylol; $2\frac{1}{2}$, 5, 10, 15, 25, 50, 75, and 100 per cent xylol, 2 or 3 hours in each grade. Change the pure xylol once or twice.
5. Paraffin and xylol, 48 hours.
6. Melted paraffin in the bath, 30 to 40 minutes.
7. Imbed.
8. Section; about 10 μ is a good thickness.
9. Fasten to the slide.
10. Dissolve off the paraffin in xylol, 5 minutes.
11. Xylol and absolute alcohol, equal parts, 2 minutes; 100, 95, 85, and 70 per cent alcohol, 2 minutes each.
12. Stain in safranin (alcoholic), 6 hours or over night.
13. Rinse in 50 per cent alcohol, using a trace of HCl if necessary; then in 70, 85, 95, and 100 per cent alcohol, 2 minutes each.
14. Stain in gentian-violet dissolved in clove oil (or in clove oil with a little absolute alcohol), 10 minutes.
15. Treat with pure clove oil until the gentian-violet stain is satisfactory.
16. Rinse in xylol, 1 minute.
17. Mount in balsam.
18. Label.

That the paraffin method is tedious and complicated is universally recognized. Many substitutes have been tried, but without enough success to justify even a reference.

CHAPTER X

THE CELLOIDIN METHOD

The celloidin method has almost disappeared from botanical microtechnic, because material too hard for imbedding in paraffin can be cut without any imbedding at all, and material too delicate to be cut without a supporting medium can be imbedded in paraffin. But these two categories do not cover all the ground; celloidin still has its advantages. Years ago, a piece of rotten wood from an ancient Egyptian mummy case was brought to the writer for identification. It could be rubbed into powder in the fingers, and had to be handled gently to keep it from falling to pieces. It was cut very successfully in celloidin. Stems too hard for paraffin may be cut in celloidin when it is desired to preserve cell contents. Celloidin is still very valuable for most of the sections used in medical schools, because the sections can be prepared in great numbers and each student can take a section, add a drop of balsam and a cover, and have a preparation of his own ready to study. Where serial sections are necessary, as in most morphological and cytological work, the method is too tedious to be worth even a trial, unless the sections cannot be cut in any other way. Besides, most of the more valuable stains color the celloidin matrix, and if the matrix be removed, the more delicate elements may be displaced or even lost.

Celloidin and collodion are forms of nitro-cellulose. They are inflammable, but do not explode. Schering's celloidin, which is only a collodion prepared by a patented process, is in general use for imbedding. Granulated and shredded forms of celloidin are on the market, but the tablets are more convenient. Directions for making the various solutions accompany the celloidin. To make a 2 per cent solution, add to 1 tablet enough ether-alcohol to make the whole weigh 2,000 g. To make a 4 per cent solution, add another tablet, and to make a 6 per cent solution, add an additional tablet, and so on.

The collodion method was published by Duval[1] in 1879. Celloidin was recommended by Merkel and Schiefferdecker[2] in 1882.

[1] Duval, *Journal de l'anatomie*, 1879, p. 185.

[2] Merkel and Schiefferdecker, *Archiv für Anatomie und Physiologie*, 1882.

The principal features of the method are as follows: Material is dehydrated in absolute alcohol; treated with ether-alcohol; infiltrated with celloidin; imbedded in celloidin; hardened in chloroform or alcohol; after which it is cut, stained, and mounted.

Eycleshymer, who brought the celloidin method to a high degree of efficiency, published in 1892 a short account, which may be summarized as follows: Put the celloidin tablet, or fragments, into a wide-mouthed bottle, and pour on enough ether-alcohol (equal parts ether and absolute alcohol) to cover the celloidin. Occasionally shake and add a little more ether-alcohol until the celloidin is all dissolved. The process may require several days. The solution should have the consistency of a very thick oil. Label this solution Number 4. Solution Number 3 is made by mixing 2 parts of solution Number 4 with 1 part of ether-alcohol. Solution Number 2 is made by mixing 2 parts of Number 3 with 1 part of ether-alcohol. Solution Number 1 consists of equal parts of ether and absolute alcohol.

After dehydrating, the material is placed successively in solutions 1, 2, 3, and 4. For an object 2 mm. square, 24 hours in each solution is sufficient; for the brain of a cat, a week is not too long.

A paper tray may be used for imbedding. Pour the object, with the thick solution, into the tray and harden in chloroform for 24 hours; then cut away the paper and place the block in 70 per cent alcohol for a few hours. The material may be left indefinitely in a mixture of equal parts of 95 per cent alcohol and glycerin.

Before cutting, the object is mounted upon a block of wood. A block, suited to the microtome clamp, is dipped in ether-alcohol, which removes the air and insures a firmer mounting. Dip the end of the block of wood in solution Number 3, and the piece of celloidin containing the object in solution Number 1. Press the two firmly together, and place in chloroform until the joint becomes hardened.

Set the blade of the microtome knife as obliquely as possible. Both the object and the knife should be kept flooded with 70 per cent alcohol, and the sections, as they are cut, should be transferred to 70 per cent alcohol.

Stain in Delafield's haematoxylin for 5 to 30 minutes. Wash in water for about 5 minutes, and then decolorize in acid alcohol (2 to 5 drops of hydrochloric acid to 100 c.c. of 70 per cent alcohol) until the stain is extracted from the celloidin, or at least until the celloidin retains only a faint pinkish color. Wash in 70 per cent alcohol (not acid) until the characteristic purple color of the haematoxylin replaces the red due to the acid. Stain in eosin (preferably a 1 per cent solution in 70 per cent alcohol) for 2 to 5 minutes. Dehydrate in 95 per cent alcohol for about 5 minutes. Absolute alcohol must not be used, unless it is desirable to remove the celloidin matrix. Eycleshymer's clearing fluid (equal parts of cedar oil, bergamot oil, and carbolic acid) clears readily from 95 per cent alcohol. Mount in balsam.

If serial sections are necessary, arrange the sections upon a slide, using enough 70 per cent alcohol to keep the sections moist, but not enough to allow them to float. Cover the sections with a strip of toilet paper, which can be kept in place by winding with fine thread. After the sections have been stained and cleared, remove the excess of clearing fluid by pressing rather firmly with a piece of blotting-paper. Then remove the toilet paper and mount in balsam.

With occasional slight modifications, we have used the method as presented by Eycleshymer in his classes. Instead of the graded series of celloidin solutions, we use a 2 per cent solution, which is allowed to concentrate slowly by removing the cork occasionally, or by using a cork which does not fit very tightly. The material is imbedded when the solution reaches the consistency of a very thick oil. If the material is to be cut immediately, we prefer to imbed it and fasten it to the block at the same time. The blocks should have surface enough to accommodate the objects, and should be about ¼ inch thick. White pine makes good blocks; cork is much inferior. Place the block for a moment in ether-alcohol and then dip into the 2 per cent celloidin the end of the block which was left rough by the saw. With the forceps remove a piece of the material from the thick celloidin and place it upon the block, taking care to keep it right side up. Dip the block with its object first in thick celloidin, then in thin, and after exposing to the air for a few minutes drop it into chloroform, where it should remain for about 10 to 20 hours. It should then be placed in equal parts of glycerin and 95 per cent alcohol, where it may be kept indefinitely. If the material is hard, like many woody stems, it will cut better after remaining in this mixture for a couple of weeks.

The following schedules, beginning with the celloidin sections in 70 per cent alcohol, will give the student a start in the staining:

Delafield's Haematoxylin and Eosin.—

1. 70 per cent alcohol, 2 to 5 minutes.
2. Delafield's haematoxylin, 5 to 30 minutes.
3. Wash in water, 5 minutes.
4. Acid alcohol (1 c.c. hydrochloric acid+100 c.c. of 70 per cent alcohol) until the stain is extracted from the celloidin, or at least until only a faint pinkish color remains.
5. Wash in 70 per cent alcohol (not acid) until the purple color returns.
6. Stain in eosin (preferably a 1 per cent solution in 70 per cent alcohol), 2 to 5 minutes.
7. Dehydrate in 95 per cent alcohol, 2 to 5 minutes. Do not use absolute alcohol unless you wish to dissolve the celloidin, which is not necessary with this staining.
8. Clear in Eycleshymer's clearing fluid, usually 1 to 2 minutes, but sometimes 5 to 10 minutes.
9. Mount in balsam.

Safranin and Delafield's Haematoxylin.—

1. 70 per cent alcohol, 2 to 5 minutes.
2. Safranin (alcoholic), 6 to 24 hours.
3. Acid alcohol (a few drops of hydrochloric acid in 70 per cent alcohol) until the safranin is removed from the cellulose walls.
4. Wash in 50 per cent alcohol, 5 to 10 minutes to remove the acid.
5. Delafield's haematoxylin, 2 to 5 minutes.
6. Wash in water, 5 minutes.
7. Acid alcohol, a few seconds.
8. Dehydrate in 95 per cent alcohol, 2 to 5 minutes, then in absolute alcohol, 2 to 5 minutes, which will partially dissolve the celloidin.
9. Clear in clove oil, which will complete the removal of the celloidin.
10. Be sure that the sections are free from fragments of celloidin and then mount in balsam.

Jeffrey's improvements in the celloidin method have been described in considerable detail by Plowman.[1] Sections of hard stems and roots cut by this method could hardly be surpassed, and they are perfectly adapted to the requirements of photomicrography. The following is a brief abstract of Plowman's paper:

1. **Preparation of Material.**—Dead and dry material should be repeatedly boiled in water and cooled to remove air. An air-pump

[1] A. B. Plowman, The Celloidin Method with Hard Tissues, *Botanical Gazette*, 37:456–461, 1904.

may be used in addition. Living material may be fixed in a mixture of picric acid, mercuric chloride, and alcohol:

Mercuric chloride, saturated solution, in 30 per cent alcohol.......	3 parts
Picric acid saturated solution, in 30 per cent alcohol..............	1 part

Fix 24 hours, and wash by passing through 40, 50, 60, 70, and 80 per cent alcohol, allowing each to act for 24 hours.

2. **Desifilication, etc.**—Silica and other mineral deposits are removed by treating with a 10 per cent aqueous solution of commercial hydrofluoric acid. The material is tranferred to this solution from water or from the 80 per cent alcohol. The process may require 3 or 4 days, with one or two changes of the acid and frequent shaking of the bottle. An ordinary wide-mouthed bottle, coated internally with hard paraffin, should be prepared, since the acid is usually sold in bottles with narrow necks. The bottles are easily prepared by filling them with hot paraffin and simply pouring the paraffin out. Enough will stick to the bottle to protect the glass from the acid. Wash in running water 3 or 4 hours.

3. **Dehydration.**—Use 30, 50, 70, 90, and 100 per cent alcohol, allowing 12 hours in each grade.

4. **Infiltration with Celloidin.**—There should be ten grades of celloidin: 2, 4, 6, 8, 10, 12, 14, 16, 18, and 20 per cent. Transfer from absolute alcohol to the 2 per cent celloidin. (We should prefer a previous treatment with ether-alcohol.) The bottle should be nearly filled, and the stopper should be clamped or wired in place. Put the bottle on its side in a paraffin bath at 50° to 60° C. for 12 to 18 hours. Cool the bottle quickly in cold water, taking care that the water does not get into the bottle. Pour out the 2 per cent solution (which, as well as all other solutions, may be used repeatedly), and replace it with the 4 per cent, and proceed in the same way with the other grades. When the 20 per cent solution is reached, a further thickening is gained by adding a few chips of dry celloidin from time to time until the mixture is quite stiff and firm. Remove each block with the celloidin adhering to it and harden it in chloroform for 12 hours. Then transfer to a mixture of equal parts of glycerin and 95 per cent alcohol, where the material should remain for a few days before cutting.

Cutting, Staining, and Mounting.—Although 10 μ is usually thin enough, sections are readily cut as thin as 5 μ by this method.

Remove the celloidin before staining by treating 10 to 15 minutes with ether; then wash in 95 per cent alcohol and transfer to water, and then to the stain. Stain to a fairly dense purple in an aqueous solution of Erlich's haematoxylin; wash in dilute aqueous solution of calcium or sodium carbonate, and then in two changes of distilled water. Add a few drops of alcoholic solution of equal parts of Grübler's alcoholic and aqueous safranin, and stain to a rich red. A dilute stain acting 1 to 2 hours is better than a more concentrated stain acting for a shorter time. Transfer directly to absolute alcohol, clear in xylol, and mount in balsam.

Haidenhain's iron-haematoxylin is a very satisfactory stain for photographic purposes.

The celloidin method has its disadvantages as well as its advantages. It is extremely slow and tedious, and it is rarely possible to cut sections thinner than 10 μ, while, on the other hand, it gives smoother sections.

Succulent tissues, which are usually damaged by the paraffin method, are easily handled without any injury in celloidin. The fact that the method may be used without heat is often a further advantage. Stems and roots which cannot be handled at all in paraffin cut well in celloidin, and much larger sections can be cut than in paraffin, but most material of this kind can be cut without any imbedding.

When material is to be imbedded, use celloidin as a last resort. Use paraffin when you *can*, celloidin when you *must*.

CHAPTER XI

THE CELLULOSE ACETATE METHOD

We believe that the cellulose acetate method is destined to prove as successful with hard woody tissues as the Venetian turpentine method has proved with unicellular and filamentous forms; but, it must be confessed that investigators have been unsuccessful in their experience with cellulose acetate. Nevertheless, refractory material, like oak and even harder woods, has yielded thin transverse sections. Cellulose acetate does not injure the finer details of structure, and, on that account, will be superior to hydrofluoric acid, when students have solved the present uncertainties in regard to it. We have such confidence in the value of the method that we are quoting, in full, Mrs. Williamson's short account in the *Annals of Botany* for January, 1921, which is the only published information upon the subject:

A NEW METHOD OF PREPARING SECTIONS OF HARD VEGETABLE STRUCTURES

In order to prepare sections of hard vegetable structures it is essential that some method should be devised by which the structure is not only embedded but softened, so that sections can be cut easily and smoothly. After various methods had been tried, the cellulose acetate method successfully used by Dr. Kernot for embedding and sectioning the fabric of aeroplane wings was used. It was discovered that this method not only embedded hard vegetable structures, but also softened them so that sections are easily obtained. It proved best to use cellulose acetate of French manufacture made from pure cellulose, as the viscosity is more uniform than that of English manufacture, which is obtained from the cellulose of wood.

In the preliminary experiments pieces of oak and beech, cut into half-inch cubes, were passed through strengths of alcohol, then placed in pure acetone for two hours and finally into a 12 per cent solution of cellulose acetate in acetone. There they were left for two months and excellent sections were obtained. Further experiments showed that the passage through alcohols was unnecessary. In the final experiments the pieces of wood were placed in water and the air removed from them, after which they were put into pure acetone for 1 to 2 hours and finally into the solution of cellulose acetate. It was found that the length of time of immersion in the solution of cellulose acetate necessary for softening the tissues varied with the hardness of the wood, the minimum time for soft woods being two days; for woods such as oak and beech, at least six days are required. Experiments were tried with sâl (*Shorea robusta*) and Pyingadu (*Xylia dolabriformis*), one of the Indian ironwoods, which is extremely hard. After fourteen days in the cellulose acetate solution it was possible to obtain transverse sections of these hard woods. The cellulose acetate solution is therefore capable of softening even the hardest wood in a relatively short time.

In order to stain sections—either hand or microtome—obtained by this method, it is necessary to wash them in pure acetone for 1 to 2 minutes to remove the cellulose acetate, wash in alcohol 1 to 2 minutes, and pass on to the stains selected. Various staining methods for cell walls—such as anilin chloride, methylene blue, and Congo red, ammoniacal fuchsin and Kleinenberg's haematoxylin, etc.—were tried with success. A comparison with stained sections of untreated wood revealed no differences. Delicate tissues in the wood and hyphae of fungi infecting the wood also stain well and are unaffected by the treatment.

A satisfactory method of preparing sections of hard vegetable structures is therefore supplied by the use of a 12 per cent solution of cellulose acetate in pure acetone for softening and embedding.—H. S. WILLIAMSON, Imperial College of Science and Technology.

Correspondence with Mrs. Williamson indicates that the various brands of cellulose acetate behave differently. Cellulose acetate obtained from wood is unsatisfactory. We found that cellulose acetate made from photographic films was also unsatisfactory. Mrs. Williamson used a cellulose acetate sold by Cellon (Richmond) Ltd., 22 Cork Street, London, England, and manufactured by the Société Chemique des Usines du Rhone. The time may be shortened by keeping the temperature at 40° C.

In making the solution, use 12 g. of cellulose acetate to 100 c.c. of pure acetone.

It would be worth while to try to adapt this method to the paraffin method.

CHAPTER XII

SPECIAL METHODS

In the preceding chapters it has been our object to present the principal methods of preparing plant material for microscopic examination in such a way that good preparations may be obtained as soon as possible. The schedules afford an easy introduction to the various methods, but it is hoped that the methods have been so presented that the student has become grounded in the fundamentals, because some modification is likely to be necessary in nearly every piece of research. A few methods designed to meet special difficulties are given in this chapter and others are mentioned in the second part of the book, in connection with various laboratory types.

VERY LARGE SECTIONS

It is sometimes desirable to cut very large sections. Sections as large as a cornstalk may be cut freehand or in celloidin. A section of a stem of *Zamia* 5 or 6 cm. in diameter is difficult to handle by the usual methods. If a large microtome, such as is used in cutting complete sections of large brains, is available, the piece of stem is easily held for the cutting. Some of the medium-sized sliding microtomes now have a rigid clamp which will grip a block 3 cm. square. The lower part of the piece can then be trimmed to fit the clamp, leaving the upper part round, so that sections across a stem 6 or 7 cm. in diameter may be cut without much difficulty. With a rather soft stem, like *Zamia*, the surface must be flooded with 95 per cent alcohol after each section, if it is desirable to cut thin sections. From stems 3 cm. in diameter, sections can be cut at about 20 to 30 μ. If the section is not more than 3 or 4 cm in diameter, it can be mounted on a 50×75 mm. slide. Sections 6 or 7 cm. in diameter can be mounted on lantern slides; if large covers are not available, use another lantern slide for a cover. It will be easier to get neat mounts if the cover is cut down so as to leave a margin 2 or 3 mm. wide. It is not easy to mount a thick section between 2 lantern slides of the same size, on account of the balsam which oozes out at the edges. Such preparations may be used directly as lantern slides. Large sections of the stem of a tree fern make good mounts without any staining. Large sections of cornstalk are rather hard to cut, because the rigid bundles tear through the soft parenchyma. Flooding with 95 per cent alcohol facilitates the process. A slight hardening is sufficient, so that about 4 or 5 sections can be cut in a minute. The corn stem, and similar structures, can be cut in paraffin by Miss Langdon's method, described on page 93.

STONY TISSUES

Sections of the stony tissues of hickory nuts, walnuts, peach stones, and similar refractory substances cannot be cut by ordinary methods.

With a fine saw, saw sections about 1 mm. in thickness. Rub a section between 2 pieces of fine sandpaper until it is not more than half a millimeter in thickness. Then rub it between 2 smooth hones, keeping the hones wet with water. When the section is thin enough, wash it thoroughly in water, using a pipette to rinse off any particles of dirt. Dehydrate in absolute alcohol, clear in clove oil, and mount in balsam. The long, narrow pores show better without any clearing. In this case, dry the section thoroughly, heat a few drops of balsam on the slide to drive off the solvent, put the section into the balsam, and add a cover. The air caught in the long, narrow pores will make them appear as black lines. Sections of most nuts show excellent detail without any staining. Thin sections, however, may be stained in the usual way.

PETRIFACTIONS

Paleobotany has made such rapid progress during the last ten years that scarcely any problem involving the anatomy of living vascular plants can be investigated intelligently without some knowledge of Mesozoic and Paleozoic forms. Material, especially that of Paleozoic Pteridophytes and Gymnosperms, is becoming available, and consequently it is increasingly necessary for laboratories to have apparatus and technic for cutting rock-sections.

The outline of the process of cutting a rock-section is very simple:

1. Saw the rock into two pieces.
2. Polish the cut surface.
3. Fasten the cut surface to a piece of glass with hot shellac.

4. With the saw, make another cut, as close to the glass as possible, so as to leave a thin section firmly fastened to the glass.
5. Grind and polish until the section is as thin as possible, or as thin as you want it.
6. Wash all polishing powder off with water.
7. Dry completely and, either with or without moistening in xylol, mount in balsam.

A word of suggestion in regard to these various points may not be amiss.

1. Most rock-sections are cut with a rather expensive and quite complicated instrument, called a "petrotome." The saw is of the circular type, is made of tin or other soft metal, has no teeth, but has diamond dust driven into the margin. A rigid clamp holds the object, and the saw, constantly cooled by a stream of water, gradually cuts through the specimen. If the piece to be cut is more than 5 or 6 cm. in diameter, a band saw is better; and if the piece is 10 or 20 cm: in diameter, the band saw is necessary.

2. The cut surface is most easily polished on a revolving brass plate, kept wet and liberally powdered with fine carborundum. When the surface has become even and smooth, the specimen is ready for the next step.

3. Fasten the polished surface to the glass slide upon which the section is to be mounted. Plate glass 3 or 4 mm. thick is best for sections larger than 3 or 4 mm. square. Gradually heat the slide until it is quite hot. Melt upon the slide the thin flakes of white shellac used by painters; heat the object and press the polished surface *very firmly* into the melted shellac. Canada balsam, from which the xylol has been driven off by heating, can be used instead of shellac. Much of the Paleozoic material is in the form of coal balls. After the ball has been cut in two, it is often difficult to hold the hemispherical piece in a clamp, especially if the piece is small. In such cases, it is better to fasten the polished surface to a convenient piece of marble, about 2.5 cm. thick, and 5 or 6 cm. square. The marble is easily held in the clamp. As soon as the slide, or marble, and object are cool, the next cut can be made.

4. Fasten the object in the clamp and saw as close to the glass, or marble, as possible, thus leaving a thin section cemented to the slide or marble. If marble has been used, the section is removed by heating or by dissolving it off with xylol. It can then be fastened

to the glass slide for grinding and polishing. Anyone who can handle tools should soon be able to cut a section 1 mm. thick. A skilled technician can cut sections as thin as 0.5 mm.

5. The second grinding must be very careful and accurate. Do the polishing on the revolving disk. The glass slide allows one to note how the process is progressing.

6. When the section becomes thin enough, or even before if it begins to crack, wash off the powder.

7. It is usually a good plan to use rather thick balsam for mounting, even if it should be necessary to heat it a little to make it flow well.

By this method, sections of fossils 10×15 cm. have been cut thin enough for examination with a 4 mm. objective. Sections 3 or 4 mm. square have been cut thin enough for satisfactory examination with a 2 mm. oil immersion lens. Figure 26 shows that a reliable study can be made from sections cut from the solid rock.

Of course, this method can be used for such objects as walnut and hickory shells.

THICK SECTIONS

It is sometimes desirable to make very thick sections to show general topography rather than detail. A longitudinal section of the fully grown ovule of *Ginkgo* or a cycad may be cut as thick as 3 to 5 mm. so as to include the entire group of archegonia. A slab can be cut from each side of the ovule with a fine saw, and a razor can be used for smoothing. If the section is from fresh material it should be fixed, washed, etc., with about the same periods as if it were to be imbedded in paraffin. When thoroughly cleared in xylol, the section should be put into a flat museum jar of suitable size and kept in xylol. Even before the stony coat of a cycad becomes too hard to be cut with a razor such thick sections are very instructive. Stain very lightly, or not at all. Sections of *Zamia* or other cycad stems, 2 mm., or even 5 mm. thick, make instructive mounts, since they show the peculiar course of the bundles, a feature which is largely lost in thin sections.

Krause prepared large objects very effectively by dehydrating, clearing in xylol, and then transferring to cedar oil. Sections of an apple, either longitudinal or transverse, about 3 or 4 mm. thick, cleared in this way, are very instructive. Strawberries, gooseberries, and similar objects treated in this way afford a kind of study which is too often neglected.

Miss Langdon cleared 15-mm. cubes of Dioon stem in a similar way, but used equal parts of xylol and carbon disulphide instead of cedar oil. The course of the bundles could be traced perfectly.

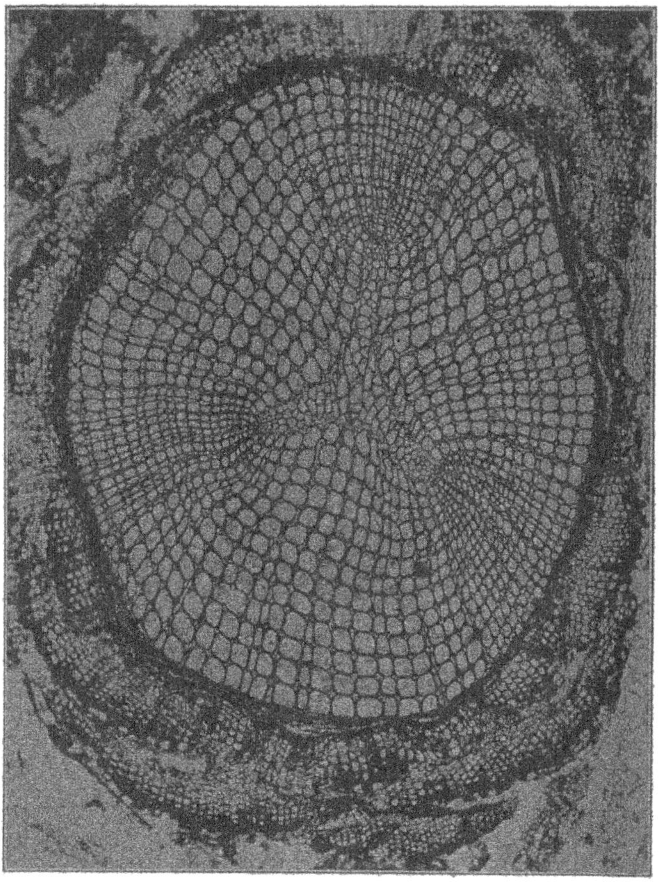

FIG. 26.—*Sphenophyllum pleurifoliatum:* transverse section of a stem, cut from a coal ball of the Upper Carboniferous. Eastman Commercial Ortho film, Wratten E (orange) filter, J. Swift and Son 1-inch lens; arc light; exposure, 1 second. Slide by Lomax, negative by Sedgwick. ×19.

LAND'S GELATIN METHOD

It is sometimes desirable to get sections of partly disorganized material. A matrix is necessary to hold the parts in place, but dehydration may make the tissue unnecessarily hard to cut.

Soak ordinary gelatin (which can be obtained at the grocery) in water until no more is taken up. Then drain off the excess water and liquefy the gelatin by heating. Place the material—previously soaked in water—in the melted gelatin and keep it there for several hours. Place also in the gelatin some small blocks of hard wood to serve as supports in the microtome. The material to be sectioned is oriented in a gelatin matrix on the supporting blocks, cooled until the gelatin sets, and then placed in strong formalin to harden the gelatin. In cutting, flood the knife with water.

If the material is to be stained, stain it in bulk before putting it into the gelatin, since the gelatin stains very deeply. Of course, the gelatin could be dissolved with hot water, or hot water and acetic acid, but all the advantage of a matrix would be lost.

It would be worth while to try this method thoroughly with soft, succulent tissues and with hard tissues which become still harder if dehydrated.

SCHULTZE'S MACERATION METHOD

Various solutions are used to separate a tissue into its individual cells. These solutions dissolve or weaken the middle lamella so that the cells are easily shaken or teased apart. Schultze used strong nitric acid and potassium chlorate. Put the material, which should be in very small pieces, into a test-tube; pour on just enough nitric acid to cover it, and then add a few crystals of potassium chlorate. Heat gently until bubbles are evolved, and let the reagent act until the material becomes white. Four or five minutes should be sufficient. The fumes are disagreeable and are very injurious to microscopes. Pour the contents of the tube into a dish of water. After the material is thoroughly washed in water, it may be teased with needles and mounted, or it may be put into a bottle of water and shaken until many of the cells become dissociated.

After a thorough washing in water, the material may be stained. The large tracheids of ferns, dissociated in this way and stained in safranin or methyl green, make beautiful preparations.

JEFFREY'S MACERATION METHOD

A method which I saw at Toronto and which gives much better results was credited to Professor Jeffrey. Wood is cut or split into sections about 300 μ thick, which are then boiled and cooled until free from air. The macerating fluid consists of equal parts of 10 per cent nitric acid and 10 per cent chromic acid. The time will vary with different woods, but is likely to be about 24 to 48 hours if the temperature is about 35° C. When properly macerated the cells may be shaken apart or are very easily teased apart. Before staining, the material should be very thoroughly washed to remove the acid. A study of such material is very valuable in modern anatomical work.

Maceration methods which act in a few minutes are likely to be so violent that fine details will not be preserved.

PROTOPLASMIC CONNECTIONS

As a rule, protoplasmic connections are not likely to be seen in an ordinary preparation. It used to be thought that the rather large protoplasmic strands seen at the sieve plates of the pumpkin and other Cucurbitaceae were exaggerated examples of protoplasmic continuity; but, as a matter of fact, the large strands do not extend entirely through the plate. The real continuity, through the middle lamella, is scanty and hard to demonstrate.

Very satisfactory material for the demonstration of the connecting strands is furnished by the seeds of the Japanese persimmon, *Diospiros Kaki*. Fix in formalin alcohol (10 c.c. formalin to 50 c.c. of 70 per cent alcohol) or in a mixture of equal parts 70 per cent alcohol and glycerin (Fig. 27).

Good views of the strands are most abundant in sections cut parallel with the flat surface of the seed. No imbedding is either necessary or desirable. Clamp the seed in the microtome directly, or fasten it to a wooden block with cellulose acetate (easily made by dissolving a photographic film in acetone; of course the emulsion should be removed), with celloidin, or even with glue. Cellulose acetate seems to be the best. Cut sections 8 to 12 μ thick; place them in ether several hours to remove any fatty substances; remove the ether with absolute alcohol; transfer to 95 per cent alcohol, then to 50 per cent alcohol, and then to water. Stain in iron-alum haematoxylin. The following schedule for *Diospiros Kaki* will introduce the method.

1. Fix in formalin alcohol or in glycerin alcohol 4 or 5 days.
2. Cut sections 8 to 12 μ thick.
3. Ether, several hours.
4. Absolute alcohol, 95 per cent and 50 per cent alcohol, 10 minutes each.
5. Wash in water, 5 minutes.
6. Iron alum, 4 per cent, 8 to 24 hours.
7. Wash in water, 30 minutes.
8. Stain in ½ per cent haematoxylin 24 hours.
9. Wash in water. At this stage, examine carefully, because it may not be necessary to reduce the stain in iron-alum. If the connections are not too deeply stained, simply dehydrate, clear, and mount in balsam.

After the first 5 stages, a strong stain in Delafield's haematoxylin, 10 to 24 hours, followed by a very weak hydrochloric acid, has given

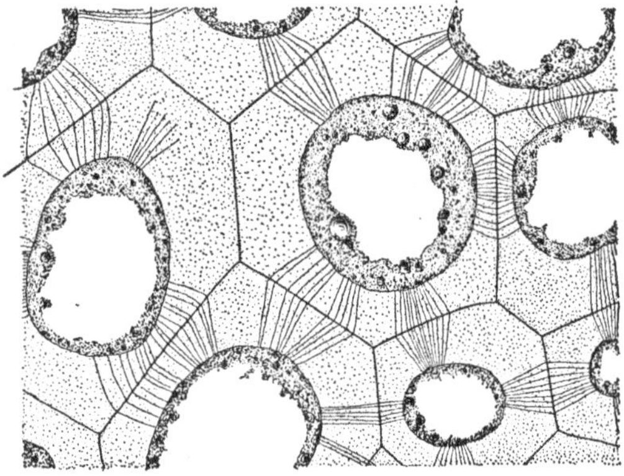

FIG. 27.—*Diospiros Kaki:* section of endosperm prepared as described for this species. ×780

good results. A sharp stain in crystal violet, differentiated with orange in clove oil, often fails, but sometimes succeeds; and, when successful, the connections stand out beautifully.

The endosperm of *Phytelephas* (vegetable ivory), of dates and many other palms, and probably most hard endosperms, will show the connections by the methods just described; but in many cases it is necessary to resort to special methods in order to demonstrate the continuity. In these special methods a reagent is used which causes the membranes to swell before the stain is applied. It is only by such an exaggeration that the more delicate connections can be shown.

Put thin sections of fresh material into a mixture of equal parts of sulphuric acid and water; and allow the reagent to act for 2 to 10 seconds. Wash the acid out thoroughly in water and stain in anilin blue. According to Gardiner, this stain should be made by adding 1 g. of the dry stain to 100 c.c. of a saturated solution of picric acid in 50 per cent alcohol. The staining solution is then washed out in water, and the sections are mounted in glycerin. The sections may be dehydrated, cleared in clove oil, and mounted in balsam. The anilin blue may be used in 50 per cent alcohol acidulated with a few drops of acetic acid.

Chloroïodide of zinc may be used instead of sulphuric acid. Treat the fresh sections for 2 hours with the iodine and potassium-iodide solution used in testing for starch; then treat about 12 hours with chloroïodide of zinc. Wash in water and stain in anilin blue. Examine in glycerin.

Meyer's pyoktanin method is one of the best. The reagents are as follows:

1. Iodine, potassium iodide solution: iodine 1 part, potassium iodide 1 part, water 200 parts.
2. Sulphuric acid 1 part, water 3 parts; this mixture to be saturated with iodine.
3. Pyoktanin coeruleum 1 g., water 30 c.c. This pyoktanin is a very pure methyl violet obtained from E. Merck in Darmstadt.

Put sections of the date seed into a watch glass full of the first solution, and allow it to act for a few minutes; then mount in a drop of the solution. The connections will be only very faintly stained, showing a slightly yellowish color. At the edge of the cover, add a drop of the second solution. The preparation will darken a little. Then allow a small drop of the third solution to run under the cover. Allow the stain to act for about 3 minutes. Then plunge the whole preparation into water. The action should be stopped before the entire section has become blue. Now wash the section quickly. If there are annoying, granular precipitates, remove them with a soft brush. Mount in glycerin. The membrane should be a clear blue, while the protoplast and connections should be a blue black.

The following is Strasburger's modification of Meyer's method, and shows the connections with great distinctness:

Permanent preparations may be secured by the following method:

1. Treat the fresh sections with 1 per cent osmic acid, 5 to 7 minutes.
2. Wash in water 5 to 10 minutes.
3. Treat with a solution of iodine in potassium iodide (0.2 per cent iodine and 1.64 per cent potassium iodide), 20 to 30 minutes.
4. Transfer to 25 per cent sulphuric acid, which should act for at least half an hour; 24 hours may be necessary.
5. Bring the sections into 25 per cent sulphuric acid which has been saturated with iodine. Add a drop of Meyer's pyoktanin solution (1 g. pyoktanin coeruleum as sold by E. Merck in Darmstadt in 30 c.c. of water).

In about 5 minutes the sections will be stained sufficiently and can be examined in glycerin. If there are annoying precipitates, remove them with a soft brush.

According to Meyer, the swelling is an advantage only when the walls are very thin. When the walls are thick, the connections show better without any previous swelling.

Try the following method with the seeds of *Diospiros, Latania Chamerops, Phoenix,* or *Phytelephas:* Soak in water and cut thin sections. Extract the oily and fatty substances with xylol; wash in 95 per cent, or in absolute alcohol; stain in anilin blue (Hoffman's blue 1 g. dissolved in 150 c.c. of 50 per cent alcohol) for a few minutes. Examine in glycerin. This method succeeds very well with seeds of the date, which is sold at all groceries.

Permanent preparations may be secured by the following method:

1. Fix in 1 per cent osmic acid, or in absolute alcohol, 5 to 10 minutes.
2. Stain for 24 hours in Delafield's haematoxylin.
3. Wash for a few minutes in acid alcohol (5 drops of hydrochloric acid in 50 c.c. of 70 per cent alcohol).
4. Wash for a few minutes in ammonia alcohol (5 drops of ammonia to 50 c.c. of 70 per cent alcohol).
5. Dehydrate in absolute alcohol, clear in xylol, and mount in balsam.

STAINING CILIA

The cilia of the large spermatogoid of *Ginkgo* and the cycads stain beautifully in iron-alum haematoxylin, which not only stains the cilia but even differentiates the free portion from the part between the blepharoplast and the surface.

The cilia of sperms of Bryophytes and Pteridophytes stain better with gentian-violet or crystal-violet. The periods are long; not less than 30 minutes, and often several hours will be required.

The cilia of the motile spores of Thallophytes may often be demonstrated by allowing a drop of the iodine solution used in testing for starch to run under the cover.

Zimmermann gives the following method: Bring the objects into a drop of water on the slide and invert the drop over the fumes of 1 per cent osmic acid for 5 minutes. Allow the drop to dry. Then add a drop of 20 per cent aqueous solution of tannin, and after 5 minutes wash it off with water. Stain in a strong aqueous solution of fuchsin (or carbol fuchsin) for 5 minutes. Allow the preparation to dry completely, and then add a drop of balsam and a cover. The cilia should take a bright red.

Zimmermann also found the following method satisfactory for the cilia of the zoöspores of algae and fungi: Fix by adding a few drops of 1 per cent osmic acid to the water containing the zoöspores; then add an equal amount of a mixture of fuchsin and methyl violet. The fuchsin and methyl violet should be 1 per cent solutions in 95 per cent alcohol. In a few seconds the cilia stain a bright red.

CHONDRIOSOMES

During the past fifteen years the terms chondriosomes, mitochondria, *Chondriokonten*, *Chondromiten*, etc., have become increas-

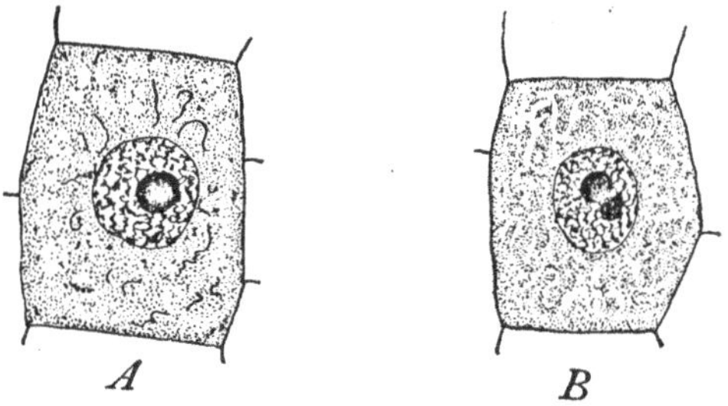

FIG. 28.—Cells from the periblem of the root-tip of *Allium cepa:* *A*, chondriosomes; *B*, canaliculi: fixed in Bensley's solution and stained in iron-alum haematoxylin. ×1,200.

ingly frequent in botanical literature. These "chondriosomes," as we shall call them, are minute structures, probably present in most cells, but not differentiated by the most usual methods and generally overlooked when they might be seen. Most of them are as small as bacteria and bear a superficial resemblance to coccus, spirillum, and bacillus forms (Fig. 28*A*).

Many fixing agents either destroy the mitochondria or make it almost impossible to demonstrate them. Fixing agents containing alcohol or any considerable percentage of acid are to be avoided. Benda's solution, followed by Haidenhain's iron-alum haematoxylin, will give good results. A solution recommended by Bensley is good also for plant material.

Bensley's Solution.—

Osmic acid 2 per cent	1 part
Corrosive sublimate ($HgCl_2$) $2\frac{1}{2}$ per cent	4 parts

Add 1 drop of glacial acetic acid to 10 c.c. of this solution. Fix for 24 to 48 hours and wash thoroughly in water. On the slide, bleach with hydrogen peroxide; wash in water; treat with the iodine solution used in testing for starch; then wash in water. The slide is now ready for staining. We recommend the usual Haidenhain's iron-alum haematoxylin.

Bensley recommends the following method which we have found rather uncertain, but which, when successful, yields magnificent preparations: On the slide, bleach for 2 or 3 seconds in a 1 per cent aqueous solution of permanganate of potash; then treat with a 5 per cent aqueous solution of oxalic acid until the preparation becomes white (a few seconds); wash in water, and then stain as follows:

1. Copper acetate (neutral) saturated solution in water, 5 to 10 minutes.
2. Wash in water.
3. $\frac{1}{2}$ per cent haematoxylin, 5 to 10 minutes.
4. Wash in water.
5. Potassium bichromate (neutral) 5 per cent solution in water until the preparation blackens, usually 30 seconds or less.
6. Differentiate in Weigert's ferricyanide solution.

Borax	2.0 g.
Ferricyanide of potassium	2.5 g.
Water	200.0 c.c.

7. Wash in water and proceed as usual.

CANALICULI

By using special methods, Bensley has obtained views of the protoplasm of plants, quite different from those seen in ordinary preparations. In the cell of a root-tip a series of small canals, or vacuoles, appears, which is much more definite and extensive than the usual display of vacuoles and which appears before any vacuoles can be recognized in preparations made in the usual way (Fig. 28*B*). Being a zoölogist, he has called these vacuoles "canaliculi."

VASCULAR BUNDLES IN LIVING TISSUES

In studying venation, and in tracing the course of vascular bundles in large ovules and in other organs, it is often an advantage to use a stain. If a stem of *Impatiens* be cut under water, and the cut surface be then placed in a dilute aqueous solution of eosin, the eosin will rise in the vessels, making them very prominent. The outer bundles of the large ovules of cycads are very easily studied by this method. The inner bundles also may be seen by opening the seed and removing the endosperm.

If such preparations could only be cleared, they would be still more valuable, but the effect is due to the presence of the staining fluid in the vessels, and any subsequent treatment diffuses or destroys the stain. Perhaps a little experimenting might obviate the difficulty.

STAINING LIVING STRUCTURES

Some stains will stain living structures. Cyanin, methyl blue, and Bismarck brown have been recommended for this purpose. The solutions should be very dilute, not stronger than 1:10,000 or 1:500,000. The solutions should be very slightly alkaline, never acid. It is claimed that such solutions never stain the nucleus, and that if the nucleus stains at all, it is an indication that death is taking place.

Campbell succeeded in staining the living nuclei in the stamen hairs of *Tradescantia* by using dilute solutions of dahlia and of methyl violet (0.001 to 0.002 per cent in water). Dividing nuclei were stained.

For determining the stage of development of fresh material it is often necessary to use a stain. For this purpose stronger stains may be used, since it is unimportant whether the tissue is killed or not. An aqueous solution of methyl green or eosin can be recommended. With 1 per cent solutions, diluted one-half with water, mitotic figures can be recognized with ease.

CHAPTER XIII

PHOTOMICROGRAPHS AND LANTERN SLIDES

While a work like the present book is hardly the place for any extended treatment of photomicrography or the making of lantern slides, a few simple directions will help the beginner and enable him to prepare most of the photomicrographs and lantern slides which may be necessary in the classroom. It is assumed that the student knows how to handle an ordinary camera and knows how to do his own developing.

In the preparation of this chapter, I am indebted to Dr. Paul J. Sedgwick for suggestions, criticism, and considerable revision.

PHOTOMICROGRAPHS

For a simple beginning, no apparatus is needed except an ordinary camera and a microscope. Try low powers first and proceed gradually to the higher magnifications. Remove both front and back lenses from the camera, leaving the lens barrel and the shutter; also, remove the eyepiece from the microscope. Bend the microscope to the horizontal position and place the lens of the camera close to the ocular end of the microscope and shut out all light at this point by winding black cloth around the end of the microscope and the barrel of the camera lens. Take great care to have a perfectly straight optical axis through the microscope and camera.

While the camera and microscope can be adjusted so as to secure a perfect optical axis by simply putting both instruments on the table and raising one or the other—according to the size of the camera—by placing a board under it, such an adjustment is extremely unsatisfactory, since the least jar may disturb it, and inserting the plate-holder is almost sure to disarrange something. It will save time if you prepare a board to keep both instruments in position. Select a clear board 1 inch thick, about 1 foot wide, and 5 feet long. On the top of this board, screw two pieces $\frac{3}{4}$ inch thick, $1\frac{1}{2}$ inches wide, and 5 feet long, so as to form a guideway for the camera (Fig. 29*A*). If the camera is so small that it must be raised to bring it into the optical axis of the microscope, fit to the guideway a board of the necessary thickness, and fasten the camera to this board. It is absolutely necessary that the preparation to be photographed and the ground glass of the camera should be perfectly parallel. The board will save time in securing this parallelism. Cut through the board a slot $\frac{1}{4}$ inch wide and extending to within 6 inches of each end. By this means the camera can be clamped with the screw used to fasten it to a tripod. Also, a piece of metal or hard wood may be placed over the horseshoe base of the microscope and with a bolt, preferably one with a butterfly nut, the microscope may be held firmly in place. This board, with the long slot, will be useful in making lantern slides.

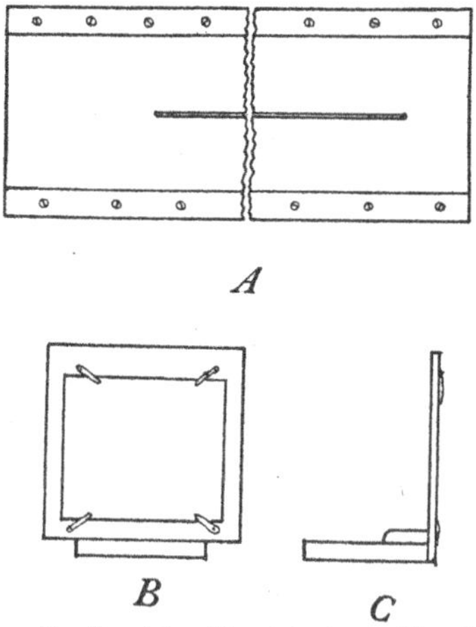

FIG. 29.—*A*, board for photomicrographic and lantern-slide work; *B*, end view with clips to hold negatives; *C*, side view of block to be used on board when making lantern slides.

As an illuminant, direct sunlight, diffuse daylight, a gas-mantle lamp, an acetylene lamp, a Mazda bulb, an arc light, or any strong light may be used. Remove the mirror from the microscope and allow the light to come directly into the optical axis. This mirror will not be needed in any photomicrographic work.

Let us suppose that we are to make a photomicrograph of a vascular bundle and that we are using a 16-mm. objective. If only a part of the bundle is shown on the ground glass, remove the ocular of the microscope. If the illumination is very uneven and shows a "flare spot," look at the inside of the tube of the microscope. Probably it was not blackened and the "flare spot" was due to reflections. Obviate the difficulty by putting a piece of black paper inside the tube. Any modern microscope should have the tube well blackened inside. Move the light back and forth and sidewise to get the best illumination. An easy way to determine whether the light is centered is to close down the iris diaphragm below the condenser until the lighted circle on the ground glass is very small. Then shift the light or the condenser, for it is possible that the condenser is also off center, or both until the illuminated circle is centered. After the light source, the condenser, and the objective have been brought into line, rack the condenser back and forth until the beam of light is focused on the object to be photographed. Then make certain that the cone of light at the point where it enters the objective has a diameter great enough to fill the aperture of the objective. The ordinary form of Abbé condenser is not likely to be satisfactory with objectives of 16-mm. focus, and should not be used at all with objectives of longer focus.

Focus the object upon the ground glass. Even with a 16-mm. objective, the ordinary ground glass is rather coarse for accurate focusing. Always examine the image with a focusing lens. A brilliant view may be obtained by fastening a thin cover-glass to the ground glass with a small drop of balsam. At this spot the image may be examined very critically. Of course, as in any photography, the ground side of the glass should be nearest the object, occupying exactly the place which is to be occupied by the emulsion side of the plate. Do not focus indiscriminately, but be sure that the image is sharp at the level of the ground side of the glass. It is a good plan to make a cross upon the ground glass with a pencil or pen, and then add a drop of balsam and a cover-glass. Focus on this mark and fix the focusing glass at this level. The cheap tripod lenses are good for this purpose.

The time of exposure will vary with the magnification, the intensity of the light, and the speed of the plate. The exposures will be much longer than in ordinary photography. It is better to use artificial light, since one can more quickly learn to estimate the length of exposure when the intensity of the light is constant. A slow plate, even the very slow contrast plate, is likely to prove most satisfactory for the beginner. With a Welsbach lamp, a contrast plate of the same speed as a lantern slide, and a 16-mm. objective used without an ocular, or Abbé condenser, try an exposure of 30 seconds. Develop the negative in whatever solution is recommended in the directions which come with every box of plates. If the negative is too weak, make a longer exposure; if too dense, shorten the exposure. A little experience with your apparatus will soon enable you to estimate the length of exposure with some certainty. We frequently use lantern-slide plates for tests and for small photomicrographs. The Cramer lantern slides and contrast plates of larger sizes have the same speed and, consequently, one can determine the length of exposure by using a cheap lantern slide. In making tests, it will save both time and money to expose for 5 seconds and then push in the dark slide so as to cover a part of the plate; then expose 5 seconds longer and push the slide in a little farther, etc. In this way you can make 4 or 5 exposures on a lantern-slide plate showing exposures of 5, 10, 15, 20, and 25 seconds, the first exposure being 5 seconds and the last, 25 seconds. By making several exposures on 1 plate, with the first probably underexposed and the last probably overexposed, you can note which exposure is best and will waste only 1 plate in getting the time. A print from such a negative is valuable, since it enables one to judge very accurately the printing quality of the various exposures.

A slow contrast plate such as mentioned above will prove satisfactory for photographing some subjects and because of ease in manipulation it offers certain advantages for the beginner; but it is not the most satisfactory in a majority of cases. It is necessary to use a plate which is color corrected when one is using ray filters to increase or decrease the contrast between differently stained portions of a preparation or to secure greater detail than could be obtained without a filter. We have used the Cramer Iso plates and the Eastman Standard Orthonon plates a great deal. They are orthochromatic, sensitive to a large part of the spectrum, and can be used with the various filters, with the exception of the reds. When it is necessary to use a red filter, we suggest the Wratten Panchromatic plates.

Professional films are rapidly replacing plates for many phases of photographic work. The Eastman Commercial Ortho film was used in the making of all the photomicrographs in this edition credited to Dr. Sedgwick. This Commercial Ortho film is thick, rather slow, and gives a combination of detail and contrast which is very desirable. It is heavy enough to lie flat in the Professional Film holders made for Professional films. It is strongly orthochromatic and may be used with all the filters commonly employed in photomicrography except for the reds. If it is necessary to use a red filter in order to make a red-stained portion of a preparation appear light, the Eastman Commercial Panchromatic film may be used. This film works as well as the Commercial Ortho, has the additional advantage of being sensitive to red, but is more difficult to handle because it is fogged by the usual red darkroom lamps. Panchromatic films, like panchromatic plates, should be developed in an unlighted darkroom, or if it seems necessary occasionally to examine the negative during development, the special Wratten safelight made for this purpose should be employed. The Eastman Kodak Company publishes an interesting booklet entitled "Professional Film" which they distribute free of charge.

With fast plates and without filters, a strong light will allow exposures of a small fraction of a second but we have had no success under such conditions. The ordinary yellow filters used in out-of-door work will be worth trying, but if one is contemplating doing any considerable amount of serious work we would suggest procuring a set of filters made up especially for photomicrography. The set of 9 filters known as the Wratten "M" filters made by the Eastman Kodak Company is very useful. With these filters used singly and in combination it is possible to utilize selected portions of the spectrum and thus to increase or decrease the contrast either between the specimen and background or between the variously stained portions of the specimen. For the beginner, perhaps the most confusing part of the procedure is in choosing the proper filter. The yellow and orange filters will probably be the ones most frequently used, for when they are used with a good orthochromatic or panchromatic plate or film the result will be an approximately correct rendering of the color values. Some workers rarely use any filters except the yellow and orange but very frequently other filters will give better results.

It is not possible to state just which filters will be best for each set of conditions but certain rules will aid in determining. To obtain the greatest amount of contrast between the specimen and background use a filter or combination of filters which transmits only that portion of the spectrum completely absorbed by the specimen. If the result is unsatisfactory because detail within the specimen has been sacrificed in favor of contrast between the background and specimen, try a filter or combination of filters whose spectral transmission is not exactly limited to the part of the spectrum absorbed by the object. To eliminate contrast between the object and its background use a filter or a combination of filters transmitting that portion of the spectrum also transmitted by the object. After a little experience one will be able to estimate the probable photographic result by making a visual examination with first one and then another of the filters in place.

An example of how the filters were chosen in making one particular photomicrograph of a stem section may be suggestive. The specimen was well stained in safranin and anilin blue. If the photograph had been made on an ordinary uncorrected plate and without a filter the result would have been unsatisfactory. The red of the safranin would have taken as pure black while the parts of the specimen stained in anilin blue would hardly appear in the photograph because the plate was not sensitive to red and very much oversensitive to blue giving the blue almost the value of white light. Therefore we used a color-corrected plate and a combination of the Wratten "B" and "E" filters which gave the maximum contrast for the anilin blue since the spectral transmission band of the combination is entirely included within the limits of the absorption band of the anilin blue. While these filters did not give the maximum contrast for the safranin-stained portion of the specimen, the contrast was sufficient and the result was more satisfactory than if the safranin stain had photographed as pure black because all detail in the vessel walls would then have been lost.

A booklet, "Photomicrography," published by the Eastman Kodak Company, includes a table showing the spectral absorption bands of some of the stains used in botanical microtechnic. It also gives data on the transmission spectra of the Wratten "M" filters.

The Abbé condenser, which should not be used at all with low powers, is very useful with objectives of 8-mm. focus and all higher powers, especially if the condenser is achromatic. If the condenser is not achromatic, it is sometimes a good plan to remove it and in its place put a 16-mm. objective, or, for very high powers, even an 8-mm. objective. The condenser may be fastened into the condenser sleeve by an improvised ring or collar. Zeiss makes a collar for this purpose. Spencer microscopes also have such collars.

In addition to the Abbé condenser, there should be another, placed between the microscope and the light. For this purpose, the large condenser from a projection lantern may be used. For magnifications higher than 100 diameters still another condenser will be useful. Place it between the last-named condenser and the microscope.

With so many condensers, the heat may damage preparations: so place between the last-named condenser and the one next the light a cooler filled with water or a solution of alum.

With all these accessories, an additional iris diaphragm is desirable. Place it between the middle one of the three condensers and the miscroscope, but quite close to the middle condenser. To make an efficient adjustment of all these parts requires patience, practice, and judgment.

It will save time and patience if the position of the object to be photographed be marked in ink on the slide by vertical and horizontal lines, or by a circle around it. Even with these lines, it is none too easy to get the object into the desired position on the ground glass. Remove the ground glass and let the image fall on a piece of white cardboard a short distance back of the camera. If the curtains are pulled down, the position of the object in the field and the focusing of the condensers will be comparatively easy.

The desirability of a rigid, straight, and accurate optical bed will soon be realized. If one is intending to do much photomicrographic work, the heavy, graduated optical bed is almost a necessity. However, if time is no object and patience is abundant, good photomicrographs at a magnification of over 1,000 diameters can be made with no apparatus except a good camera, a good microscope, and a good lamp.

The relative positions of the various parts, as we have used them in making the illustrations for this book, are indicated below:

Ground glass	Camera bellows	Shutter	Microscope	Abbé condenser Filter	Filter Diaphragm Condenser	Cooler Condenser	Light

Some data which may be helpful will be found in the legends under some of the photomicrographs.

Easily available and inexpensive help will be found in three booklets published by the Eastman Kodak Company, of Rochester, New York: "Photomicrography"; "Wratten Light Filters"; "The Photography of Colored Objects."

After a little practice the student will read with profit the more extended works on this subject, among which are the following: *The A, B, C of Photomicrography*, by W. H. Walmsley (Tennant & Ward, New York); *Photomicrography*, edited by J. Spitta (Scientific Press, London, England); *Lehrbuch der Microphotographie*, by Dr. Richard Neuhaus (Harold Bruhn, Brunswick, Germany).

LANTERN SLIDES

Lantern slides are now so universally used in the lecture-room that every teacher should be able to make them. Three general classes of lantern slides, as far as the technic of making them is concerned, will be described: (1) lantern slides by contact, (2) by reducing or enlarging, and (3) by copying illustrations.

1. Lantern Slides by Contact.—This method is very simple. Imagine that the lantern-slide plate is a piece of printing-out paper, and proceed just as in making a print on paper. Remember that dust on the negative or plate causes spots in the print, and that spots so small as to be almost unnoticeable in an ordinary print will be greatly magnified when they appear on the screen. Brush both negative and plate very gently with a soft clean brush before making the print. If the negative is $3\frac{1}{4}\times4\frac{1}{4}$ inches, it can be placed in a printing frame of that size, and the lantern slide placed upon it with the two films in contact, just as in printing paper. If there is no small printing frame, use a 4×5, a 5×7, or even an 8×10 frame. In such cases, put in a piece of *clean* glass free from scratches or bubbles, and lay the negative upon it. Lantern slides may be printed from a portion of a 4×5 or some larger negative by simply placing the lantern-slide plate over the desired spot. Take great care not to scratch the negative.

A much more satisfactory method is to use a box 10 inches square and 16 inches high, inside measurement. The box should be painted white inside, and the top should be a strong piece of glass; but there should be also a wooden lid, hinged at the back and covered inside with some thick, soft cloth. Inside the box, about 4 inches from the bottom, place two electric bulbs, one red and the other white. To diffuse and tone down the light, a piece of glass, ground on one side should be placed a little more than halfway between the light and the glass top. If the light is still too strong, place a piece of white paper above the ground glass. It is most convenient to have the white light operated by a switch. With a 40-watt light, exposures will vary from 1 to 30 seconds, as the negatives are very thin or very dense. With properly exposed negatives, the exposures will vary 2 to 5 seconds.

If you are printing in a printing frame with an average negative at a distance of 3 feet from a gas-mantle lamp, try an exposure of 3 seconds; if the negative is weak, shorten the exposure; if strong, lengthen it. If you always use the same light at the same distance, you should soon be able to estimate the exposures for negatives of various densities.

If a negative is uneven, the distance from the light may be increased so as to lengthen the exposure to several seconds, thus giving time to shade the weak parts, just as in case of prints on paper. If a negative is harsh and shows too much contrast, hold it closer to the light and shorten the exposure; if weak and lacking in contrast, hold it farther away and increase the time of exposure.

Be careful not to underdevelop. A lantern slide looks stronger in the developer, and even in the hypo, than it really is.

2. **Reducing and Enlarging.**—If a slide is to be made from a 4×5 or larger negative, there must be a reduction. A camera is necessary. A $3\frac{1}{4}\times4\frac{1}{4}$ camera is large enough. If any larger size is used, the plate-holder must be "kitted" down to $3\frac{1}{4}\times4$, the standard size of lantern slides in America. In using the larger cameras, mark upon the ground glass the exact size and location of the lantern-slide plate. Fasten the negative in some convenient place where the light may shine through it: diffuse daylight is good. Then arrange the camera just as in taking any ordinary picture. The board shown in Figure 29 will be just as useful in making lantern slides as in making photomicrographs. At one end of the board fasten a frame which will hold an 8×10 negative and also hold kits for smaller negatives (Figs. 29*B* and *C*). The long slot in the board will allow the camera to be fastened at the proper distance. If buildings, trees, or shadows are in the way, tilt the board so as to have a clear sky for a background.

Be very careful in focusing; it is best to examine, with a pocket lens, the image on the ground glass. In general, use a rather small stop, F16 or even F32. If reducing from an average 5×7 negative, in good daylight, with an F16 stop, try 2 or 3 seconds. If enlarging from a negative somewhat smaller than a lantern slide, try 8 or 10 seconds. Other things being equal, the best lantern slides are made by reduction of larger negatives and the poorest by enlargement from smaller negatives.

The superiority of the larger negative is easily demonstrated. With a 5×7 camera, make a negative of some elaborately ornamented building, making the building just cover the plate. Then, with a $3\frac{1}{4}\times4\frac{1}{4}$ camera, make a similar negative, so that the building just covers the plate. Make lantern slides from both negatives. While the building, as it appears on the lantern slide, and on the screen, is of the same size in the two cases, the one from the 5×7 negative will show much finer detail. The same principle holds true for all kinds of plant subjects. The small cameras are easy to carry, make good views of interesting bits of scenery, on a tripod, well stopped down, will do some fairly good scientific work; but for real scientific investigation, take a 5×7. If you are fond of work, or can afford to have someone else carry the heavy load, take an 8×10.

3 **Copying Illustrations.**—It is often desirable to get lantern slides from photographs, maps, or pictures in books. Here, it is necessary to make a negative and then make the lantern slide from the negative. In such cases make a $3\frac{1}{4}\times4$ negative and print the lantern slide by contact. A lantern-slide plate is good for such copying. The exposure will depend upon the light, the character of the print, and the amount of reduction or enlargement. Other things being equal, the exposure will always be longer in case of enlargement than in case of reduction. If an average 5×7 photograph is to be copied in good diffuse daylight, with an F16 stop and a lantern-slide plate, try 15 seconds.

For making a negative of a line drawing, about natural size, with a Cramer lantern slide for a plate, at F16, dull day with no shadows, try 15 seconds.

For a diagrammatic line drawing, 16×20 inches, F16, good light, Cramer lantern slide for a plate, try 10 seconds.

If one prefers to use film, Eastman Process film is very good for copying line work. With the proper exposure and correct development it gives excellent results, the lines appearing clean, sharp, distinct, with the background perfectly black in the negative. Use the developer suggested by the manufacturer. In copying maps and line drawings where dead blacks and pure whites are desired, expose fully, but do not overexpose, and develop until the image shows plainly on the back of the plate. In such cases, the usual developers will not give as good results as a developer modified to give dead blacks upon a perfectly transparent background. For this purpose, Dr. Land's formula is the best we have tried.

Land's Contrast Developer.—

Hydrokinon	20 g.
Sodium sulphite (dry)	60 g.
Sodium carbonate (dry)	140 g.
Potassium bromide	12 g.
Water	1,000 c.c.

If kept tightly stoppered, with no air space between the liquid and the cork, this developer keeps almost indefinitely. When some is taken out for use, the rest should be put into a smaller bottle, so that there shall be no air between the liquid and the cork.

Another formula, developed by Dr. Land to meet the trying conditions of the tropics, is also useful for lantern slides.

Land's Tropical Developer.—

Hydrokinon	8 g.
Metol	3 g.
Sodium sulphite (dry)	30 g. (60 g. if crystals are used)
Sodium carbonate (dry)	30 g. (90 g. if crystals are used)
Potassium bromide	2 g.
Water	1,000 c.c.

This formula will develop an underexposed plate when the usual developers fail. With this developer, the image flashes into sight with surprising suddenness, but do not become startled and remove the slide too soon, lest you fail to secure details.

It is not necessary to furnish the usual formulas for developers and fixing solutions, since these are furnished with every box of plates. We have found the Cramer plates very satisfactory for all kinds of photographic work. The firm will send gratis to anyone who requests it Cramer's manual on *Negative-making and Formulas* (G. Cramer Dry Plate Co., St. Louis, Mo.).

Warm Tones.—A pyro-ammonia developer for warm tones is recommended in Harrington's *Photographic Journal* for June, 1914:

		Metric	Apothecaries'
A.	Pyro	31 g.	(1 oz.)
	Sodium sulphite crystals	62 g.	(2 oz.)
	Citric acid	2.6 g.	(40 g.)
	Water	237 c.c.	(¼ pt.)
B.	Ammonia	31 g.	(1 oz.)
	Water	237 c.c.	(½ pt.)
C.	Ammonia bromide	31 g.	(1 oz.)
	Water	237 c.c.	(½ pt.)

The solutions A, B, and C keep well separately, but not when mixed. When wanted for immediate use, mix A, 1.8 c.c. (30 minims), B, 3.7 c.c. (60 minims), and C, 1.8 c.c. (30 minims), and add 30 c.c. (1 oz.) of water.

A correctly exposed plate will develop in about 2½ minutes, and the tone should be a warm black. Brown tones are secured by increasing the quantity of C, while A and B remain the same.

Reducing Overexposed Negatives and Lantern Slides.—In case of overexposure, the negatives or lantern slides can be saved by reducing. The reducing solution should be applied as soon as the negative is well fixed in hypo. If a negative which has been washed and dried is to be reduced, it should be soaked in water for half an hour before using the reducing solution.

The following is a good solution for most purposes:

		Metric	Apothecaries'
A	Water	473 c.c.	(16 oz.)
	Hyposulphite of soda	31 g.	(1 oz.)
B	Water	473 c.c.	(16 oz.)
	Red prussiate of potassium	31 g.	(1 oz.)

Solution B must be protected from the light. Cover the bottle with black paper and keep it in the dark when not in use.

Mix only for immediate use 8 parts of A to 1 of B and use in rather subdued light. A darkroom is not necessary, but avoid bright light.

When the negative or lantern slide becomes satisfactory, wash it in water as thoroughly as if it had just come from the ordinary hypo fixing solution.

Ordinarily, if it is possible to do so, we believe it is better to make a new, correctly exposed negative or lantern slide rather than to reduce or intensify an incorrectly exposed one.

Intensifying Underexposed Negatives and Lantern Slides.—Even if a negative or lantern slide has been considerably overexposed, it can be reduced quite satisfactorily; if much underexposed, little can be done for it; if only slightly underexposed, it may be greatly improved by the following solution:

		Metric	Apothecaries'
A	Bichloride of mercury........	2 g.	(31 gr.)
	Water.......................	100 c.c.	(4 oz.)
	Bromide of potassium........	2 g.	(31 gr.)
B	Sulphite of soda crystals......	10 g.	(154 gr.)
	Water.......................	100 c.c.	(4 oz.)

The solutions keep indefinitely and may be used three or four times.

Apply the intensifier after fixing in hypo and washing in water. If the negative or slide has been allowed to dry, soak it in water for half an hour before intensifying.

Place the negative or slide in A, rocking the tray as in developing, until it becomes gray or even white. Wash in water for 1 minute and then transfer to B and leave until the dark color can be seen on the back of the negative or slide. Wash in water as thoroughly as after fixing in hypo.

Some use a saturated aqueous solution of the bichloride of mercury, without the bromide of potassium; and, instead of solution B, use water to which ammonia has been added—about 1 part ammonia to 40 parts water. Excellent sepia tones may be secured in this way. Wash well in water.

After the plate has been thoroughly washed in water, wipe it gently with a tuft of cotton. The cotton must, of course, be thoroughly wet; it is better to hold the plate under a stream of water while wiping. *This should always be done before placing a negative or slide in the rack to dry, after a washing in water.*

Toning Lantern Slides.—A lantern slide may sometimes be made more effective by judicious toning. The hints given here merely introduce the student to the possibilities of the subject.

Light Sepia to Red Tones.—Overexpose up to four or five times the length of exposure for a normal slide in black and white; develop thoroughly; fix and wash as usual; then tone in the following solution:

	Metric	Apothecaries'
A. Potassium ferricyanide	6 g.	(90 gr.)
Water	295 c.c.	(10 oz.)
B. Copper sulphate	7 g.	(110 gr.)
Potassium citrate	65 g.	(1,000 gr.)
Water	295 c.c.	(10 oz.)

(The metric and U.S. measures are practically rather than arithmetically equivalent.)

When needed for use, pour some of A into an equal quantity of B, stirring or shaking constantly. Put the slide into the solution in a tray and rock just as if developing a plate. The solution is a strong reducer. The tone should change from black to warm, then to sepia, and may finally become quite red. The time may vary from 1 to 20 minutes, according to the density of the slide and the tone desired. The finished product must not be too dense, for a slide, toned in this way, may seem rather weak and yet appear surprisingly strong on the screen.

After toning, wash in water for about 20 minutes.

Moonlight Tints.—Some excellent formulas, recommended by Anderton, will be of service:

	Metric	Apothecaries'
Ferric ammonia citrate (10 per cent solution)	15 c.c.	(½ oz.)
Potassium ferricyanide (10 per cent solution)	15 c.c.	(½ oz.)
Glacial acetic acid (10 per cent solution)	148 c.c.	(5 oz.)

The following gives a more greenish-blue tint:

Uranium nitrate (10 per cent solution in water)	3.6 c.c.	(1 dram)
Ferric ammonia citrate (10 per cent solution in water)	3.6 c.c.	(1 dram)
Potassium ferricyanide (10 per cent solution in water)	7.2 c.c.	(2 drams)
Nitric acid (10 per cent solution in water)	7.2 c.c.	(2 drams)

Both solutions intensify considerably, so that slides to be toned should be rather weak. After toning, wash 20 minutes in water.

Green Tones.—

		Metric	Apothecaries'
A.	Potassium bichromate (10 per cent solution in water)	60 drops	
	Potassium ferricyanide (10 per cent solution in water)	30 c.c.	(1 oz.)
	Water	120 c.c.	(4 oz.)
B.	Cobalt chloride	3.9 g.	(60 gr.)
	Ferric sulphate	3.9 g.	(60 gr.)
	Hydrochloric acid	15 c.c.	(½ oz.)
	Water	120 c.c.	(4 oz.)

Bleach in A, wash 10 minutes in water, tone in B, and then wash 20 minutes in water.

Staining Lantern Slides.—Some of the stains used in staining microscope slides will give a pleasant tone to lantern slides. Light green gives a clear, moonlight effect. Magdala red gives a transparent, rosy tint. Sepia and other tones could doubtless be imitated by this easy method.

Clearing Lantern Slides.—Sometimes a slide will seem perfectly clear, just as it comes from the fixing bath, especially from an acid fixing bath; usually, however, it will be better to transfer the slide from the fixing bath to a weak solution of acetic acid—just enough acid to give the solution the taste of weak vinegar—and then rock for a minute before washing.

The following clearing fluid may be used in the same way:

	Metric	Apothecaries'
Alum	20 g.	(1.3 gr.)
Iron sulphate	20 g.	(1.3 gr.)
Citric acid	20 g.	(1.3 gr.)
Water	500 c.c.	(17 oz.)

Coating Lantern Slides.—After the slide has become thoroughly dry, a coat of balsam or shellac will add much to its brilliancy. Dilute the Canada balsam with xylol until it becomes almost as thin as water; balance the slide on the thumb and first, second, and third fingers, holding it as level as possible; pour the balsam over it, letting the balsam flow evenly over the whole surface; then tilt the slide and pour the balsam back into the bottle. Put the slide in the rack to dry.

Mounting.—Add a suitable mat and a clean lantern-slide cover. Remember that the effect of a first-class lantern slide may be impaired or even ruined by an inartistic mat. Bind the slide and cover together with a lantern-slide binding strip. Paste on the label, or, if you prefer, paste the label on the mat before binding, so as to have it protected by the cover. Lay the slide down so that the positions are just as they were in the original, and then paste the "thumb mark" in the lower left-hand corner.

Some efficient and easily available help in making lantern slides and in general photography will be found in some booklets published by the Eastman Kodak Company, Rochester, New York: "Lantern Slides; How to Make and Color Them"; "Elementary Photographic Chemistry;" "Color Plates and Filters"; "About Lenses."

PART II

SPECIFIC DIRECTIONS

In the preceding chapters the principles and methods of technic have been described in a general way. It is difficult, especially for a beginner, to apply general principles to specific cases, and, besides, the types which he might select for the preparations might not form a symmetrical collection. Consequently, a series of forms has been selected which will not merely serve for practice in microscopical technic, but will also furnish the student with preparations for a fairly satisfactory study of plant structures from the algae up to the angiosperms. It is not at all our purpose to discuss general morphology, but rather to answer, by means of sketches and specific directions, the multitudinous questions which confront the instructor in the laboratory. For those who have had a thorough training in general morphology the following suggestions will be in some degree superfluous. Those who are beginning the study of minute plant structure are referred to the standard textbooks for descriptions of the plants mentioned here.

The directions for collecting and growing laboratory material constitute an important feature of this part of the book.

With a few exceptions, the order in which the forms are presented is that given in Engler's *Syllabus der Pflanzenfamilien.*

CHAPTER XIV

MYXOMYCETES AND SCHIZOPHYTES

MYXOMYCETES

With the exception of a few forms like *Fuligo* (often found on oak stumps and on oak bark in tanyards), the myxomycetes are small, and are usually overlooked by collectors (Fig. 30). A careful examination of rotting logs in moist woods will usually reveal an abundance of these delicate and beautiful organisms. Various species may be found in spring, summer, and autumn. The plasmodia are most abundant just after a warm shower. A couple of days of dry weather will then bring sporangia in abundance. The specimens should be pinned to the bottom of the box for safe carrying. An excellent collecting-box can be made from an ordinary paper shoebox. On the bottom of the box place a thin piece of soft pine, or a piece of the corrugated paper so commonly used in packing; or, better still, a sheet of cork. At each end nail in a piece of pine half an inch thick and an inch high. Upon these end pieces place a thin piece of pine, thus making a second bottom, which, of course, should not be fastened. A second pair of ends with a third pine bottom nailed to them may rest upon the second bottom. The three bottoms will give a considerable surface upon

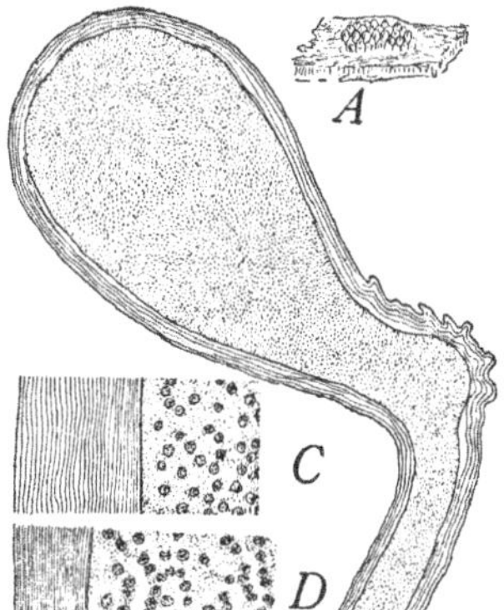

FIG. 30.—*Trichia*, a myxomycete. *A*, habit of a group of sporangia growing on rotten wood. Natural size. *B*, single sporangium with some of the rotten wood. The dots represent fairly well the size and distribution of the nuclei at this stage. ×70. *C*, *D*, and *E*, successive stages, much later than *B*, showing condensation of the wall, origin of elaters (*D*), and mature sporangium (*E*). Preparation stained in safranin, gentian-violet, orange. ×320.

which the material may be pinned. For most purposes, the specimens are simply allowed to dry, and are then fastened with glue or paste to the bottom of a small box.

Plasmodia and young sporangia may be fixed in chromo-acetic acid or Flemming's fluid. Sections are easily cut in paraffin, and should not be more than 5 μ in thickness; for nuclear details, sections should not be thicker than 3μ. The safranin, gentian-violet, orange combination is good for a study of the general development and for some cytological features, but iron-alum haematoxylin is better for nuclear details.

Spores of most myxomycetes will germinate as soon as they are thoroughly ripe, and, during the first year, germination is more prompt than in case of older spores. Fresh spores may germinate in half an hour; the time may extend to several hours; spores two or three years old may germinate in three or four days, or may not germinate at all. We have never succeeded in germinating spores which were more than three years old. The longevity is doubtless different in different species. In most cases, spores will germinate in water, if they will germinate at all. For small cultures, the hanging-drop method, described on page 77, may be used.

Plasmodia may be raised by sowing spores on moist, rotten bark or wood and placing the culture under a bell jar, where the moist, sultry condition favorable to their growth is easily imitated. Plasmodia may be got upon the slide by inclining the slide at an angle of about 15°, with one end of the slide at the edge of the plasmodium, and allowing water to flow very gently down from the upper end of the slide to the lower. The proper flow of water could be secured by dropping water from a pipette, but a less tedious plan is to arrange a siphon so as to secure a similar current. The plasmodium will creep up the slide against the current, furnishing an excellent illustration of rheotropism. Enough plasmodium for an illustration may be formed in two or three hours. Examined under the microscope, the preparation should give an excellent view of the streaming movements of protoplasm.

The following is another method for getting the plasmodia upon the slide: Place the slides upon a pane of glass and upon each slide place a small piece of plasmodium-bearing wood. Cover with a bell jar. Wet blotting paper or a small dish of water included under the jar will help to create the warm, sultry atmosphere necessary. The slides may be covered with the plasmodium in a few hours. Permanent preparations may be made by immersing the slide in chromo-acetic acid, then washing and staining without removing the plasmodium from the slide. Acid fuchsin is a good stain for bringing out the delicate strands of the plasmodium. Iron-alum haematoxylin, followed by acid fuchsin or erythrosin, brings out both nuclei and cytoplasmic strands.

Some of the foregoing methods are taken from an article by Professor Howard Ayers in the January and February (1898) numbers of the *Journal of Applied Microscopy*. Other methods, with directions for various experiments, are given in the same article.

SCHIZOPHYTES (*Fission Plants*)

BACTERIA (*Schizomycetes, Fission Fungi*)

The methods of modern bacteriological technic are so numerous and so specialized that we must refer to laboratory manuals for instruction in this subject. The method given here will merely enable the student to study the form and size of those bacteria which are more easily demonstrated.

Foul water at the outlets of sewers and such places will usually afford an abundance of *Coccus*, *Bacillus*, *Spirillum*, and *Beggiatoa* forms. Place a drop of water on a slide, heat it gently until the water evaporates, then stain with fuchsin or methyl violet, dehydrate, clear in xylol, and mount in balsam.

The hay infusion is a time-honored method for securing bacteria for study. Pour hot water on a handful of hay, and filter the fluid through blotting paper. Place the fluid in a glass dish, and cover with a piece of glass to keep out the dust. When the fluid begins to appear turbid, bacteria will be abundant. The active movements are easily observed in a mount from the turbid water. As the bacteria pass into the resting condition, they form a scum on the surface of the water. Usually, the first to appear is a somewhat rod-shaped form, the *Bacterium termo* of the older texts. *Spirillum* and *Coccus* forms often appear later.

Fine preparations may be obtained by inoculating a mouse with *Anthrax*, and then cutting paraffin sections of favorable organs. For making mounts of a dangerous form like *Anthrax*, secure properly fixed material from a bacteriologist. Stain in gentian-violet or crystal-violet. The following schedule gives good results with *Anthrax* and many other bacteria:

1. Gentian-violet, 5 minutes.
2. Rinse in water a few seconds.
3. Gram's solution (iodine 1 g., potassium iodide 2 g., water 300 c.c.) until the color is almost or quite black; this will generally require 1 or 2 minutes.
4. 95 per cent alcohol until the color has nearly disappeared.
5. Rinse in water and examine. If the bacteria are well stained, a counter-stain may be added.
6. Light green or erythrosin, 5 seconds; or Bismarck brown, 5 or 10 seconds.
7. 95 and 100 per cent alcohol, dehydrating as rapidly as possible. Not more than 5 or 10 seconds can usually be allowed.
8. Xylol, 1 to 5 minutes.
9. Balsam.

After the rinsing in water of stage 5, the preparation may be dehydrated rapidly in 95 per cent and 100 per cent alcohol, and then stained for 5 or 10 seconds in orange dissolved in clove oil. From the clove oil, transfer to xylol and mount in balsam (Fig. 31).

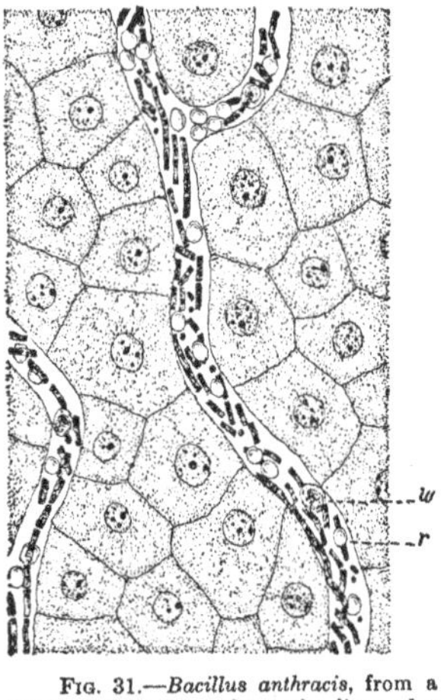

FIG. 31.—*Bacillus anthracis*, from a paraffin section cut from the liver of a mouse and stained in crystal-violet; *w*, white blood corpuscle; *r*, red blood corpuscle. ×580.

With sections not more than 5 μ thick, excellent results can be obtained by staining in iron-alum haematoxylin.

The following rapid method gives fairly good results:

1. Place on a clean cover a drop of water containing the bacteria and dry completely in a flame or on a hot plate.
2. Stain 2 to 5 minutes in gentian-violet or methyl violet.
3. Rinse quickly in water.
4. Dip into 95 per cent alcohol to reduce the stain.
5. Remove most of the alcohol by touching a corner of the cover with filter paper and then dry completely by passing through a flame.
6. Mount in balsam.

Leptothrix may often be obtained by scraping the inside of the cheek. *Beggiatoa*, one of the sulphur bacteria, with oscillating movements like *Oscillatoria*, is often found in foul water. Its presence may be indicated by whitish patches on the bottom.

The Bacteria are the only plants in which a nucleus has not been conclusively demonstrated, and some claim that a nucleus is present even in Bacteria. In determining the presence or absence of a nucleus in Bacteria, the crude method, just given, would be of no value, and even the most critical methods of the bacteriologist, who mounts the organisms whole, would be entitled to only scant consideration. The presence or absence of a nucleus will have to be determined by a study of thin, well-stained sections of perfectly fixed material.

CYANOPHYCEAE. BLUE-GREEN ALGAE (*Schizophyceae Fission Algae*)

The blue-green algae include unicellular, colonial, and filamentous forms. They occur everywhere in damp or wet places. On the vertical faces of rocks where there is a constant dripping of water, brilliant blue-green forms are abundant. In the Yellowstone National Park the brilliant coloring of the rocks is due in large measure to members of this group. Many forms occur as brownish or greenish gelatinous layers on damp ground or upon rocks, or even upon damp wooden structures in greenhouses. Other forms float freely in water or on the surface of the water.

Oscillatoria.—For most purposes it is best to study *Oscillatoria* in the living condition. It is readily found in watering-troughs, in stagnant water, on damp earth, and in other habitats. The commonest forms have a deep blue-green or brownish color. It is very easy to keep *Oscillatoria* all the year in the laboratory. Simply put a little of a desirable form into a gallon glass jar half filled with water. By adding water occasionally to compensate for evaporation, the culture should keep indefinitely. In a jar with a tightly fitting cover we have kept such a culture for years without renewing the water.

For the purposes of identification and herbarium specimens the material may simply be placed on a slip of mica and allowed to dry. When wanted for use, add a drop of water and a cover, and the mount is ready for examination. After the examination has been made remove the cover, allow the preparation to dry, and then return it to the herbarium.

Good mounts may be made by the Venetian turpentine method. Species of medium size are more satisfactory for a study of the nucleus than the very large species. Fix in a chromo-acetic-osmic solution (1 g. chromic acid, 3 c.c. acetic acid, and 1 c.c. 1 per cent osmic acid). Stain in iron-alum haematoxylin, and follow the Venetian turpentine method. While the nuclei are easily seen in such preparations, still better views can be secured from sections of paraffin material fixed in the same solution or in Flemming's weaker solution. The section should be about 3 μ thick. After staining with haematoxylin, stain lightly with orange dissolved in clove oil. In paraffin sections the scattered condition of the material as it appears in thin sections is very annoying. As soon as the material is thoroughly washed in water, arrange it so that the filaments will all have the same general direction. This will enable you to get longitudinal

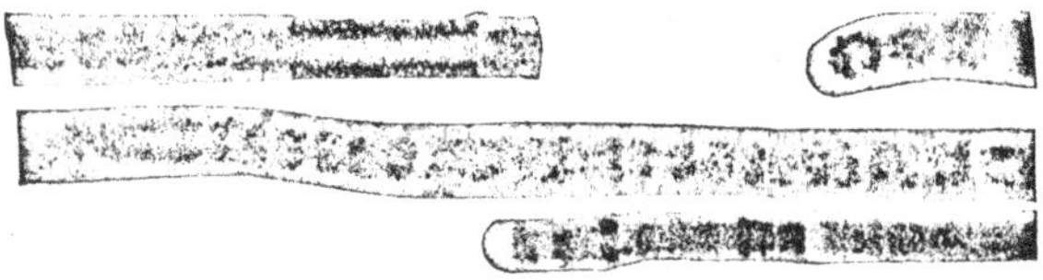

FIG. 32.—*Oscillatoria:* photomicrograph from a paraffin section 3 μ in thickness and stained in iron-alum haematoxylin. ×373.

and transverse sections. As you begin with the alcohols, use a Petri dish and lay a slide over the material, and keep it there until you imbed in paraffin. This will keep the filaments from spreading out too much, and you will be able to get as much on one slide as you would be likely to get on a dozen slides without such precaution.

Oscillatoria, as it appears in section, is shown in Figure 32.

Tolypothrix.—This form occurs as small tufts, either floating in stagnant water or attached to plants and stones. Some species grow upon damp ground. It furnishes an excellent example of false branching (Fig. 33). Like all small filamentous algae, it may be dried on mica for herbarium purposes. Venetian turpentine mounts and paraffin sections are prepared as in *Oscillatoria*. *Tolypothrix* is even better than *Oscillatoria* for a study of the nucleus.

Scytonema is a similar form which is fairly common. It is often found as a feltlike covering on wet rocks.

In staining forms like *Tolypothrix* and *Scytonema*, which have a thick sheath, take care not to obscure the cell contents by staining the sheath too deeply. If the sheath is not stained at all, you may not be able to see the nature of the false branching. Iron-alum haematoxylin, with orange in clove oil for the sheath, is good for sections. Magdala red, with light green for the sheath, is good for Venetian turpentine mounts.

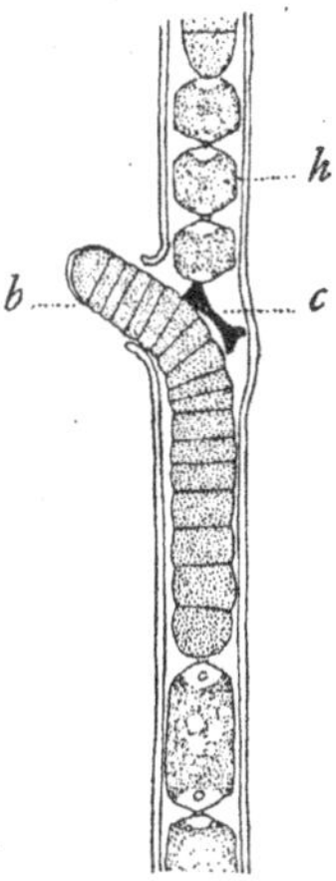

FIG. 33.—*Tolypothrix*, showing "false branching": *h*, heterocyst; *c*, concave cell; *b*, end of false branch with beginning of new sheath. ×620.

Nostoc.—*Nostoc* is a cosmopolitan form. It occurs on damp earth or floating freely in water. In a fruit can or a battery jar, *Nostoc* is easily kept year after year in the laboratory. Young specimens are generally in the form of gelatinous nodules, but in older specimens the form may be quite various. It is very easy to make sections, since the gelatinous matrix cuts well and holds the filaments together. Chromo-acetic acid is a good fixing agent. Stains which stain the gelatinous matrix make the preparations look untidy, but they show that each filament of the nodule has its own gelatinous sheath. Small nodules may be stained in bulk and be got into Venetian turpentine. Crushed under the cover, they make instructive preparations.

Rivularia.—This form is readily found on the underside of the leaves of water-lilies (*Nuphar*, *Nymphaea*, etc.), but is also abundant on submerged leaves and stems of other plants. It occurs in the form of translucent, gelatinous nodules of various sizes. Chromo-acetic acid gives beautiful preparations, but good results can also be secured from formalin or picric-acid material.

The most instructive preparations for morphological study can be obtained by the Venetian turpentine method. Stain in iron-haematoxylin and very lightly in erythrosin, the latter stain being used merely to outline the sheath. When ready for mounting, crush a small nodule under a cover-glass. The paraffin method is easily applied, since the gelatinous matrix keeps the filaments in place. Any form of similar habit may be prepared in the same way.

Gloeotrichia.—*Gloeotrichia* (Fig. 34), in its later stages, is a free-floating form. In earlier stages it is attached to various submersed aquatic plants. The nodules, when young, are firm like *Nostoc*, but as they grow older and larger they become hollow and soft. The older forms become so much dissociated that they lose their characteristic form and merely make the fixing fluid look turbid. Allow a drop of such material to spread out and dry upon a slide which has been slightly smeared with albumen fixative. Leave the slide in 95 per cent alcohol 2 or 3 minutes to coagulate the albumen fixative, and then stain in safranin. If the background appears untidy, stain for 24 hours, or longer; you can then extract the stain from the background, and still leave the long spore and some of the other features of the filament well stained. A touch of cyanin will bring out the sheath. Cyanin and erythrosin is a good combination if the material is clean. The firmer nodules may be treated like *Nostoc* or *Rivularia*.

FIG. 34.—*Gloeotrichia:* photomicrograph from a preparation stained in cyanin and erythrosin: negative by Dr. W. J. G. Land.

Wasserblüthe.—Many genera of the Cyanophyceae occur as scums, often iridescent, on the surface of stagnant or quiet water. Some of the commonest forms are *Coelosphaerium* and *Anabaena* (Fig. 35). Some of the Chlorophyceae also occur as *Wasserblüthe.* Where the material is very abundant, it may be collected by simply skimming it off with a wide-mouthed bottle, but where it is rather scarce, it is better to filter the water through bolting silk and finally rinse the algae off into a bottle, adding enough formalin to the water in the bottle to make a 5 per cent solution. The material may be kept here indefinitely, but after a few hours it is ready for use. If the forms are small, like *Anabaena*, smear a slide lightly with Mayer's albumen fixative, as if for paraffin sections, add a drop of the material and allow it to dry over night or for 24 hours; then immerse the slide in strong alcohol for a few minutes, and then proceed with the staining. Cyanin and erythrosin form a good combination for differentiating the granules. Delafield's haematoxylin, used alone, stains some granules purple and others red. Iron-alum haematoxylin is excellent for heterocysts. With patience, these *Wasserblüthe* forms may be stained in iron-haematoxylin and brought into Venetian turpentine, from which they will yield much better preparations than can be secured by the drying-down method.

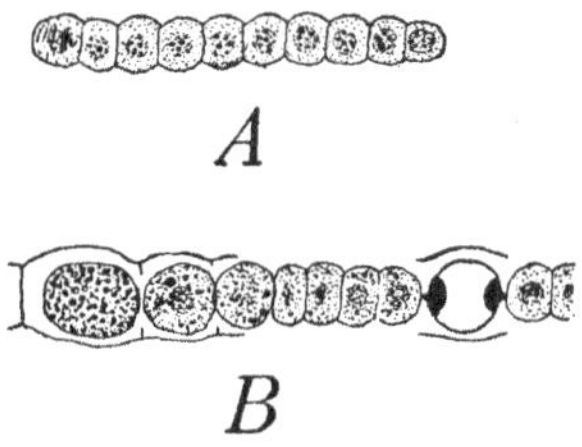

FIG. 35.—*Anabaena:* *A*, hormogonium showing well-defined nuclei; *B*, older filament showing a spore and a heterocyst.

Sometimes *Anabaena*, mixed with *Gloeothece* or *Gloeocapsa*, occur floating in gelatinous masses which hold together fairly well, so that it is easy to fix in the chromo-acetic–osmic solution recommended for *Oscillatoria*, stain in iron-alum haematoxylin, and follow the Venetian turpentine method.

With such material we have tried a more expeditious method with excellent results. After staining in haematoxylin, we have used a series of alcohols, $2\frac{1}{2}$, 5, 10, 15, 20, 30, 40, 50, 70, 85, 95, and 100 per cent, allowing only 3 or 4 hours for the entire series. Then use mixtures of clove oil and absolute alcohol, beginning with 1 part clove oil to 4 parts alcohol, followed by equal parts clove oil and alcohol, then 3 parts clove oil to 1 of alcohol. At this point, stain in orange dissolved in clove oil. Drain off the stain and transfer to pure clove oil. Then place the material in thin balsam, about 1 part of the balsam used for mounting to 3 parts of xylol. Here the material may be kept indefinitely. Mounts may be made in balsam from this stock. Figure 35 was drawn from material prepared in this way.

CHAPTER XV

CHLOROPHYCEAE. GREEN ALGAE

For experiments in most phases of botanical microtechnic, no group of plants affords better material than the green algae, since the killing, fixing, and staining can be watched directly; the effect of the change from one solution to another can be observed; and even the behavior during infiltration with paraffin can be determined with considerable accuracy.

Since the Chlorophyceae furnish our best illustrations of the evolution of the plant body, the origin and development of sex, and also the beginning of alternation of generations, they occupy a prominent place in any well-planned course in the morphology of plants; and, if they were better known, the ease with which the reactions of the individual cell may be observed would make them valuable to the physiologist.

They are found in both fresh and salt water, but are most abundant in fresh water. The ponds, ditches, and rivers of any locality will yield an abundance and variety both of the unicellular and the multicellular members of this group. Most of the forms are independent, but there are epiphytic, endophytic, and saprophytic species. The larger forms and those which grow in tufts or mats are readily recognized in the field. Many of the smaller forms are attached to other water plants. Drain the water plants and then squeeze them over a bottle. The sediment is likely to contain a variety of unicellular and other small algae.

Many of the genera are easily kept in the laboratory. It is not necessary to have very large aquaria. Ordinary glass battery jars holding about a gallon are good for most forms. Jars holding 2 gallons will be as good or better. For some cultures which are to be kept for a long time, like *Scenedesmus*, small glass jars, or dishes, with ground-glass tops are desirable. For a limited amount of material, quart or 2-quart fruit cans are very efficient. Put about an inch of pond dirt in some, clean sand in others, and in still others use a gravel bottom. Many forms grow well without any soil or sand in the bottom of the jar.

When possible, use the water in which the algae were growing, since very few take kindly to a sudden change of water. If the material has been brought to the laboratory in a very small quantity of water, fill the jar about two-thirds full with tap water. Let the water run for 2 or 3 minutes before you fill the jar, since the water standing in the pipes is injurious, or even fatal, to most algae. Add water occasionally, only a little at a time, to compensate for evaporation. If the water has evaporated until the jar is about one-third full and you fill it nearly to the top with tap water, you are likely to kill some of the most desirable forms.

It is a mistake to put too much material into a jar. A wad of *Spirogyra* half as large as one's finger is as much as should be put into a gallon jar. As it grows to ten or twenty times that amount, it is not necessary to keep throwing it out, since it will gradually accommodate itself to conditions. However, do not let the jar become choked with the material.

Cultures may be started even in the winter. Bring in some mud over which algae were growing the previous summer or autumn; put it into a jar and fill it two-thirds full of tap water. Also bring in sticks, leaves, and stones from good alga localities and put them into jars of tap water. Cultures may be started either by taking mud and sticks from under the ice or by taking them from places which have entirely dried up during the summer or autumn. A few such jars will be likely to yield a variety of material.

If you have a good jar of *Oedogonium,* or some other desirable form, do not throw it out if the alga should disappear. Remember that temporary disappearances occur in nature. Allow the culture to become dry and then set it aside where it will be protected from dust. After a few months, pour on tap water and it is very likely that you will soon have a good jar of *Oedogonium.* Many algae behave similarly; some, like *Volvox,* appear for a short time and then disappear for a long time; some, like *Cladophora,* may last the whole year and grow so luxuriantly that the excess material must be removed; and some, like *Ulothrix,* we have not been able to cultivate at all in the laboratory.

Some very useful hints on collecting and growing fresh-water algae for class work will be found in an article by Dr. J. A. Nieuwland in the *Midland Naturalist,* **1**:85, 1909.

Professor Klebs has shown that various phases in the life-histories of many algae and fungi may be produced at will. By utilizing his results, the fruiting condition may be induced in many of the common laboratory types. Knop's solution will be needed in most cases. A stock solution which can be diluted as required may be made as follows:

Potassium nitrate, KNO_3	1 g.
Magnesium sulphate, $MgSO_4$	1 g.
Calcium nitrate, $Ca(NO_3)_2$	3 g.
Potassium phosphate, K_2HPO_4	1 g.

Dissolve the first, second, and fourth ingredients in 1 liter of distilled water, and then add the calcium nitrate. A precipitate of calcium phosphate will be formed. For practical purposes this may be called a 0.6 per cent solution. Whenever a dilute solution is made from the stock solution, the bottle must be shaken thoroughly in order that a proper amount of the precipitate may be included in the diluted solution. To make a 0.1 per cent solution, add 5 liters of distilled water to 1 liter of the stock solution; for a 0.3 per cent solution, add 1 liter of distilled water to 1 liter of the stock solution, etc.

The addition of a liter of a 0.2 per cent solution to 4 or 5 liters of water will often produce a more thrifty growth. Directions for inducing reproductive phases are given in connection with the various types. With a good supply of glass jars, plenty of Knop's solution, a reasonable control over temperature, and the teacher's usual amount of patience, most laboratory types can be studied in the living condition at all seasons of the year.

Collecting algae need not be so laborious as most botanists make it. Forms like *Spirogyra*, *Zygnema*, *Cladophora*, *Vaucheria*, and *Hydrodictyon* may be rolled up in wet newspaper and carried in a botany can. They suffer less from lack of water than from lack of air. Large quantities of material can be brought in and transferred to water after reaching the laboratory. Even after 24 hours in the wet paper, such forms seem to suffer no damage.

Permanent preparations are needed to show details which are not so evident in the fresh material. The unicellular and filamentous members, together with such forms as *Volvox*, are best prepared by the Venetian turpentine method. The structure is so much more complicated than in the Cyanophyceae that it demands far more care and skill to make good preparations.

In many, probably in most of the green algae, nuclear and cell division takes place at night. This is definitely known to be the case in *Spirogyra, Zygnema, Closterium, Ulothrix,* and others. Mitosis is most abundant about midnight, or an hour before midnight, and continues up to three or four o'clock in the morning. The most extensive work on the time of day at which nuclear division occurs is a paper by G. Karsten, "Ueber embryonales Wachstum und seine Tagesperiode, *Zeitschrift für Botanik,* **7**:1–34, 1915. Although the paper is in German, the numerous tables can be understood by those who are unfamiliar with the language. The paper contains a bibliography of the subject.

Chromo-acetic acid, with or without osmic acid, is a good killing and fixing agent for the entire group. We prefer the following formula: Chromic acid, 1 g.; glacial acetic acid, 3 c.c; 1 per cent osmic acid, 1 c.c.

With any fixing agent, it is worth while to place a few filaments in the mixture and watch the effect under the microscope. If plasmolysis occurs with the chromo-acetic mixture, weaken the chromic or strengthen the acetic until the suitable proportions are determined. In the previous edition, the usual method was to weaken the chromic acid. While this avoided any shrinking of the cell contents, the fixing was not very thorough, and material often suffered during staining or other subsequent processes. An extensive series of experiments, especially with coenocytic forms which are notoriously difficult to prepare, proved that it is better to keep the chromic acid up to 1 per cent and strengthen the acetic acid, if necessary. The function of the osmic acid is to make the killing almost instantaneous. This high percentage of acetic acid, excellent for algae and fungi and many other forms, is not so good for many tissues of higher plants, because the proportion of acetic acid is too great. About 24 hours in any of the chromic series and a 24 hours' washing in water will be sufficient for members of this group. Only a few of the most commonly studied will be mentioned.

With Marine Forms use sea-water in making up the fixing agents and in washing, but use fresh water in making up alcohols and for the 10 per cent glycerin.

Volvox.— *Volvox* is found in ponds and ditches, and even in shallow puddles. The most favorable place to look for it is in the deeper ponds, lagoons, and ditches which receive an abundance of rain water. It has been claimed that where you find *Lemna,* you are likely to find *Volvox;* and it is true that such water is favorable, but the shading is unfavorable. Look where you find *Sphagnum*, *Vaucheria*, *Alisma*, *Equisetum fluviatile*, *Utricularia*, *Typha*, and *Chara*. Dr. Nieuwland reports that *Pandorina*, *Eudorina* and *Gonium* are commonly found in summer as constituents of the green scum on wallows in fields where pigs are kept. The flagellate, *Euglena*, is often associated with these forms. If you have a culture in the laboratory, do not throw it out when the culture disappears, because new coenobia are likely to develop from the oöspores.

For collecting, use "bolting cloth" or "bolting silk" of the finest mesh available. With a piece of thin cloth about 15 cm. square, laid over an ordinary coffee strainer, you can pour through about 4 liters of water in a minute. In this way you will secure all the *Volvox* in about a barrel of water in half an hour. *Eudorina* may be collected in the same way. Smaller members of the *Volvox* family like *Pandorina*, *Gonium*, and *Chlamydomonas* are too small to be held by the cloth; but if material is very abundant, the water goes through faster than the organisms and you will soon have many times as much material in a bottle as you could get by dipping. Many small organisms are effectively collected in this way, even when they are so small that most of them pass through the cloth.

Material of *Volvox* and all the Volvocaceae may be fixed in the corrosive sublimate–acetic mixture, used hot—85° C. If material is to be stained and mounted whole, use the aqueous mixture; if it is to be imbedded and cut, use the alcoholic. For mounting whole, stain in iron-alum haematoxylin, or in Magdala red and anilin blue, following the Venetian turpentine method. A few bits of broken cover-glass, placed among the colonies, will prevent crushing.

Many fixing agents cause the protoplasmic connections between the cells to be withdrawn. Figure 36*A*, showing the protoplasmic connections in nearly the living condition, was drawn from material fixed in the hot aqueous corrosive sublimate–acetic acid solution; *B*, fixed in a 1 per cent chromo-acetic solution, does not show the connections; *C* to *E* show details drawn from sections.

For paraffin sections, the material, preferably in sufficient abundance to make a layer half an inch deep in the bottom of a bottle as large as one's finger, is infiltrated with paraffin in the usual way. In imbedding, simply pour the contents of the bottle out so as to form a thin layer on a piece of glass. If material is so abundant that you can afford to lose some of it, simply cool the paraffin in the bottle or dish. If the paraffin sticks, as it probably will, break the dish and cut the paraffin into suitable blocks. Or, the paraffin may be poured into the usual imbedding dish, keeping it just warm enough to let the organisms settle; then cool and cut into blocks. Safranin, gentian-violet, orange, are very good for olden stages, especially for pyrenoids and starch. Iron-alum haematoxylin is better for nuclei and the younger stages of oögonia, antheridia, and new colonies.

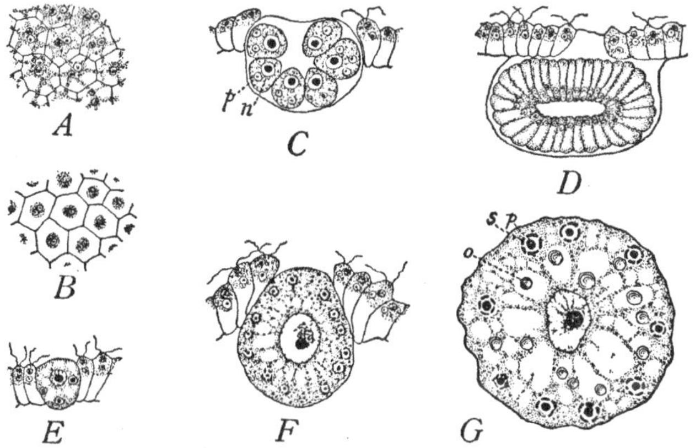

Fig. 36.—*Volvox:* surface view of several cells, fixed in the hot aqueous corrosive sublimate-acetic acid mixture, stained in iron-alum haematoxylin, and mounted in Venetian turpentine; *B*, fixed in chromo-acetic acid, but otherwise treated like *A*; *C–G* fixed in 1 per cent chromo-acetic acid imbedded in paraffin and cut 5 μ; *C*, *E*, and *F* stained in iron-alum haematoxylin; *D*, and *G* stained in safranin gentian-violet orange; *C*, new colony showing pyrenoids, *p*, and nucleus, *n*; *D*, a nearly mature antheridium; *E*, young egg; *F*, egg before fertilization; *G*, egg after fertilization, showing oil globules, *o*; pyrenoids, *p*; starch cut off from pyrenoid, *s*. All ×780.

Figure 37 shows that even such delicate forms as *Volvox* can be imbedded in paraffin without shrinking.

The Power's Methods.—Professor J. H. Power's mounts of *Volvox* and other members of the Volvocaceae have been the delight and despair of both botanists and zoölogists for many years. They have never been surpassed and, probably, never equaled. Professor Powers has kindly given me an outline of his methods; but, as in other phases of technic, judgment, skill, and patience must be furnished by the student.

For fixing, Professor Powers uses the aqueous potassium iodide solution used in testing for starch. This solution may be weaker than the usual formula so that it has a light-brown color. From 10 to 24 hours is sufficient for fixing, but material may be left here for several days. Wash thoroughly in tap water which has stood long enough to give off all of its excess of air; otherwise bubbles will form on the colonies, causing them to float and hindering subsequent processes. Change after change of water should be made rapidly, using large amounts of water and decanting just as soon as the colonies have settled. From 1 to 3 hours' washing should be sufficient to remove the brown color of the iodine.

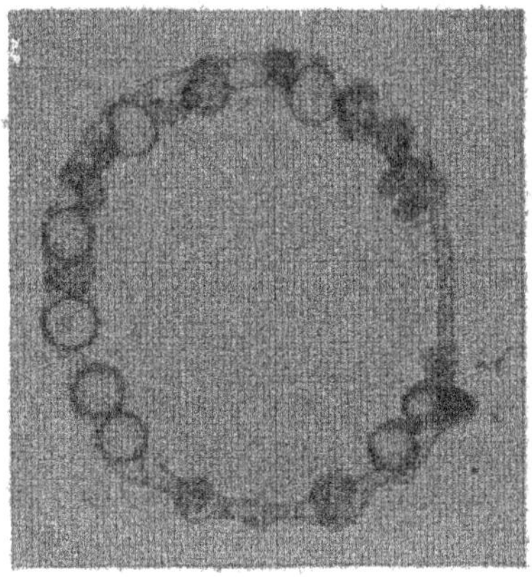

FIG. 37.—*Volvox:* photomicrograph of a section from material fixed in chromo-acetic acid and stained in Delafield's haematoxylin; from a preparation and negative by Dr. W. J. G. Land.

Stain in Mayer's Carmalum.[1] Use a pure carminic acid in making the stain, 1 g. carminic acid, 10 g. alum, and 200 c.c. distilled water. Dissolve with heat, filter, and add a crystal of thymol to keep out fungi. After staining, follow the Venetian turpentine method, taking care to wash the glycerin out completely. The 10 per cent turpentine should not be allowed to concentrate too rapidly.

Material fixed in weak osmic acid is even better for protoplasmic connections and cilia. About 4 or 5 drops in 50 c.c. of distilled water is sufficient. From 6 to 24 hours is long enough for fixing.

After either of these fixing agents, following the washing in water, material may be preserved in a nearly saturated solution of alum; or in a dilute aqueous carmalum, with a crystal of thymol to prevent mold. Staining for 3 weeks in a weak carmalum stains the cells

[1] Lee *Vade-Mecum* (8th ed. p. 137).

but not the matrix. About 3 months will be necessary to stain the matrix enough to make it a background for the cells.

An aqueous solution of nigrosin gives an effect like that of iron-alum haematoxylin. Rose benzol and Lee's pyrogallic acid method were also useful.

Diatoms.—Living diatoms are often found clinging in great numbers to filamentous algae, or forming gelatinous masses on various submerged plants. *Cladophora* is frequently covered with *Cocconeis*, an elliptically shaped diatom; *Vaucheria* is often covered with small forms. Other algae will pay for examination, especially if they look brown. If stones in the water have a brown, slippery coating, you can be sure of diatoms. Sometimes the brown coating on sticks and stones is so abundant that it streams out with the current. If rushes and stems of water plants have a brown, gelatinous coating, you are likely to find millions of specimens of the same diatom. The surface mud of a pond, ditch, or lagoon will always yield some diatoms. They can be made to come out from the mud by putting a black paper around the jar and letting direct sunlight fall upon the surface of the water. The diatoms, in a day or even less, will come to the top in a scum which can be easily secured.

Since diatoms form an important part of the food of molluscs, tunicates, and fishes, the alimentary tracts of these animals often yield deep-water forms which are not easily secured in any other way.

Fresh-water diatoms appear in greatest abundance in spring, are comparatively scarce in summer, and reappear in autumn, though not so abundantly as in the spring.

Marine forms can be secured by scraping barnacles, oyster shells, and other shells. The big *Strombus* shell from the West Indies, which we use to keep the door open, will yield a good collection if you get it before it is cleaned.

The silicious shells of diatoms are among the most beautiful objects which could be examined with the microscope (Fig. 38). To obtain perfectly clean mounts requires considerable time and patience, but when the material is once cleaned, preparations may be made at any time with very little trouble. Diatom enthusiasts have devised numerous methods for cleaning them, and separating the various forms from one another, but we shall give here only a few simple, practical methods.

Material for mounts of frustules of living forms may be obtained by skimming off the brownish scum found on ponds, by squeezing out water weeds, by scraping sticks and stones which are covered at high water, or from the mud of filter beds and pumping-works, or in other places. The material is put in a dish of water, and after it has settled the water is decanted. This is repeated until the water will clear in about half an hour. The sediment is then treated with an equal bulk of sulphuric acid, after which bichromate of potash is added until all action ceases. After a couple of hours the acid is washed out. To separate the diatoms, place the sediment in a glass dish with water, and when the water becomes clear give the dish a slight rotary motion. This will bring the diatoms to the top, when they may be removed with a pipette and placed in alcohol. To mount, place a number in distilled water, evaporate a few drops of the mixture on a cover-glass, which is then mounted on a slide in balsam.[1]

It is better to use a very slight smear of Mayer's albumen fixative to prevent the diatoms from floating to the edge of the cover.

Many scouring soaps and silver polishes contain large quantities of fossil diatoms, and the diatomaceous earths are particularly rich. Diatomaceous earths from Cherryfield, Maine, and from Beddington, in the same region, are the richest we have seen. The deposits at Richmond, Virginia, have long been famous. In some of our western states there are deposits 300 feet thick, with 80 per cent of silica, the silica being the valves of diatoms.

Break up a small lump of such material and boil it in hydrochloric acid. An evaporating dish or a test-tube is convenient for this purpose. Let the diatoms settle, pour off the acid, and then wash in water. As soon as the diatoms settle, the water should be poured off. The washing should be continued until the hydrochloric acid has been removed. When the washing is complete, pour on a little absolute alcohol, and after a few minutes pour off the alcohol and add equal parts of turpentine and carbolic acid. The material will keep indefinitely in this condition, and may be mounted in balsam at any time. In making a mount, put a little of the material on a slide and allow it to become dry, or nearly dry, and then add the balsam and cover. If the balsam should be added too soon, the diatoms are likely to move to the edge of the cover.

We have had excellent results with the following method: After washing in water, keep the diatoms in 5 per cent formalin. To

[1] From a review of Dr. Wood's paper on "Diatoms," *Journal of Applied Microscopy*, March, 1899.

make a mount, smear the slide very slightly with Mayer's albumen fixative, add a little of the material, and heat just enough to coagulate the albumen. When perfectly dry, add a drop of balsam and a cover. Or, after coagulating the fixative, dip in absolute alcohol and then in xylol before mounting in balsam. Without the alcohol and xylol,

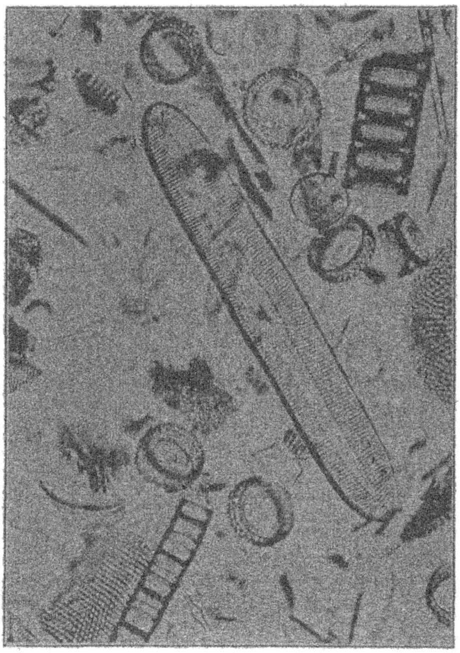

FIG. 38.—Diatoms: diatomaceous earth from Cherryfield, Maine, a Pleistocene deposit, showing the great variety of forms usually found in such material; photomicrograph by Miss Ethel Thomas from a preparation by Rev. E. L. Little. ×400.

some air is sure to be caught and it may accentuate some markings; but, in general, we prefer to use the alcohol and xylol.

To show the cell contents, diatoms must be fixed and stained. If they are clinging to filamentous algae, the algae with the diatoms attached should be put into chromo-acetic acid (24 hours) and then washed in water for 24 hours. Stain in iron-haematoxylin and proceed by the Venetian turpentine method. When ready for mounting, the diatoms can be scraped off from the algae or other substratum. Other stains may be used.

When the material is in gelatinous masses it may be fixed in chromo-acetic acid, with or without a little osmic acid, and imbedded in paraffin. There will, of course, be some difficulty in cutting, but the knife often breaks the frustules very cleanly, so that good sections may be secured. It might be worth while to try a weak solution of hydrofluoric acid to dissolve the silicious shells.

Desmids.—The desmids are unicellular, free-floating or suspended algae. They are not found in salt water and are more abundant in soft water than in hard. Deep pools, quiet ponds, and quiet margins of small lakes are good collecting-grounds. Collections of other fresh-water algae often contain some desmids. It frequently happens that a single desirable desmid appears during examination of field collections. In such a case, remove it with a fine pipette, and get it into a drop of water on a clean slide, invert it over a bottle of 1 per cent osmic acid for 2 minutes, leave the slide exposed to the air until almost all the water has evaporated, and then add a drop of 10 per cent glycerin. In a few hours (6 to 24) put on a cover and seal. It requires more time, care, and patience than it is worth to attempt staining in such a case.

Sometimes desmids occur in great abundance. They may then be treated like the filamentous algae, except that more care must be taken not to lose them when changing fluids. Four or five drops of 1 per cent osmic acid to 50 c.c. of water fixes well, and material from this solution may be placed directly into 10 per cent glycerin and mounted by the Venetian turpentine method. It looks almost as if stained in iron-alum haematoxylin. The iodine solution used in testing for starch gives good results and may be followed by any stains. The larger desmids stain beautifully in iron-alum haematoxylin.

The Venetian turpentine method, with Magdala red and anilin blue, will give beautiful preparations. A deep stain with Magdala red and a rather light stain with anilin blue is better for the pyrenoids and nucleus, while a light stain in the red and a deep stain in blue is better for the chromatophores. When the material is sufficiently abundant, paraffin sections may be made as directed for *Volvox*.

Lutman has found that *Closterium* divides at night. If mitotic figures are wanted they are likely to be obtained if the material is fixed about midnight.

Zygnema.—*Zygnema* is one of the commonest algae of the ponds, swamps, and ditches. The mats are very slippery to the touch. In the field it resembles *Spirogyra*, but is distinguished by the two characteristic chromatophores which are readily seen with a good pocket lens. Sometimes conjugation can be induced by bringing the material into the laboratory and placing it in open jars with plenty of water and not too much light.

The special chromo-acetic-osmic-acid solution is good for fixing. Stain in iron-alum haematoxylin and also in Magdala red and anilin blue. Follow the Venetian turpentine method, and in mounting put material from both stains on each slide.

Textbooks describe "stellate chromatophores" in *Zygnema*. A good preparation should show that the chromatophores have an even outline, with no trace of a stellate form. The boundary between the chromatophore and the protoplasm—often of a stellate form—in which the chromatophore is imbedded, should be seen in well-fixed material with either of the foregoing stains. The large starch grains, extending from the pyrenoid almost to the border of the chromatophore, are better differentiated by the Magdala red and anilin blue; the nucleus and pyrenoid are better stained by the iron-alum haematoxylin. Careful staining should bring out the features shown in Figure 39. The chromatophores do not stain as readily as those of *Spirogyra*, and consequently it is necessary to use stronger stains or more prolonged periods. Use the Venetian turpentine method.

For a detailed study, imbed in paraffin and cut thin sections. After washing in water, arrange the filaments so that most of them will have the same general direction; then, in running up through the alcohols, keep the filaments from spreading too much by placing a slide on the material. After imbedding, the material can be cut into blocks about a centimeter square. If sections thinner than 5 μ are wanted, cut out smaller paraffin blocks.

Spirogyra.—Probably no alga has been more studied by pupils, teachers, and investigators than *Spirogyra*. Nearly all of the numerous species belong to the low, quiet waters of ponds and ditches, where they often form large, flocculent green mats nearly covering the surface of the water. A few species occur in running water. The mats are very slippery to the touch—a character which assists in recognizing the genus in the field. In the larger species the characteristic spiral chromatophores can be seen with a good pocket lens, thus completing the identification, as far as the genus is concerned. Mats in which zygospores have been formed are likely to show a pale, or even a brownish, color, due to the brownish walls

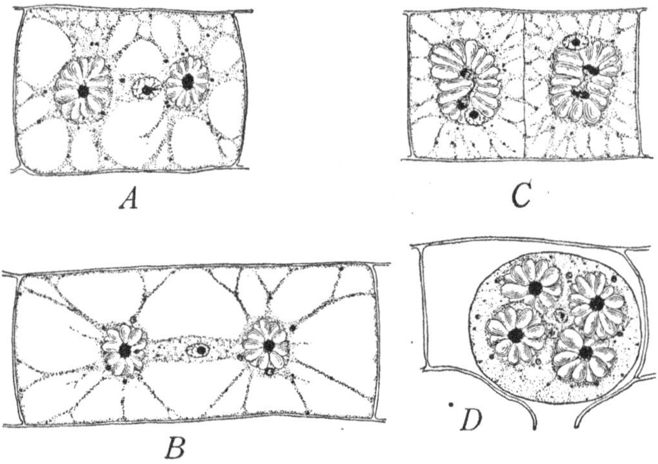

FIG. 39.—*Zygnema:* fixed in the special chromo-acetic-osmic solution: *A* and *B* stained in Magdala red and anilin blue; *C* and *D* stained in iron-alum haematoxylin. All show the nuclei, the chromatophores differentiated from the cytoplasm, and the large starch grains arranged radially about the pyrenoids. In *C*, cell division has just taken place and the pyrenoids and chromatophore in each new cell are dividing. In *D*, a young zygospore, the two nuclei have not yet fused. ×790.

of the zygospores. This color, however, is not always, or even usually, due to zygospores, but is more often due to the death and degeneration of the plants. Mats in early stages of conjugation and those with young zygospores show as bright a green as vigorously growing material.

Spirogyra is not easy to keep in the laboratory. The small species keep better than the larger ones. Put only a small amount of the material in a jar and use rain water. If it is necessary to use tap water, let the water run for a minute before taking the water for the culture. Most metals are poisonous to *Spirogyra*, even the small amount taken up by the water while standing in the water pipe being detrimental.

The species found in running water will usually conjugate within a week when brought into the laboratory and placed in rain water or tap water. Species belonging to quiet waters, when brought into the laboratory and placed in a 0.2 per cent Knop's solution, are likely to undergo rapid cell division and growth. After the alga has remained in such a culture for a few days or for a week, conjugation may be induced by transferring to rain water or tap water, and

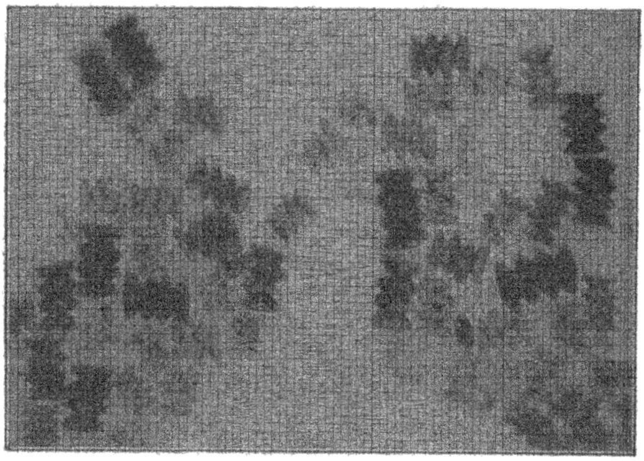

FIG. 40.—*Scenedesmus:* photomicrograph from a preparation by Dr. Yamanouchi, mounted whole and stained as described in the text; Cramer contrast plate; 4-mm. objective; ocular ×4; yellowish-green filter; camera bellows, 1 meter; arc light; exposure, 6 seconds. ×675.

keeping the culture in bright sunlight. Conjugation may begin within 3 or 4 days. Variations in temperature between 1° and 15° C. have little influence upon conjugation.

The special chromo-acetic-osmic solution fixes well. Stain some material in iron-alum haematoxylin and some in Magdala red and anilin blue. Use the Venetian turpentine method, and on each slide mount material stained in both ways. With Magdala red and anilin blue the spiral chromatophore takes the blue and its pyrenoids the red. If the material contains figures, stain in iron-haematoxylin. This will stain the figures, but will hardly touch the chromatophore or cell wall, thus allowing an unobstructed view of the figures. While figures occur occasionally in the daytime, collect your material at night, preferably near midnight.

Spirogyra is easily imbedded and cut.

Scenedesmus.—*Scenedesmus* (Fig. 40) is found everywhere as a constituent of the fresh-water plankton. It is more abundant in stagnant water. It often appears in considerable quantity in laboratory cultures, where it may be kept for years in a tightly closed glass jar without renewing the water, the lid being removed only when material is needed.

The form is so small that in living material little more than the general form can be distinguished. Excellent mounts are easily and quickly made. Smear a very thin layer of albumen fixative upon the slide, and add a drop of water containing the *Scenedesmus.* The drop may be inverted for 1 or 2 minutes over the fumes of 1 per cent osmic acid. No washing is necessary, and good mounts may be made without any fixing whatever. Allow the drop to dry completely. It is better to leave it for 24 hours before proceeding. The usual difficulty with this form, and with many others, is that the background stains and so makes the mounts untidy. The following method by Yamanouchi will produce beautiful preparations (Fig. 40):

1. Dry on the slide, 24 hours.
2. 10 per cent alcohol over night to remove chlorophyll.
3. Safranin (alcoholic), 4 days.
4. Water, 5 minutes.
5. Aqueous gentian-violet, 2 days.
6. Water, a few seconds.
7. Orange G, aqueous, 3 minutes.
8. 95 per cent alcohol, a few seconds.
9. Absolute alcohol, 1 minute.
10. Clove oil, until the stain is satisfactory. Different collections of *Scenedesmus* stain very differently, but the time in clove oil is likely to be long, even as long as 6 hours.
11. Xylol, 5 minutes.
12. Mount in balsam.

Hydrodictyon.—This is popularly known as the "water-net." *Hydrodictyon* is found floating or suspended in ponds, lakes, or slow streams. The young nets are formed within the segments of the older nets. Examine segments 4 or 5 mm. in length for the formation of young nets. The old nets may reach a length of 10 cm. Cultures are easily kept in the laboratory. If material which has been growing in a 0.5 to 1 per cent Knop's solution be brought into tap water or pond water, zoöspore formation may begin within 24 hours. Nets brought from the nutrient solution into a 1 to 4 per cent cane-sugar solution produce zoöspores for a few days.

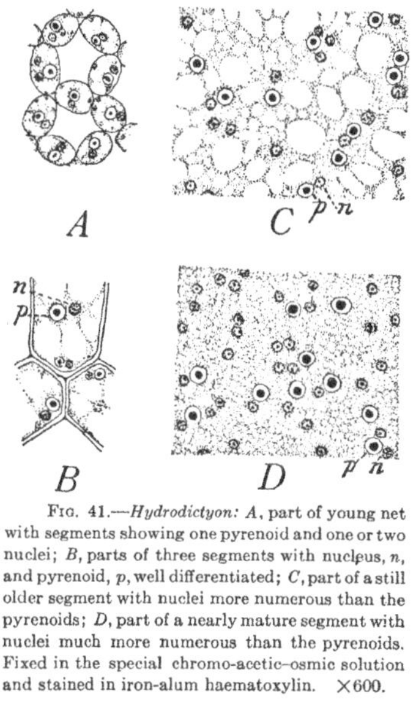

FIG. 41.—*Hydrodictyon: A*, part of young net with segments showing one pyrenoid and one or two nuclei; *B*, parts of three segments with nucleus, *n*, and pyrenoid, *p*, well differentiated; *C*, part of a still older segment with nuclei more numerous than the pyrenoids; *D*, part of a nearly mature segment with nuclei much more numerous than the pyrenoids. Fixed in the special chromo-acetic–osmic solution and stained in iron-alum haematoxylin. ×600.

Nets of all sizes should be selected for study. The segments are coenocytic, and the nuclei of the older segments are hard to differentiate, except in stained preparations. Only one nucleus will be found in the young segments, but in the older segments the nuclei become very numerous.

For fixing, use the special chromo-acetic–osmic formula. This should not produce plasmolysis in nets of any age. Iron-alum haematoxylin will differentiate the nuclei and pyrenoids, which may look alike with less precise stains. Use the Venetian turpentine method for mounting whole young nets and parts of older nets. Fine scissors should be used freely, because any attempt to arrange the material with needles will make it look as if the whole method of preparation were wrong. Parts of nets mounted whole are shown in Figure 41.

For details of the formation of starch and for the finer details of the development of zoöspores and gametes, *Hydrodictyon* should be imbedded and cut.

Pleurococcus.—This form, which is used everywhere as a laboratory type of the unicellular green algae, is found on the bark of trees, where it is more abundant on the north side and near the ground.

It is also found on stones and fences, and in moist situations generally. It is easily secured in nearly all localities and at all seasons.

The life-history of *Pleurococcus* is variously described in textbooks, but it is very doubtful whether there is any mode of reproduction except by cell division. The zoöspores and gametes which are sometimes described probably belong to other forms which are occasionally associated with *Pleurococcus*, especially when growing in very moist situations. The life-history was examined very critically by the great algologist, Wille, who not only concluded that cell division is the sole mode of reproduction, but showed how investigators, even those relying upon cultures, had made their mistakes. Wille's paper was published in 1913 in *Nyt Magazin for Naturvidenskaberne.*

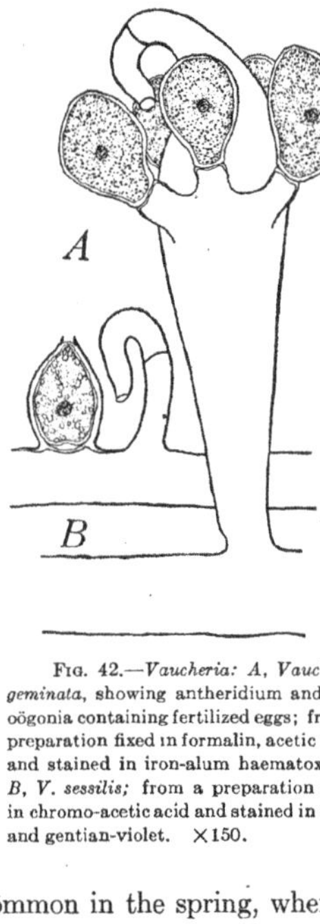

FIG. 42.—*Vaucheria: A, Vaucheria geminata*, showing antheridium and five oögonia containing fertilized eggs; from a preparation fixed in formalin, acetic acid, and stained in iron-alum haematoxylin; *B, V. sessilis;* from a preparation fixed in chromo-acetic acid and stained in eosin and gentian-violet. ×150.

A study of the living material is sufficient for any general course. The bright-green cells, scraped off and mounted in a drop of water, show the rather thick wall, the chromatophores, and usually the nucleus. A drop of iodine will bring out the nucleus, if it does not show already, and will also stain the pyrenoid, if the cell contains one. A mount in Venetian turpentine, stained in Magdala red and anilin blue, shows the nucleus very clearly.

Vaucheria.—This form can always be obtained in greenhouses, especially in the fernery, where it forms a green felt on the pots. The greenhouse form is likely to be *Vaucheria sessilis.* Another species, *V. geminata*, is very common in the spring, when it may be found in ponds and ditches (Fig. 42). *Vaucheria* is also found in running water, but in this situation is almost certain to be sterile. In the vicinity of Chicago, *V. geminata* appears late in March or early in April and within a few weeks begins to fruit abundantly. The fruiting continues for 4 to 8 weeks, and then the alga may disappear until later in the season, when some of the oöspores germinate.

Vaucheria sessilis is found at all seasons in the greenhouses, but it is usually in the vegetative condition. Klebs found that the formation of oögonia and antheridia can be induced in *V. repens* (a variety of *V. sessilis*) within 4 or 5 days by putting the material into a 2 to 4 per cent cane-sugar solution in bright sunlight. The sex organs will not be formed in weak light or in darkness.

The formation of zoöspores may be induced in the following way: Cultivate in a 0.1 to 0.2 per cent Knop's solution for a week, then bring the material into tap water, and keep the culture in the dark. Zoöspores may appear within 2 days. Bright light or a temperature higher than 15° C. will check the production of zoöspores. A 2 per cent cane-sugar solution kept in the dark is also likely to furnish zoösporic material. If no zoöspores are formed when the solution is kept in the dark, the nutrition has been too weak: strengthen the nutrient solution and keep the culture in the light for a few days; then put the culture in the dark, and zoöspores should appear. The formation of zoöspores may continue for a couple of weeks.

Aplanospores of *V. geminata* are formed in nature when the plant is growing upon damp ground. The aplanospores may also appear in a 4 per cent cane-sugar solution.

In fresh 0.5 per cent Knop's solution in bright light, cultures remain in the vegetative condition, and the result is the same in weak light if the nutrient solutions are seldom changed. Such cultures may be kept indefinitely by changing the nutrient solution whenever a whitish scum appears on the surface.

Vaucheria is not easy to fix. Solutions which give fine results with *Spirogyra* and *Zygnema* may be ruinous to *Vaucheria*. We have secured the best results with a formalin-acetic solution (10 c.c. formalin, 5 c.c. glacial acetic acid, and 90 c.c. water). Chromic-acid solutions, even with 4 or 5 per cent acetic acid, cause some plasmolysis. If the chromic acid is weakened enough to prevent plasmolysis, the fixing is not thorough enough to prevent shrinking during subsequent processes.

Iron-alum haematoxylin is the best stain. Magdala red and anilin blue give beautiful results, *occasionally*, but preparations are almost sure to fade. Eosin is good for topography, but will not show the nuclei.

Use the Venetian turpentine method. In mounting, use small scissors freely. You cannot untangle a mat of *Vaucheria* so as to give good views.

For the development of the oögonium and antheridium, for fertilization and for the structure and development of the various spores, thin sections are necessary. Imbed in paraffin. For nuclear details, use iron-haematoxylin; for cytoplasm, use safranin, gentian-violet, orange.

Cladophora.—This genus is found in both salt and fresh water. The fresh-water forms are usually attached to sticks or stones in quiet or running water. The mats feel rough and crisp and, even under a pocket lens, show the characteristic branching by which the form is easily recognized. The absence of a mucous coat makes *Cladophora* a convenient host for numerous parasitic algae, among which diatoms belonging to the genera *Cocconeis* and *Gomphonema* are particularly abundant.

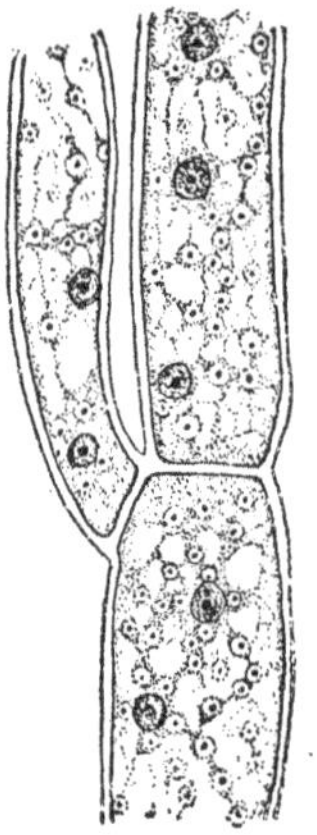

FIG. 43.—*Cladophora:* fixed in chromo-acetic acid and stained in iron-alum haematoxylin.

For laboratory cultures, select the forms found in quiet water, but for preparations, forms growing where the waves dash hard are better, since you can get a fine display of branches under a small cover. Forms growing in still water or in gently flowing water may look like unbranched filaments under an ordinary cover. The special chromo-acetic–osmic solution is excellent for *Cladophora.* Iron-alum haematoxylin, followed by the Venetian turpentine methods, gives the best results for nuclei and pyrenoids. Magdala red and anilin blue are better for the cell wall and chromatophores (Fig. 43).

Ulothrix.—Where the problem of the origin and evolution of sex is studied, *Ulothrix* is an indispensable type. *Ulothrix zonata* is found in springs, brooks, and rivers, occurring in bright-green masses attached to stones in riffles, especially in sunny places. It is abundant on stones and piles along the beaches of lakes. Nuclear division takes place at night, most abundantly about midnight, and is followed by a rapid development of zoöspores and gametes, which continue to be discharged throughout the forenoon. In the afternoon the material is largely vegetative. Another species is found in stagnant ponds, ditches, and even in watering-troughs and rain-barrels. It is difficult to keep in the laboratory the forms which are found in rapidly flowing water. However, if they are brought in still attached to stones and placed under a stream of tap water, they may live for a couple of weeks and may produce zoöspores every morning. The production of zoöspores may continue for a few days, if the material is merely put into a jar of water; in a 2 to 4 per cent cane-sugar solution the production of zoöspores continues a little longer.

No form is better than *Ulothrix* for illustrating to a class the difference between zoöspores and gametes. Even when gametes are not conjugating, their more rapid movement is noticeable; and when conjugating, the awkward, jerky movements of the pair contrasts sharply with the graceful movements of the zoöspores.

Fix in the special chromo-acetic–osmic solution and stain some material in iron-alum haematoxylin, which will differentiate the nuclei and pyrenoids; and stain some in Magdala red and anilin blue, which is better for the cell wall and chromatophore. Mount on each slide material from both lots, and the preparation will then afford a rather complete study.

Oedogonium.—Most species are found in quiet waters, especially in ponds and ditches. The best fruiting material is often attached to submerged twigs, rushes, and various plants, where, to the naked eye, it forms only a fuzzy covering. Some species form floating masses, bearing some resemblance to *Spirogyra*, but they are not so slippery.

The special chromo-acetic–osmic solution is good for the filaments; the iodine solution used in testing for starch, or a weak aqueous solution of osmic acid (4 or 5 drops to 50 c.c. of water) is good for zoöspores and androspores.

For cell division and the peculiar method of forming the new cell wall, stain in Magdala red and anilin blue. Iron-alum haematoxylin is better for most of the other phases; but the fertilized eggs stain very deeply. Consequently, stain some material lightly, for the fertilized eggs; and some more deeply for young eggs, chromatophores, and other phases. Mount in Venetian turpentine.

For details of blepharoplasts and the development of the various motile forms, material should be imbedded and sectioned.

Nanandrous species have antheridia only in the dwarf males; and species with antheridia in the ordinary filaments have no androspores or dwarf males in the life-history. A species with dwarf males is shown in Figure 44.

In studying *Oedogonium diplandrum*, Klebs found that a change from a lower to a higher temperature would induce the production of zoöspores. A culture which had been kept in a cold room with a temperature varying from 6° to 0° C., when brought into a warmer room with a temperature varying from 12° to 16° C., produced an abundance of zoöspores within 2 days. Light does not seem to have any influence upon the formation of zoöspores in this species, but light is necessary for the formation of antheridia and oögonia.

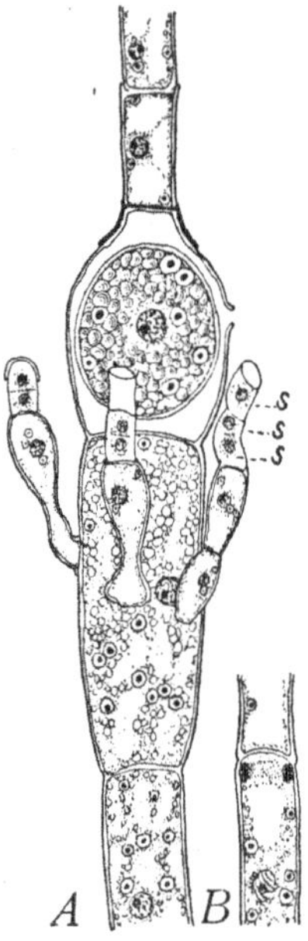

FIG. 44.—*Oedogonium:* an undescribed species collected by Dr. Elda Walker. In *A*, on the left, is a dwarf male still showing the cap cell; on the right, the cap cell has come off and the antheridium at the top has discharged one of the sperms, but the other, *s*, still remains; in the antheridium below are two young sperms. *B*, upper part of a cell, showing the cellulose ring which will form the side wall of the new cell. The nucleus is in the spirem stage of division. ×280.

We have secured an abundance of oögonia and antheridia by keeping the material for 4 or 5 days in a very weak Knopf's solution and then tranferring to distilled water. The oögonia appeared in 3 or 4 days. The method seems to succeed with some species, especially those which occur floating or suspended in the water, but we have not succeeded with species which form a fuzzy covering on grasses and twigs under water. Sterile material sometimes fruits when brought into the laboratory and placed in open jars with plenty of water and not too much light.

Coleochaete.—*Coleochaete* is epiphytic upon the stems and leaves of submerged plants. *C. scutata*, which is the most common species, has a flat, orbicular thallus generally less than 1 mm. in diameter. *C. pulvinata* has a hemispherical thallus and might be mistaken for *Rivularia*, unless examined with a lens.

For most purposes it is better to mount the whole plant, *in situ* on the leaf upon which it is growing. If it is growing on stems or petioles, strip off the epidermis and mount small pieces. In this way you get the very young stages, from the fertilized egg up to the adult plant. The young stages are likely to be overlooked and even when noticed are likely to be lost if you attempt to remove *Coleochaete* from its host. If you think you must remove it from the host, complete the staining before the removal.

Fix in the special chromo-acetic–osmic solution and stain sharply in Delafield's haematoxylin. If you overstain and reduce with hydrochloric acid (3 or 4 drops in 100 c.c. of water), the *Coleochaete* will stand out against the host, which is not likely to stain so deeply as the alga. A slight tinge with orange in clove oil may increase the contrast between the alga and its host.

After staining, the material may be run up through a series of alcohols, 5, 10, 20, 35, 50, 70, 85, 95, and 100 per cent, about 15 minutes in each grade. If orange in clove oil is used, apply it here and transfer to xylol. Mount in balsam. If no orange is used, transfer to 25, 50, 75, and 100 per cent xylol, and mount in balsam. After staining, the Venetian turpentine method may be used instead of the rather long series of reagents.

Sections are easily cut and, especially in forms with a flat thallus, show features which might escape if one depended entirely upon plants mounted whole. Cut out small pieces of leaf or stem abundantly covered with *Coleochaete*, imbed in paraffin, and cut host and guest together.

Chara.—*Chara* is found in ponds, lagoons, and ditches. Once seen, it is always readily recognized. In the ponds and lagoons along the southern shores of Lake Michigan it fruits so abundantly that the whole pond shows an orange color due to the immense numbers of antheridia. In the lagoons of the Chicago parks *Chara* is so abundant that it must be dredged out every summer.

Chara is easily kept alive throughout the year in the laboratory. A 2-gallon glass jar with an inch of pond dirt, sand, and gravel at the bottom, and nearly filled with tap water, is all that is needed for a successful culture. If the jar is to be covered, it should not be more than two-thirds full of water. Not more than a dozen plants should be put into such a jar.

A rather strong solution should be used for fixing. The following will give good results:

Chromic acid..................................	1 g.
Glacial acetic acid...........................	1 c.c.
Water...	100 c.c.

In about 24 hours this not only fixes, but it dissolves the lime with which most species are coated.

For paraffin sections select the tip of the plant, a piece about a centimeter in length. Sections of this may show, not only the large

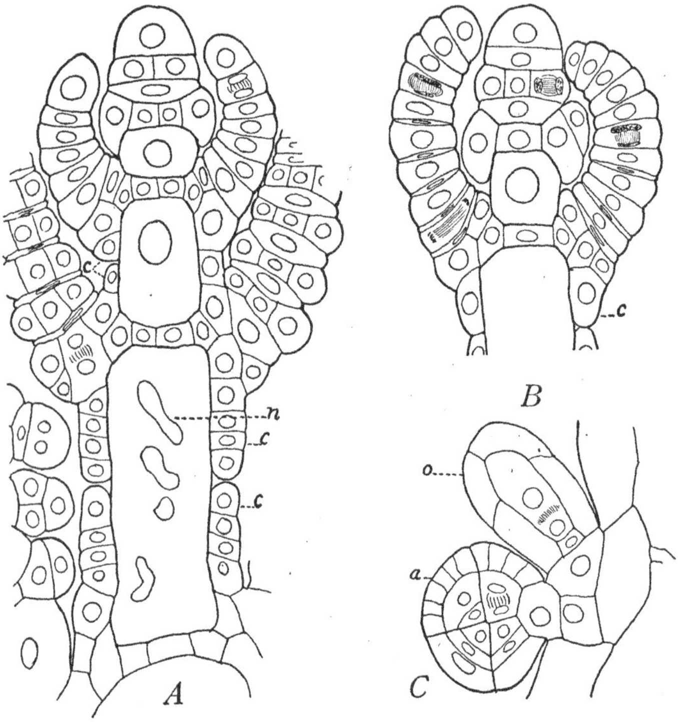

FIG. 45.—*Chara:* fixed in chromo-acetic acid and stained in safranin, gentian-violet, orange. *A*, longitudinal section of apex showing five nodes; *c*, corticating filaments; *n*, part of the fragmented nucleus. *B*, note the different condition of the nodes and first branches. *C*, young antheridium, *a*, and young oögonium, *o*. ×230.

apical cell, but also various stages in the development of antheridia and oögonia (Fig. 45). For the development of the plant body from the apical cell and also for early stages in the development of oögonia and antheridia, the safranin, gentian-violet, orange combination is excellent; for later stages, especially in the development of the antheridia, iron-haematoxylin is much better.

The antheridium of *Chara* stains so rapidly that the beginner uniformly makes poor preparations. In order to get good preparations

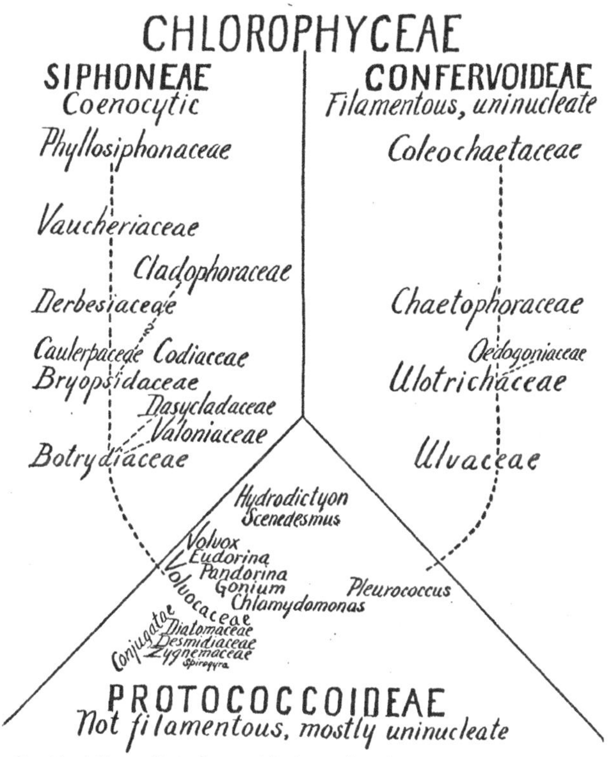

FIG. 46.—A diagram illustrating an opinion in regard to relationships in the Chlorophyceae

of the antheridium, it is necessary to disregard other structures, which will be stained lightly or not at all when the stain is just right in the antheridial filaments.

If it is desired to mount whole branches showing the antheridium and oögonium in position, use the Venetian turpentine method, staining in Magdala red alone, or in Magdala red and anilin blue. Good mounts showing shield, manubrium, capitula, and filaments may be obtained by crushing an antheridium under a cover-glass. For this it is better to stain in Magdala red alone, since any over-staining is easily corrected by exposing the preparation to direct sunlight.

As slides accumulate, the thoughtful student will feel the need of some kind of classification. Of course, one might arrange alphabetically and there would be no need for thought; but we assume that the student who has made enough slides to need a classification will want one which expresses some idea of relationship, even if the idea may be more or less faulty. The classification indicated in Figure 46 is essentially that of Oltmann, and does not differ much from that given in Engler and Prantl's *Die natürlichen Pflanzenfamilien.* If the student would compile similar diagrams in all the groups, his slides would mean, not only proficiency in technic, but an increasing knowledge of the structure, development, and relationship of plants.

CHAPTER XVI

PHAEOPHYCEAE. BROWN ALGAE

The Phaeophyceae, or brown algae, are almost exclusively marine. They include a great variety of forms, ranging from delicate filaments to coarse, leathery plants 100 feet in length. There are no unicellular members.

For fixing marine algae, fixing agents should be made up with sea-water, never with fresh water, and the washing should be done with sea-water; but fresh water should be used in making the series of alcohols. When the Venetian turpentine method is to be used, wash in fresh water before placing in the 10 per cent glycerin.

For habit work, material may be put into formalin—about 6 c.c. commercial formalin to 100 c.c. of sea-water—and kept there indefinitely. If it is desired to transport large quantities of coarse forms, the material may remain in this solution for a week and may then be removed from the liquid and packed in closed pails or tubs or any water-tight containers. After reaching its destination, the material should be put into formalin again.

For material to be mounted by the Venetian turpentine method, 6 to 10 per cent formalin (always in sea-water) is a good fixing agent. Wash in sea-water for 1 hour, then in equal parts sea-water and fresh water for ½ hour, then in fresh water ½ hour. The material is now ready for staining in aqueous stains, or for the 10 per cent glycerin, if alcoholic stains are to be used.

The following formula by Flemming will also give good results, both for the Venetian turpentine method and for the paraffin method:

Chromic acid..............................	1 g.
Glacial acetic acid..........................	0.4 c.c.
Sea-water.................................	400 c.c.

Fix 24 to 48 hours and wash 24 hours in running sea-water. A convenient washing-box can be made from an ordinary washtub. Bore a dozen ⅜-inch holes in the bottom; insert rubber tubes 6 inches long, and in the end of each tube place the glass part of a pipette. The tub may be elevated by nailing three narrow boards to the sides so as to form a tripod. Place the bottles or cans of material under the pipettes and let sea-water flow into the tub.

If such chromic-acid material is to be used at once for Venetian turpentine mounts, follow the washing in sea-water by ½ hour's washing in equal parts sea-water and fresh water (not necessarily running water) and then ½ hour's washing in fresh water. The material is now ready for an aqueous stain or for 10 per cent glycerin. If desirable to keep it for future staining, put it into 5 or 6 per cent formalin *in fresh water*.

Material for sections may be treated in the same way, but it is often better to add 2 to 10 c.c. of 1 per cent osmic acid to 100 c.c. of the chromic-acid solution. The 1 per cent osmic acid should be made up in distilled water.

Small filamentous forms, if they are to be mounted whole, are usually more satisfactory to handle if fixed in formalin. A strong solution, about 10 c.c. of commercial formalin to 90 c.c. of sea-water, is very good. Material is well fixed in a few days, but may be left here until it is needed. To make preparations, wash in fresh water and follow the Venetian turpentine method.

For habit demonstrations many of the smaller forms can be floated out and dried on paper. *Ectocarpus*, *Desmotrichum*, *Dictyota*, *Cutleria*, and even small specimens of *Laminaria* are quite useful when prepared in this way. Take a light pine board, a little larger than the standard herbarium sheet, float it in a tub of water, place on the board the paper upon which the material is to be mounted, arrange the material with a toothpick or the blunt end of a needle, dipping all or a part of the board under water whenever necessary. Cover with a piece of cheese-cloth, add a blotter or two, as in case of flowering plants, and dry under gentle pressure, changing the blotters frequently. The algae have enough mucilage to make them adhere to the paper. Coarse forms, like *Fucus*, may need to be held down by strips of gummed paper.

Sphacelaria.—The apical cell of *Sphacelaria* or the nearly related *Stypocaulon* affords an excellent study of the structure of cytoplasm. Flemming's weaker solution, with the osmic acid even a little weaker than recommended in the formula, is good for the apical cell and the mitotic figures, which are quite conspicuous. For these features it is a good plan to break off the tips so as to have only pieces 6 to 12 mm. long, which will lie flat in the paraffin. The tips should be broken off after the material has been brought into xylol. If whole tufts are imbedded, the branches diverge enough to make perfectly longitudinal sections of the apical cells rather rare. Iron haematoxylin with a faint staining in orange is a satisfactory combination.

Ectocarpus.—For general morphological study, branches should be mounted whole in Venetian turpentine. A 6 to 10 per cent formalin solution (in sea-water, of course), or the chromo-acetic acid will give good fixation. Stain some in iron-haematoxylin and some in Magdala red and anilin blue. Mount on each slide some from each lot. Sporangia usually appear earlier than the gametangia. So collections should be made at different seasons. August is rather late for sporangia, but gametangia are abundant at this time. You should have both sporangia and gametangia on each slide (Fig. 47).

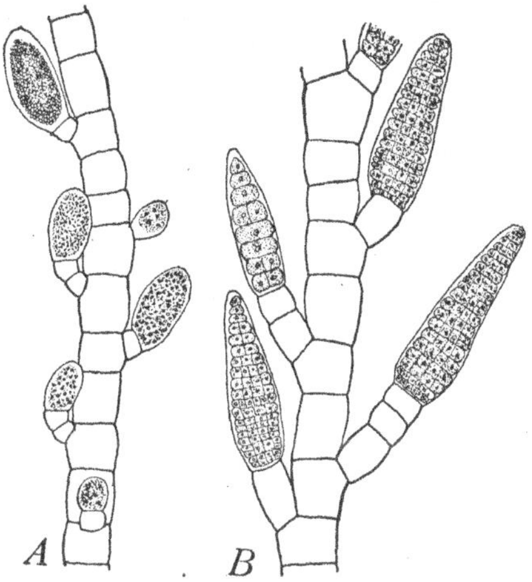

FIG 47.—*Ectocarpus:* *A*, sporangia ("unilocular sporangia") of various ages; gametangia ("pleurilocular gametangia"), three of them nearly mature. Fixed in 10 per cent formalin and stained in iron-haematoxylin. ×280.

Desmotrichum.—Forms as large as *Desmotrichum* can be handled like *Ectocarpus*, but care must be taken not to overstain.

Laminaria.—In such large forms, small portions showing the structure and development of the thallus and also the reproduction should be cut out with a razor and then placed in the fixing agent. The sporangia of *Laminaria* stain very deeply and quickly. Iron-haematoxylin is good, but be careful not to overstain. After this stain is just right, about 3 to 5 minutes in alcoholic safranin will stain the mucilaginous structures and add to the value of the preparation.

The zoöspores from the sporangia germinate immediately, forming dioecious, filamentous gametophytes bearing antheridia and oögonia. The fertilized egg at once begins to develop into the *Laminaria* plant. So little is known about this stage that it would be worth while for those who are at the seashore to germinate the zoöspores of any members of the Laminariaceae.

For habit study, small specimens up to 45 cm. in length can be mounted upon paper. They stick well and seldom need to be secured by gummed paper. Larger specimens may be allowed to dry and may then be stored away in a box. When wanted for use, wet them under the tap, or, better, in salt water; after using, let them dry and return them to the box. Specimens will stand 4 to 5 such soakings in fresh water; if a pint of salt is added to 3 or 4 gallons of water, the material may be soaked a dozen times before it passes its usefulness. If material has been fixed in formalin, it may be washed in sea-water—not very thoroughly, but enough to remove the pungent odor—and then soaked in equal parts of glycerin and water. Use only enough of the glycerin to make the specimens flexible, not enough to make them wet to handle. In this way, material of *Laminaria, Macrocystis, Nereocystis, Postelsia,* and other large forms can be kept in condition for demonstration and will last for years without any attention. When not in use, they should be kept stored in a box.

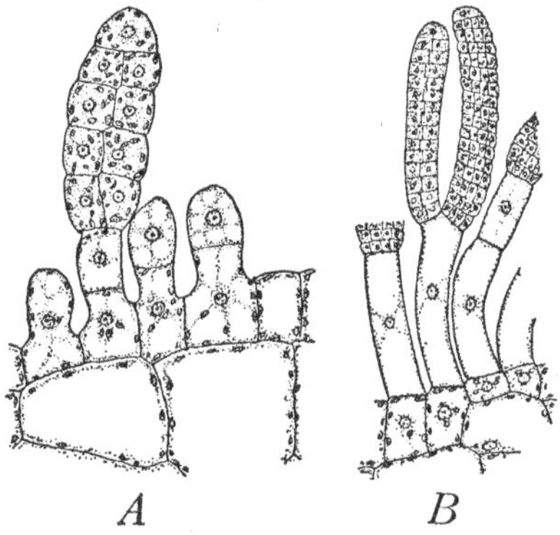

FIG. 48.—*Cutleria multifida: A,* oögonia; *B,* antheridia. Fixed in Flemming's weaker solution, cut 3μ, and stained in iron-alum haematoxylin. ×470.

Cutleria.—This alga deserves a place in any course in morphology, if the course is thorough enough to permit the study of three members of the Phaeophyceae. These three should be *Ectocarpus* (or *Pylaiella*), *Cutleria*, and *Fucus*. *Cutleria* is not found on the American coasts, but is abundant at Naples. The habits of gametophyte (known as *Cutleria*) and the sporophyte (known as *Aglaozonia*) are so different that they furnish a good illustration of alternation of generations. Beginners understand such an illustration more readily than they do an illustration like *Dictyota,* with its two generations looking so nearly alike. *Cutleria* also furnishes a good stage in the evolution of sex, about midway between isogamy and the extreme heterogamy of *Fucus.*

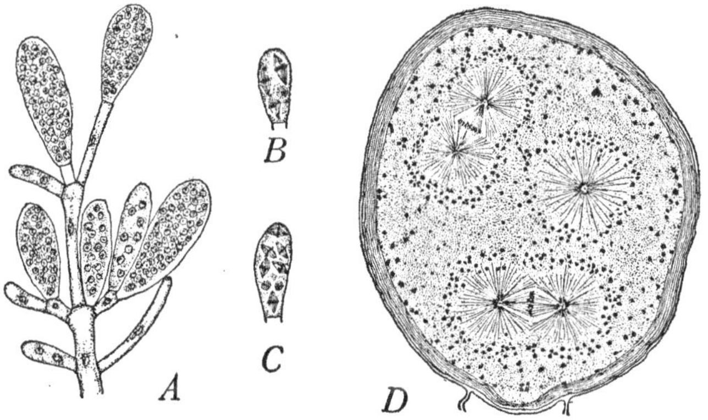

FIG. 49.—*Fucus vesiculosus: A,* antheridial branch with antheridia in various stages of development; *B,* the third mitosis in the antheridium, one figure showing a transverse view in which the chromosomes can be counted; *C,* the fourth mitosis, with two of the eight figures cut transversely but hard to count chromosomes in the figure, which is reduced one-half; *D,* the third mitosis in the oögonium, showing centrosomes and radiations. Fixed in Flemming's weaker solution, cut 3 μ, stained in iron-alum haematoxylin. $\times$480.

For habit study, both generations should be mounted upon paper. The gametophyte (*Cutleria*) sticks well, but the sporophyte (*Aglaozonia*) will need some glue or gummed paper.

For paraffin sections, fix in chromo-acetic acid. Cut 10 μ thick. For mitotic figures, some osmic acid should be added to the chromo-acetic acid and the sections should be much thinner, about 3 to 5 μ. Use iron-haematoxylin and then stain for 3 to 5 minutes in alcoholic safranin (Fig. 48).

Fucus.—Material for habit study may be dried, or preserved in formalin, or mounted on paper. In the latter case, glue or gummed paper will be necessary. Most satisfactory of all is to send to Woods

Hole, Massachusetts (George M. Gray), for living material. Fertilization occurs at all seasons, but autumn is the most favorable. In summer the material dies before it reaches Chicago, but during the rest of the year a pailful will reach Chicago, and even as far west as the Mississippi River, in good condition for showing the rotation of the egg by the sperms. The eggs and sperms form slimy masses, the antheridia being orange red, and that containing the eggs a dirty green. Mix a drop of the red with a drop of the green. The movements of the egg can be observed, and material for a study of fertilization and later stages is easily secured. In fixing fertilization and succeeding stages, it is worth while to use some of the regular

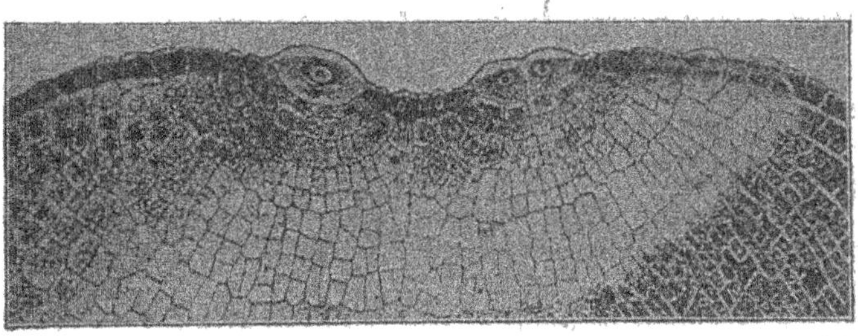

FIG. 50.—*Dictyota dichotoma:* longitudinal section through apical region showing two apical cells, the one on the right dividing again to produce another dichotomy. Stained in iron-alum haematoxylin. Photomicrograph by Miss Ethel Thomas. ×167.

Flemming's weaker solution, as well as the solution without the osmic acid.

For the growing points and conceptacles, small pieces should be cut off with a razor. If the fruiting tips be cut through lengthwise before they are cut off, the fixing will be more satisfactory. For sections of the conceptacles it is better not to cut across the whole tip, but to cut off pieces of the rind containing half-a-dozen conceptacles. Such pieces are more easily imbedded and cut. There is no difficulty in cutting such pieces in paraffin. Iron-haematoxylin is a good stain. Safranin and gentian-violet are also satisfactory, but care must be taken not to overstain since *Fucus* usually stains deeply and rapidly.

For the cytologist, *Fucus* might be used as a test object for testing proficiency in technic, just as *Pleurosigma angulatum* is used in testing an objective. The nuclear divisions in the antheridium are simultaneous, and at the sixth division, which is the last, there are 32 mitotic figures, each with 32 chromosomes which split so that 32 go to each pole. When you can make a preparation in which these chromosomes can be counted, your technic is adequate for research work in cytology. In a good preparation, the mitotic figures in the oögonium show small but brilliant centrosomes, with a great display of radiations. (Fig. 49).

Dictyota.—*Dictyota* deserves a place in the series illustrating the evolution of sex, since its large egg has lost all motility, but the difference in size of egg and sperm is not so extreme as in *Fucus*. It also furnishes an excellent example of the development of a thallus from an apical cell (Fig. 50).

Mount habit material on paper. For sections, fix in chromo-acetic acid. For figures, cut 3 to 5 μ, but for general views of apical cell and reproductive phases, cut 10 μ. Stain in iron-haematoxylin and counter-stain for 2 or 3 minutes in safranin.

CHAPTER XVII

RHODOPHYCEAE. RED ALGAE

The red algae belong almost exclusively to salt water, but a few genera are found only in fresh water, usually in running water, and a few forms occur both in salt and in fresh water. Nearly all are small forms, and for habit work can be floated out and mounted on paper. Very few will need glue or gummed paper.

For more critical habit work and for Venetian turpentine mounts, fix in 6 to 10 per cent formalin in sea-water. Material keeps indefinitely in 10 per cent formalin.

For sections, use the chromo-acetic acid with or without the addition of a little osmic acid, as recommended for the brown algae. The same method of fixing and washing should be used as for the brown algae, except that in the case of the few fresh-water forms, fresh water should be used in making the fixing agent and in washing it out. For *Polysiphonia*, and doubtless for many other forms, the period in the fixing agent should be very much shortened. Picric acid, corrosive sublimate, and absolute alcohol have been tried, but the results have not been encouraging.

Batrachospermum.—This is a green, fresh-water member of the red algae. It is not very uncommon in small streams. Fix in chromo-acetic acid (in fresh water) and use the Venetian turpentine method. Good preparations showing the nuclei may be obtained by staining in Mayer's haem-alum, or Haidenhain's iron-haematoxylin. After the material is ready for mounting, tease out a small portion, and still further dissociate the filaments by tapping smartly on the cover.

Nemalion.—Methods for preparing *Nemalion* have been described by Wolfe. Chromo-acetic acid proved to be most satisfactory for fixing. For studying fertilization, mounts were made as follows:

Young tips were crushed in water under a cover-glass and on a slide that had previously been treated with fixative; the cover was then removed, and the water on the slide allowed to evaporate. The gelatinous nature of the wall prevents the contents of the cell from being affected by this treatment, even when the albumen has hardened sufficiently to hold the filaments firmly in place.[1]

Stain in safranin and gentian-violet, and mount in balsam.

Iron-haematoxylin is recommended for paraffin sections. The sections must be very thin, 5 μ or less. "Material killed in 2 per cent formalin in sea-water and gradually transferred to pure glycerin kept its color perfectly."

For material to be mounted whole, we should recommend fixing in 10 per cent formalin and staining in iron-alum haematoxylin. Place

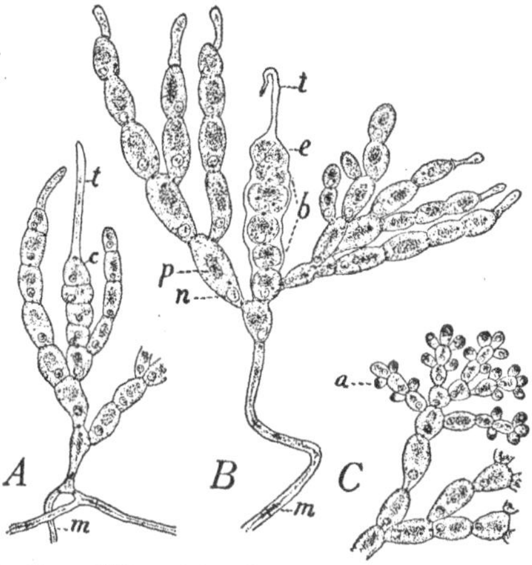

FIG. 51.—*Nemalion multifidum:* *A*, branch showing carpogonium, *c;* trichogyne, *t;* central strand, *m*. *B*, somewhat older stage showing two-cell stage immediately following fertilization, *e;* carpogonial branch, *b*. *C*, branch with antheridia, *a*. Fixed in 10 per cent formalin and stained in iron-alum haematoxylin. ×400.

the material in 10 per cent glycerin until all the water is out. Mount in glycerin jelly. To make a mount, take a small piece of the material, not more than 3 or 4 mm. long, touch it to filter paper to remove as much of the glycerin as possible, put it into the melted glycerin jelly, add a round cover, and crush by tapping on the cover. The antheridia and procarps are in the slender filaments, and the cystocarps are in the larger filaments. If several slides are to be made, it is a good plan to select slender, medium, and thicker filaments,

[1] James J. Wolfe, "Cytological Studies in Nemalion," *Annals of Botany*, **18**:607-630, 1904.

remove the surface glycerin with filter paper, and then crush the filaments between two slides. There is scarcely any danger of crushing too much. A little of the crushed material, including the various stages, can be put into the melted glycerin jelly. Add a round cover, tap gently until the jelly comes just exactly to the edge of the cover. As soon as the jelly is cool, the mount may be sealed with balsam but we prefer to leave the mounts for a day or two before sealing. Such mounts would probably keep for a year or two without sealing (Fig. 51).

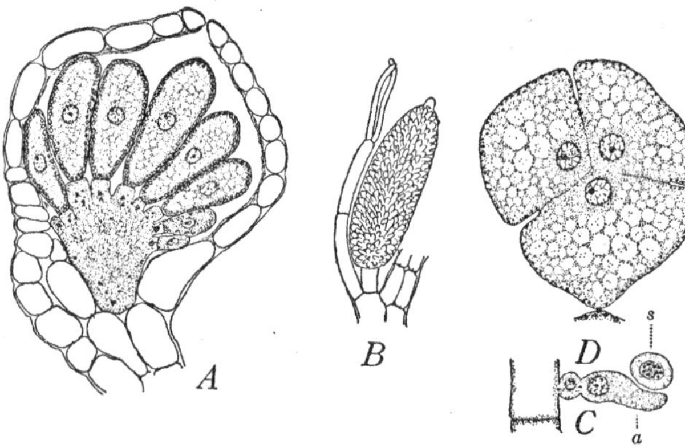

FIG. 52.—*Polysiphonia fibrillosa:* *A*, nearly mature cystocarp, showing the large cell formed by the fusion of several auxiliary cells with the pericentral cell—the carpospores are large and elongated; *B*, an antheridium—the term "antheridium" is more correctly applied to the structure shown in *C*, *a*, which cuts off one or more sperms, *s*; *D*, young tetraspores. Fixed in Flemming's weaker solution, cut 3 μ, and stained in iron-alum haematoxylin. *A* and *B* ×240; *C* and *D* ×780.

We have mounted *Nemalion* in Venetian turpentine; but by this method the material becomes hard and behaves like cartilage, so that it cannot be crushed under a cover. However, it can be crushed on a piece of glass with a scalpel.

Nemalion, stained in eosin, makes beautiful mounts, but they always fade.

Polysiphonia.—This is a very difficult form to handle, but Dr. Yamanouchi has developed an adequate method, and, by following it, anyone should be able to get good preparations.

For mounting in glycerin, glycerin jelly, or in Venetian turpentine, fix in 10 per cent formalin and stain in iron-haematoxylin.

For sections, fix in Flemming's weaker solution, but omit the osmic acid for spermatogenesis and germination of carpospores. The time should be *very short*, 5 to 40 minutes being sufficient. If material is left too long, it goes to pieces. Wash in a gentle stream of sea-water for 24 hours. Stain in iron-haematoxylin and then for 2 to 3 minutes in safranin (Fig. 52). This short stain in safranin gives a faint rosy tinge to mucilaginous structures, but does not obscure the fine nuclear detail. In the nucleus of the sperm, the chromosomes remain distinct, so that the number, 20, can be counted from the time the sperm is formed (Fig. 52*C*) up to fertilization.

With very delicate forms, like *Callithamnion* and *Griffithsia*, the washing may be in part or even wholly omitted, and the chromic acid extracted by the lower alcohols, the material being kept in the dark.

Corallina.—*Corallina* and other forms whose surface is incrusted with lime need special treatment. The following solution is good:

Chromic acid	1 g.
Glacial acetic acid	1 c.c.
Sea-water	100 c.c.

Fix 24 hours, changing the fixing agent 2 or 3 times. Wash 24 hours in sea-water.

If carefully applied, the following is a good method: Put the material into 5 per cent glacial acetic acid (in sea-water) and watch it. As soon as the vigorous effervescence begins to subside, rinse in sea-water and transfer to Flemming's weaker solution, and fix 24 hours. Iron-haematoxylin is best for figures, but for general structure the safranin, gentian-violet, orange combination gives beautiful results.

CHAPTER XVIII

FUNGI

In recent years such progress has been made in culture methods for fungi, especially for those which cause disease, that only the trained pathologist could be expected to know just what is the best medium for each particular fungus. We shall not attempt to deal with the subject of culture media, but shall simply indicate how a student, not trained in pathology, may secure material for preparations. Professor Kleb's methods make it possible to secure material of many forms in various phases of their life-histories.

In general, filamentous fungi are treated like the filamentous algae, while the fleshy forms are cut in paraffin.

PHYCOMYCETES

Rhizopus (Mucor).—This familiar mold appears with great regularity on bread. The following is a sure and rapid method for obtaining *Mucor:* Place a glass tumbler in a plate of water, put on the tumbler a slice of bread which has been exposed to the air for a day, and cover with a glass jar. The bread must not become too wet.

To obtain a series of stages in the development of the sporangium it is better to use living material. For class work, time the cultures so as to have a plenty of sporangia which have not yet begun to turn brown.

For permanent preparations, fix for at least 24 hours in 10 per cent formalin; wash ½ hour in water, and then follow the Venetian turpentine method. Eosin gives good results. Stain over night or 24 hours, treat with 2 per cent acetic acid, changing several times, and then put into glycerin, merely pouring off the 2 per cent acetic acid and not rinsing the acid out in water. When washing out the glycerin, do it with alcohol which has about 0.5 c.c. of acetic to the 100 c.c. of alcohol, and leave 1 or 2 c.c. of this slightly acid absolute alcohol on the material when you add the 10 per cent Venetian turpentine.

For permanent preparations, fix for at least 24 hours in the special chromo-acetic–osmic solution; wash 24 hours in water and follow the Venetian turpentine method. Eosin gives good results for general topography. Stain over night or 24 hours, treat with 2 per cent acetic acid in water for 5 or 10 minutes, changing several times, and then put the material into 10 per cent glycerin, merely pouring off the acid but not washing it out with water. Wash out the glycerin with alcohol containing about 0.5 c.c. acetic acid to 100 c.c. of the 95 and 100 per cent alcohol. Leave about 1 c.c. of the slightly acid alcohol on the material as you add the 10 per cent Venetian turpentine.

To bring out the nuclei, use iron-alum haematoxylin. *Rhizopus* stains very rapidly, so that an hour in iron-alum and an hour in the haematoxylin may be long enough. The sporangia stain more readily than the mycelium; consequently to show the coenocytic character of the mycelium, the action of the second iron-alum must be stopped earlier than when staining for the sporangium. Extract the stain until the nuclei of the mycelium show clearly, and then remove part of the material to wash in water. For the rest of the material, continue to extract the stain until the sporangia are satisfactory. Mount some of each lot on each slide.

The finer details of the sporangium can be seen only in thin sections. *Rhizopus* is the most easily obtained material for showing the progressive cleavage of protoplasm by vacuoles.

The zygosporic stage in the life-history is rarely met in nature or in cultures, but when once secured it may be propagated indefinitely. We have a culture which has been furnishing illustrative material for nearly twenty years. Once in a while, when a particularly good culture appears, lay aside some of it to start the next culture. The best series of stages generally appears between the fourth and seventh days. Dr. Blakeslee shows why zygospores are so infrequent. The conjugating filaments belong to different strains of mycelia which he calls "plus and minus strains," and which, for convenience, may be called "female and male strains." The more vigorous mycelium is +, and the less vigorous −. When the two strains come together, zygospores are formed along the line of meeting. If + and − strains are started at opposite sides of a dish, they will meet near the middle and form a dark line of zygospores. Through Dr. Blakeslee's generous distribution of material, the + and − strains are now available in practically all of the great universities of the world.

While *Rhizopus* may be grown on bread, it is better to use culture media in Petri dishes. While it grows well on agar media, it is hard to pick it off; and on liquid media, the growth is abnormal. Mrs. Alice A. Bailey has devised a method which is ideal for securing material. She puts about the usual amount of a potato dextrose agar in a Petri dish and pours over it a potato decoction about 2 mm. deep. The + and − strains are then added. The potato decoction is made as follows: Use 300 g. of Irish potatoes and 1000 c.c. distilled water. Peel and slice the potatoes and boil for 1 hour in the distilled water. Strain off the liquid through cheese-cloth and make up to the original quantity by adding distilled water. Flask and sterilize.

The potato dextrose agar is made as follows: to 1,000 c.c. of potato decoction, add 20 g. of dextrose and 30 g. of agar. Boil $\frac{1}{2}$ hour in a double boiler. Strain, flask, and sterilize.

The potato dextrose agar is an excellent medium and the thin layer of potato decoction keeps the material from sticking to the agar so that it can be lifted off intact. Rinse it under the tap and fix in the chromo-acetic–osmic solution.

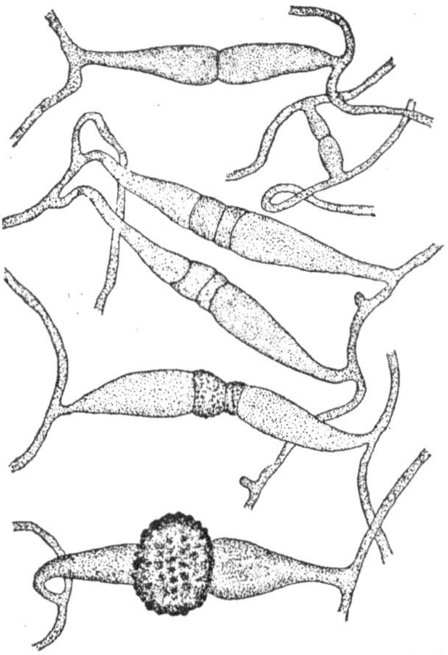

FIG. 53.—*Rhizopus nigricans:* various stages in the development of zygospores from a culture on bread; preparation stained in eosin and mounted in Venetian turpentine. ×80.

Zygospores may begin to form within 3 days, and mature zygospore may appear within 4 or 5 days. Watch the cultures and fix so as to secure a series of stages. (Fig. 53).

Paraffin sections should not be thicker than 5 μ, and 3 μ is better for nuclear detail. Iron-alum haematoxylin is best for nuclei, but safranin, gentian-violet, orange will give beautiful preparations.

In the related genus, *Sporodinia*, which is rather common in summer on fleshy fungi, especially upon *Boletus* and its allies, the zygosporic condition is not infrequent, because *Sporodinia* does not have + and − strains. *Rhizopus* behaves like a dioecious plant, while *Sporodinia* behaves like a monoecious one. The very damp atmosphere and the nutrition necessary for the formation of zygospores may be provided in the laboratory in the following way: Put a little water in a glass battery jar and place filter paper around the inside of the jar so that it will take up water and thus keep the sides of the jar moist. Place a small beaker or dish, without any water in it, in the bottom of the jar, and in the beaker place a small piece of bread dampened with the juice of prunes. Infect the bread with spores, or use a piece of bread upon which mycelium is already growing. Sections of the root of *Daucus carota* may be used instead of the bread. Put a piece of wet filter paper on a pane of glass and cover the jar. Begin to examine after 24 hours. The zygospores may appear in 4 or 5 days. A very full account of the methods by which the various phases of the life-history of *Sporodinia* may be produced at will is given by Klebs in the *Jahrbücher für wissenschaftliche Botanik*, **32**:1–69, 1898.

Zygorhynchus is another interesting relative of *Rhizopus*, readily distinguished by having suspensors of very unequal size. Dr. Florence A. McCormick sent us magnificent zygosporic material, raised on beef broth and fixed in 10 per cent formalin in water.

Saprolegnia.—This is an aquatic mold, very common upon insects and algae. Cultures are easily and quickly made. Bring in a quart of water from any stagnant pond or ditch, and into the water throw a few flies. After 12 to 24 hours throw the water away, rinse the flies in clean water, and put then into tap water. The water must be changed every day to keep bacteria from ruining the culture. The larvae of ants or small pieces of boiled white of egg are sometimes better than flies. Sporangia may appear within 24 hours but may be a day later. Sporangia may be produced in the greatest abundance by cultivating the mycelium for several days and then transferring it to pure water or to distilled water. As long as the nutrient solution is sufficiently strong and fresh, only sterile mycelium will be produced.

To secure oösporic material, mycelium which has been highly nourished for several days in a nutrient solution is brought into a

0.1 per cent solution of leucin, or into a 0.05 to 0.1 per cent solution of haemoglobin. Begin to examine after 24 hours.

Oögonia have been produced in great numbers by the following method: cut ordinary corn (*Zea Mais*) into small pieces and boil for 20 minutes. When cool, put pieces into a Petri dish and add enough pond water to nearly cover the pieces of corn. Oögonia may appear within 3 or 4 days.

For fixing, the following formula is excellent for material which is to be mounted whole:

Formalin	10 c.c.
Glacial acetic acid	5 c.c.
Water	85 c.c.

Fix at least 24 hours, but material may be left for months in this fixing agent.

Stain some in Magdala red and anilin blue, and some in iron-alum haematoxylin. Mount some of each lot on each slide (Fig. 54).

For sections, it is better to fix in the special chromo-acetic-osmic-acid solution.

Satisfactory material for general laboratory purposes can be secured as just described. Absolutely pure cultures can be secured only by observing all the precautions necessary in bacteriological work.

Achlya is similar and equally good for illustrative purposes. It is found on insects, fishes, dead fish eggs, and on algae. The zoöspores escape in a mass, which, for a short time, is held together by a transparent pellicle; in *Saprolegnia* the zoöspores swarm separately. In *Saprolegnia*, the new sporangia grow up through the empty ones; in *Achlya*, the later sporangia arise on lateral branches below the earlier ones. *Dictyuchus* and *Olpidium* often appear when one is

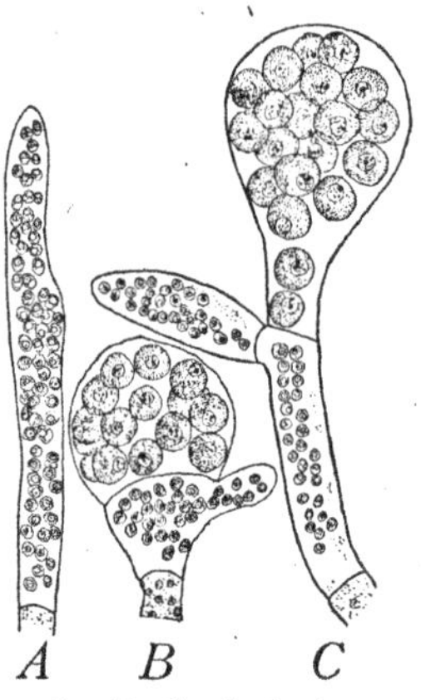

FIG. 54.—*Saprolegnia: A*, sporangium; in *B* and *C* the upper cell is an oögonium; the cells below are probably sporangia. The material was grown on flies by Miss Mary Norton at middle Granville, New York, and was fixed in formalin acetic acid and stained in Magdala red and anilin blue. ×150.

trying to get *Saprolegnia* or *Achlya*. The fixing and staining described for *Saprolegnia* will give good results with the other genera.

Albugo.—This fungus is quite common on Cruciferae, where the white "blisters" or "white rust," *Albugo candida*, form quite conspicuous patches. Affected portions of leaves and stems should be fixed in chromo-acetic acid and cut in paraffin. Sections 5 μ or less

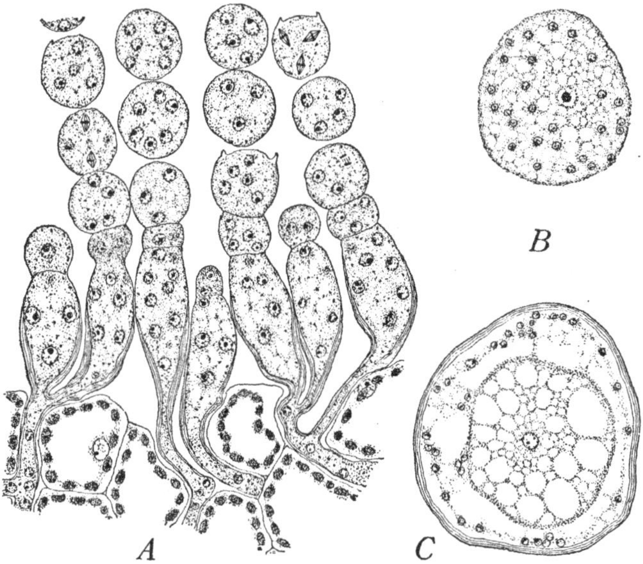

FIG. 55.—*Albugo candida:* *A*, small portion of a section of a blister showing coenocytic mycelium, conidiophores, and multinucleate conidia; *B*, a young oögonium with large coenocentrum and many nuclei: *C*, later stage, after differentiation into a central oöplasm, surrounded by the multinucleate periplasm; all of the nuclei of the oöplasm, except one, have disorganized. Fixed in the special chromo-acetic acid solution and stained in iron-alum haematoxylin and orange. ×780.

in thickness will be found most satisfactory. Stain in iron-alum and counter-stain lightly with orange (Fig. 55).

The white blisters cause little distortion, but are easily recognized by their color; the oögonia do not cause any change in color, but they cause great distortion in the pods or stems, so that these organs may reach several times their normal size. Parts only slightly distorted should be selected, as well as the extreme cases; otherwise, you will secure only old fertilized eggs, with very few of the younger stages. The oösporic phase of *Albugo bliti* is easily recognized on *Amaranthus*, especially on *A. retroflexus*, where the oöspores may be seen with the naked eye by holding the leaf up to the light. The oöspores usually occur in more or less circular patches upon the leaf. When they occur among the floral structures, there is often a slight reddish coloration. Unfortunately for the collector, it is very seldom that any red coloration in *Amaranthus* is due to the desired material.

The oösporic stage of *Albugo Ipomeae*, on the morning-glory, causes extreme distortion of the stem. For sections, it is well to cut out small pieces of the cortex, rather than to fix larger pieces of the stem.

HEMIASCOMYCETES

Saccharomyces.—Formerly it was considered rather difficult to demonstrate the nucleus of the yeast cell. With fresh-growing yeast the following method by Wager made the classical demonstration. Fix in a saturated aqueous solution of corrosive sublimate for at least 12 hours. Wash successively in water, 30 per cent alcohol, 70 per cent alcohol, and methyl alcohol. Place a few drops of alcohol containing the cells on a cover, and when nearly dry add a drop of water. After the yeast cells settle, drain off the water and allow the cells to dry up completely. Place the cover, or slide, with its layer of cells in water for a few seconds, and then stain with a mixture of fuchsin and methyl green, or fuchsin and methylin blue. Mount in glycerin or in balsam.

With modern methods, there should be no more difficulty than in demonstrating the nucleus in the Cyanophyceae or in the mycelium of the Phycomycetes. Use an abundance of vigorously budding material, so that you can afford to lose most of it and still have a plenty left: Fix in the special chromo-acetic-osmic-acid solution and stain in iron-alum haematoxylin. Use the Venetian turpentine method, or imbed in paraffin and cut sections about 5 μ.

To obtain the spore stage, put a cake of Fleischmann's yeast in a mixture of equal parts of grape juice and distilled water; add 1 g. of peptone and allow to bud freely over night at 30° C.; place the material in a plaster-of-Paris cup with a depression, and put the cup in a small Stender dish with water coming nearly to the top of the cup. In 60 to 70 hours there should be abundant spore formation.

ASCOMYCETES

This group, popularly known as the "sac fungi," contains an immense number of saprophytic and parasitic forms. The green mold on cheese and leather, the leaf curl of peach, the black knot of cherry and plum, and the powdery mildews are familiar to everyone. The few objects selected will enable the student to experiment, but he must not be discouraged if success does not crown the first attempt, for some members of the group present real difficulties.

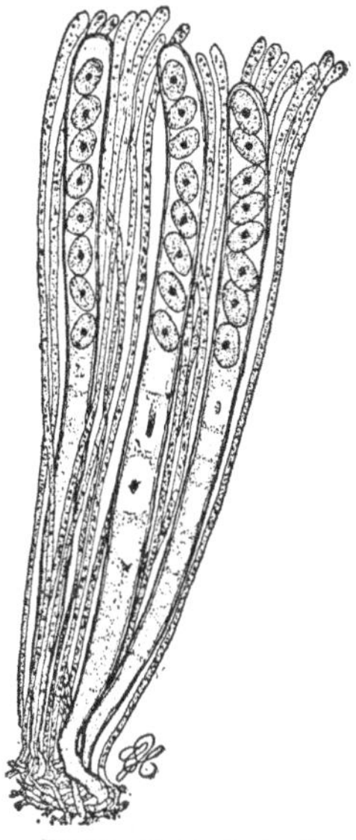

FIG. 56.—*Peziza odorata:* three asci and many paraphyses; fixed in corrosive sublimate, stained in bulk in alum carmine, teased out, and mounted in balsam. ×245.

Peziza.—The Pezizas and related forms are fleshy, and present but little difficulty in fixing, cutting, or staining. They are abundant in moist places, on decaying wood, or on the ground. The apothecia have the form of little cups, which are sometimes black and sometimes flesh-colored, but often orange, red, or green.

For general morphological work it is better to tease out fresh or preserved material. Such views as that shown in Figure 56 are easily obtained in this way. For permanent preparations showing such views, it is better to stain in bulk in alum carmine or in Delafield's haematoxylin, and then tease out the asci in glycerin or balsam. Sections showing the entire ascus should be 10 to 15 μ in thickness.

For the free nuclear division in the ascus, and also for the development of the ascospores, Flemming's weaker solution, followed by the safranin, gentian-violet, orange combination has given excellent results. The centrosomes, especially at the first division in the ascus, are sharply defined and the radiations are conspicuous. With iron-alum haematoxylin and orange, the nuclear detail and the centrosomes are better, but the spindle and radiations are not so sharply defined. For such details, sections should not be thicker than 5 μ, and 3 μ will give a clearer view.

Morchella esculenta is very good for the development of the ascus because the nuclei are very large.

For showing the ascogonium, ascogenous hyphae, and the origin of the asci, nothing is better than *Pyronema*. Fix in formalin, acetic acid (10 c.c. formalin, 5 c.c. acetic acid, and 85 c.c. water) for 24 hours or more; wash in water and stain in eosin. Or, fix in the special chromo-acetic-osmic solution and stain in iron-alum haematoxylin. In either case, use the Venetian turpentine method and tease the material so as to obtain instructive views.

Eurotium.—*Eurotium* with its conidial stage, *Aspergillus*, is a very common mold found on bread, cheese, decayed and preserved fruit, etc. In the conidial stage it is green and in the ascosporic stage yellow, reddish yellow, or reddish brown. *Aspergillus* is

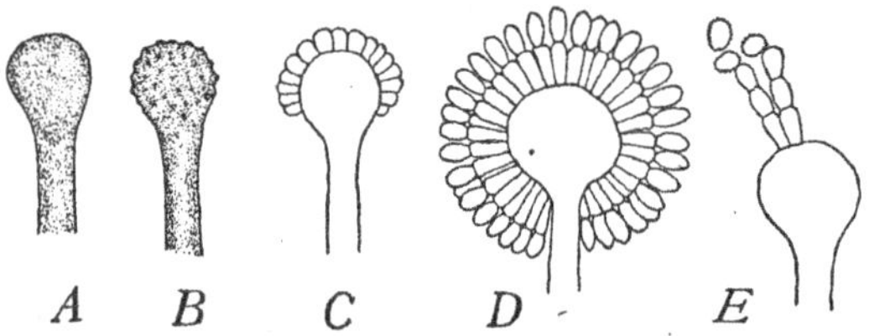

Fig. 57.—*Aspergillus:* from material growing on a hectograph pad; fixed in chromo-acetic acid, stained in eosin, and mounted in glycerin; *A–E*, successive stages in development. ×375. All such material is more satisfactory when mounted in Venetian turpentine.

almost sure to appear upon bread which is kept moderately moist, because the conidia are usually abundant in the atmosphere. If the bread be wet with a 10 per cent solution of cane-sugar or with grape juice, this stage appears sooner and in greater abundance. A temperature of 22° to 30° C. is also a favorable condition.

The perithecial stage is not found so frequently, but can sometimes be secured by examining moldy preserves. The sexual stage has been induced. Soak a piece of bread in a 20 per cent solution of grape-sugar in grape juice; upon this sow the spores and keep at a temperature of about 28° C. After 4 or 5 days, begin to examine. A 40 per cent solution of cane-sugar in the juice of prunes is also a good nutrient solution.

For class use or for permanent preparations it is best to select rather young material which shows various stages in development, from the swollen end of the hypha to the ripe spore (Fig. 57). Permanent preparations of the conidial stage, as shown in Figure 57, and also of the coiled twisted filaments which initiate the ascosporic stage, should be made by the Venetian turpentine method or by the glycerin method.

Fix in 1 per cent chromo-acetic acid (1 g. chromic acid and 1 c.c. acetic acid and 100 c.c. water) for 24 hours; wash in water 24 hours, stain sharply in eosin, transfer to 10 per cent glycerin, and follow the Venetian turpentine method.

Material may be fixed in corrosive sublimate-acetic acid (corrosive sublimate 2 g., glacial acetic acid 2 c.c., and water 100). Use it hot (85° C.). One minute is long enough. Wash in water and add, a few drops at a time, the iodine solution used in testing for starch. At first, the brownish color caused by the iodine will disappear, but after a certain amount has been added the brownish color will remain. Stain in eosin or iron-haematoxylin and follow the Venetian turpentine method.

A very rapid method for this and for similar small filamentous forms may be added. Forms as large as *Thamnidium elegans* can be mounted successfully by this method.

1. 100 per cent alcohol, 2 minutes.
2. Eosin (aqueous), 2 minutes.
3. 1 per cent acetic acid, 2 to 10 seconds.
4. Mount directly in 50 per cent glycerin and seal.

If the material gets through the first three stages without shrinking but collapses at the fourth, put it into 10 per cent glycerin and allow it to thicken, following the Venetian turpentine method.

The earlier perithecial stages are more instructive when mounted whole; but later stages, even before the formation of the asci, are very unsatisfactory by this method, and should be cut in paraffin.

Penicillium.—This green mold is found everywhere upon decaying fruit, upon bread, and upon almost any decaying organic substance. Material is even more easily secured than in case of *Aspergillus*, and *Penicillium* is an easier type for laboratory study. Such a satisfactory study can be made from the living material that it is hardly worth while to fix and stain. The very rapid method described for *Aspergillus* will furnish good mounts if permanent preparations are desired.

The Erysipheae.—The mildews are found throughout the summer and autumn on the leaves of various plants. Some of the most abundant forms are *Microsphaera alni* on the common lilac; *Sphaerotheca castagnei* on *Bidens frondosa* and other species, on *Erechtites hieracifolia*, and on *Taraxacum officinale; Uncinula necator* on *Ampelopsis quinquefolia*, and *U. salicis* on *Salix* and *Populus;*

Erysiphe commune on *Polygonum aviculare;* and *Erysiphe cichoriacearum* on numerous Compositae and Verbenaceae. *Podosphaera* may be found on the leaves of young cherry trees and apple trees, and on young shoots of older trees. The infected leaves are likely to be more or less deformed. *Phyllactinia* is sometimes abundant on leaves of *Alnus incana.* It is also found on *Celastrus, Desmodium, Typha,* and on various members of the Amentiferae. For herbarium purposes they may be preserved by simply drying the leaves under

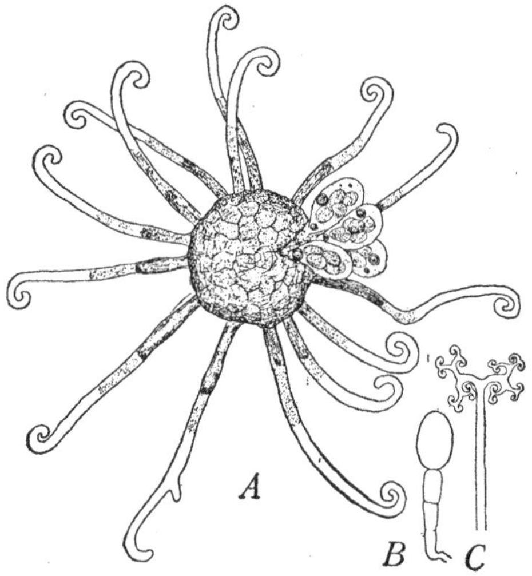

FIG. 58.—*Uncinula necator* on *Ampelopsis quinquefolia: A,* four asci containing ascospores have been forced out by pressing on the cover; stained in fuchsin and mounted in balsam; *B,* a conidiospore; and *C,* an appendage of *Microsphaera alni,* drawn from living material. ×192.

light pressure. When needed for examination the leaf should be soaked in water for a few minutes, after which the perithecia may be scraped off and mounted in water. In mounting great care must be taken not to break off the appendages. The asci may be forced out by tapping smartly on the cover (Fig. 58).

For permanent mounts of entire perithecia with appendages, fix in 5 per cent formalin 24 hours, wash in water 1 hour, stain in aqueous eosin 24 hours, remembering to keep all solutions slightly acid. Use the Venetian turpentine method. If chromic acid, corrosive sublimate, or alcohol be used for fixing, the appendages become brittle and very easily break off. However, the chromo-acetic mixtures are better if it is desired to make paraffin sections showing the developing of the perithecium with its asci and spores. For this purpose the omnipresent *Erysiphe commune* on *Polygonum aviculare* is exceptionally favorable, because, after the material has been fixed and has been brought into alcohol, the whole mycelium, with the developing perithecia, may be stripped from the leaf without the slightest difficulty, thus avoiding the necessity of cutting the leaf in order to get the fungus. Material in which the perithecia are still white or yellowish contain stages up to the formation of the uninucleate ascus; brownish perithecia show the development of ascospores, and dark-brown or black perithecia contain the mature asci with fully developed ascospores. In early stages while the perithecia are still yellow or very slightly brownish, the material can be stripped off from the leaves before fixing.

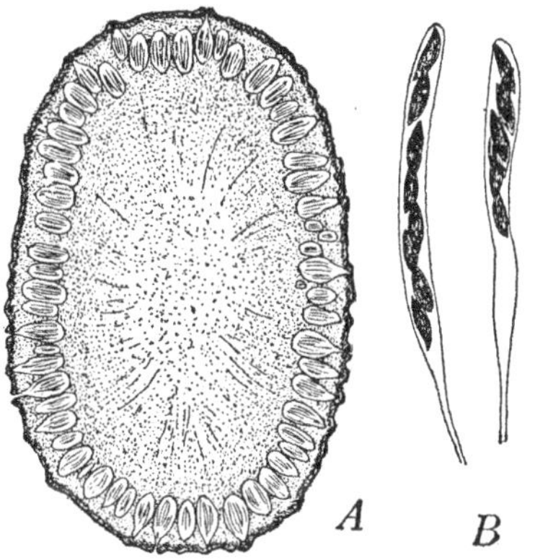

FIG. 59.—*Xylaria:* *A*, transverse section of a young stroma, showing perithecia. ×8. *B*, two asci with ascopores. ×245.

An air-pump will remove any air. Use iron-alum haematoxylin and orange, or the safranin, gentian-violet, orange combination. Sections thicker than 5 μ will be hard to stain effectively.

The Xylariaceae.—Most of these forms, in their mature condition, are black. In younger stages the color is lighter, often showing gray, brick-red, or brownish tints. *Nummularia* is common on dead branches of beech, elm, oak, locust, and other trees. It is generally flat, orbicular, or elliptical in form. *Ustilina* is a crustaceous form, rather diffuse and irregular in shape. It is most common on the roots of rotten stumps. *Hypoxylon* is more or less globose in form, and the color is brick-red, brown, or black. It is found on dead twigs and bark of various trees, especially beech, and is more abundant in moist situations. *Xylaria* (Fig. 59) is found on decaying stumps and logs, and often apparently on the ground, but really growing on twigs, wood, and bark just under the surface. When mature it is black outside and white or light-colored within. When young, it is easily cut in paraffin; in some forms the ascospores are fully formed before the stroma becomes hard enough to occasion any difficulty in cutting. When the stroma becomes black, many members of the Xylariaceae become very hard and brittle, so that sections are likely to be unsatisfactory. For general morphological study it is better to break the stroma transversely and examine with the naked eye and with a pocket lens. The asci with their spores can be teased out and mounted in water. For permanent preparations, soak the stroma for a month in equal parts of 95 per cent alcohol and glycerin; then cut sections, and, after leaving them in glycerin for a day or two, mount in glycerin jelly. It is better not to stain the old stages. For illustrative purposes, select forms which can be cut in paraffin. The method just given merely shows that such material can be cut.

LICHENS

The lichens are usually regarded as difficult forms. In younger stages they occasion no trouble, but an old apothecium or a leathery thallus often fails to cut well. By employing the gradual processes already described in chapter ix, satisfactory sections should be obtained from thalli and mature apothecia of *Physcia, Usnea, Sticta, Parmelia,* and *Peltigera. Collema* and other lichens of such gelatinous consistency, while they cut readily, show a strong tendency to wrinkle.

Cyanin and erythrosin is a very good stain for lichens. The algae stain blue and the filaments of the fungus take the red. Where the association of the alga and the fungus is rather loose, as in *Dichonema,* more satisfactory mounts can be made by staining in eosin, or haem-alum and eosin, and then teasing slightly with needles and mounting in glycerin.

BASIDIOMYCETES

This is an immense group, of which the smuts, rusts, mushrooms, toadstools, puffballs, and bracket fungi are the most widely known representatives.

The Smuts (Ustilagineae).—The smuts are abundant on wheat, oats, corn, and various other plants.

The smuts may be studied in the living material. The following method, described by Ellis, is worth remembering: A supply of smutted barley may be obtained by sowing soaked, skinned barley that has been plentifully covered by *Ustilago* spores. In such material it is easy to trace stages in the development of spores. Freehand sections of ears about 12 mm. long show the mycelium and spore clusters. If smutted ears be removed and kept floating on the water, the spores continue to develop and often germinate. For paraffin sections desirable stages should be fixed in Flemming's fluid or picro-acetic acid. Delafield's haematoxylin, followed by a very light touch of erythrosin or acid fuchsin, will give a good stain.

For a study of the germinating spores and conidia, cultures may be made in beerwort on the slide or in watch crystals. Harper's method of making preparations from such material is ingenious and is valuable in making mounts of various small plant and animal forms. A drop of the material is taken up with a capillary tube and is then gently blown out into a drop of Flemming's weaker solution (15 minutes to 1 hour was sufficient for the fungus spores). Cover a slide with albumen fixative, as if for sections. A drop of the material, without previous washing, is drawn up into the capillary tube and touched lightly and quickly to the surface of the albumen. A series of such drops, almost as small as the stippled dots in a drawing, may be applied to the slide. The fixing agent may now be allowed to evaporate somewhat, but the preparation must not be allowed to dry. As the slide is passed rapidly through the alcohols, the albumen is coagulated, and the preparation may be treated just as if one were dealing with ribbons of sections.

The Rusts (Uredineae).—*Puccinia graminis*, the common rust of wheat and oats, is familiar to everyone. The uredospores, or summer spores, known as the red rust, and the winter spores, known as the black rust, are found in unfortunate abundance, but the aecidium stage on the barberry is not necessary for the vigorous development of rust in the United States, and it is not nearly so prevalent as the red- and black-rust stages. When found, it may be so abundant that most of the leaves of the barberry are spotted with the cluster cups. It is a curious fact that wheat and oats may be quite free from the red and black rust in localities where the aecidium stage is very abundant; and that the rust stages may be most destructive where there are no barberry bushes. But no one doubts that the aecidium stage on barberry belongs in the life-history. The aecidium on barberry cuts easily in paraffin (Fig. 60). If the aecidium stage is not easily available, there are various aecidia which are just as good, or even better, for morphological study. The aecidia growing on *Euphorbia maculata* (spotted spurge) are abundant and are very easy to fix and cut. The infected plants are also very easily recognized, normal plants having the prostrate habit, while infected plants become erect and the internodes become greatly elongated. Aecidia growing on *Arisaema triphyllum* (Jack-in-the-puplit) are also easy

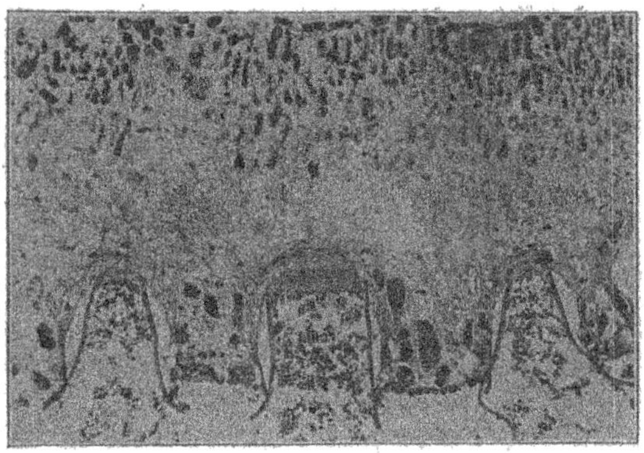

FIG. 60.—*Puccinia graminis:* photomicrograph of aecidium stage on barberry. Fixed in chromo-acetic acid and stained in cyanin and erythrosin: Eastman Commercial Ortho film, Wratten E filter (orange); arc light; Spencer 16-mm. objective N.A. .25; Bausch and Lomb projection eyepiece; exposure, ½ second. ×47. Negative by Dr. P. J. Sedgwick.

to cut. The *Aecidium* on *Hepatica* has large nuclei and affords particularly good views of the intercalary cells.

The special chromo-acetic–osmic-acid solution is recommended for fixing and iron-haematoxylin with a faint touch of orange is a satisfactory stain (Fig. 61).

It is rather difficult to get good sections of uredospores and teleutospores of *Puccinia graminis*, because the leaves of wheat and oats are refractory objects to cut (Fig. 62). For illustrative purposes, soak the leaves, scrape off the spores, and study without sectioning. For sections, select species growing on less refractory hosts.

Everyone who studies the rusts should attempt to germinate the uredospores and teleutospores. For this purpose the hanging-drop culture may be employed, as described in the chapter on temporary mounts (chap. v). The uredospores germinate readily all summer, but in most forms teleutospores will germinate only in the spring following their maturity. However, the teleutospores of "lepto" species, like *Puccinia xanthii* on *Xanthium canadense* (cocklebur), will germinate as soon as they ripen, and will serve equally well for study. If a particularly good specimen is secured, it may be preserved by the method previously described for desmids, except that in this case it might be worth while to attempt staining with Mayer's haem-alum or with eosin.

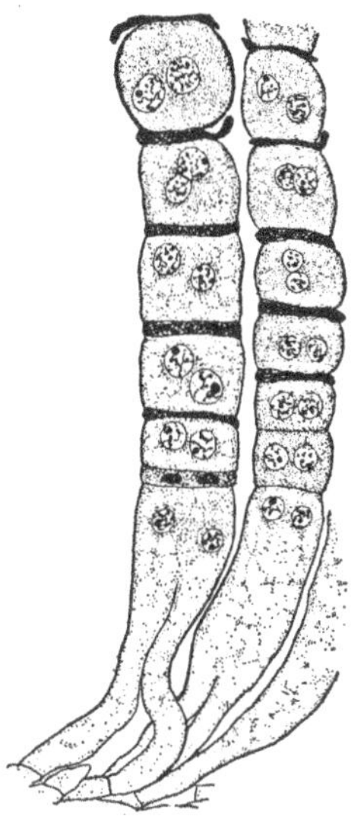

FIG. 61.—*Aecidium* on *Hepatica:* fixed in chromo-acetic acid with a little osmic acid, and stained in safranin, gentian-violet, orange; from a preparation by Dr. Wanda Pfeiffer Vestal. ×950.

Gymnosporangium, which is rather common on *Juniperus virginiana* (red cedar), forms its basidia in the "cedar-apple" stage. Bring the cedar apples into the laboratory in late winter or early spring and put some into a dish of water. The yellowish, gelatinous strands with the germinating teleutospores may appear within 24 hours. The various stages are easily recognized under a low-power dry lens without even crushing the gelatinous masses. Fix in 10 per cent formalin for 24 to 48 hours. The material may be left here indefinitely. In chromo-acetic acid the gelatinous substance goes to pieces. Stain in iron-alum haematoxylin and follow the Venetian turpentine method; or, when the thick glycerin is reached, mount in glycerin jelly. The material spreads out better and fewer basidia are torn off from the teleutospores in glycerin jelly than in the Venetian turpentine.

The interesting nuclear conditions in the life-history of a rust which has uredospores, teleutospores, and aecidiospores are not difficult to demonstrate. After the uredospore or teleutospore stage has been fixed in chromo-acetic acid and washed in water for an hour, treat with 10 per cent hydrofluoric acid in water for an hour and then continue the washing in water for 24 hours. If fixed in alcoholic corrosive sublimate and acetic acid, rinse in 50 per cent alcohol, then treat with 10 per cent hydrofluoric acid in 50 per cent alcohol, wash in two or three changes of 50 per cent alcohol. In either case, imbed in paraffin. Cut about 5 μ in thickness and stain in iron-alum haematoxylin. A rather strong stain of orange in clove oil will make it easier to trace the mycelium in the host. Uredospores and teleutospores, both young and old, are shown in Figure 62.

Various stages in the germination of the teleutospores of *Gymnosporangium* are shown in Figure 63. Reduction of chromosomes takes place at the first two divisions of the nucleus in the basidium. The sterigmata in this species are very large.

The Fleshy Fungi.—For habit study, nothing is equal to fresh material; for second choice, buy canned "mushrooms" (usually *Agaricus campestris*) at the grocery; forms not readily available in field or grocery may be preserved in formalin alcohol (6 c.c. of formalin to 100 c.c. of 50 per cent alcohol). When formalin is used in water, the fungi become too soft. Larger forms of the mushroom, puff-ball, and bracket types may be dried in an oven. The circulation of air should be good and the temperature should be kept at about 50° C. After drying, the fungi should be poisoned.

For sections, Gilson's fluid deserves more recognition than it has received. It is particularly good for soft forms, like *Tremella*.

Gilson's Fluid.—

95 per cent alcohol	42 c.c.
Water	60 c.c.
Glacial acetic acid	18 c.c.
Concentrated nitric acid	2 c.c.
Corrosive sublimate (saturated solution in water)	11 c.c.

Fix about 24 hours and wash in 60 or 70 per cent alcohol.

Coprinus micaceus is particularly good for a study of gills, basidia, and the formation of basidiospores, because it is so small that a single section may show a fine series of stages. Gills which are becoming brownish at the tip, but which are still white toward the top of the cap, will show a splendid series of stages. For fixing, cut out pieces of the gills 1 cm. long and 3 mm. thick. Such material fixes well in Flemming's weaker solution. Cut paraffin sections perpendicular to the gills. To show the four basidiospores, sections should

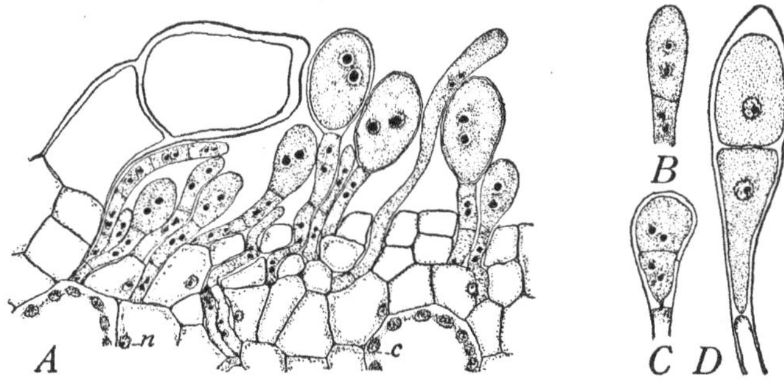

FIG. 62.—*Puccinia graminis: A*, uredospores on oats, showing binucleate mycelium and binucleate uredospores with binucleate stalk cells; *c*, chloroplasts, and *n*, nuclei of host cells; *B*, *C*, and *D*, stages in the development of the teleutospore; the two nuclei of the binucleate cells in *C* fuse to form the uninucleate cells of *D*. ×780.

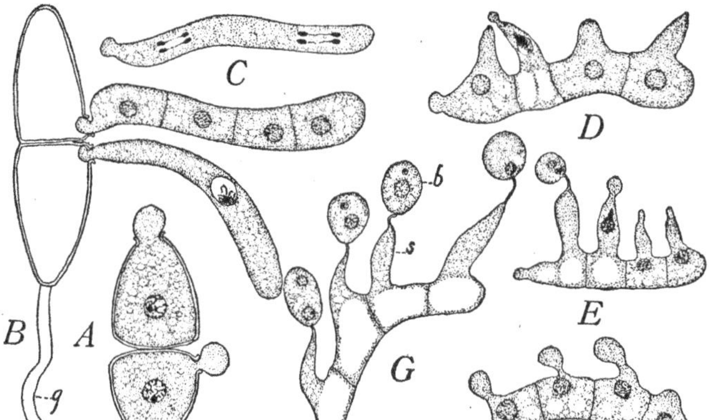

FIG. 63.—*Gymnosporangium: A*, beginning of germination of teleutospore; *B*, later stages, the lower basidium showing the nucleus in the synapsis stage of the heterotypic mitosis, and the upper basidium showing the four-cell stage following the homotypic mitosis; *C*, the two figures of the homotypic mitosis; *D*, *E*, *F*, and *G*, later stages showing the formation of basidiospores, many of the basidiospores being binucleate—*g*, gelatinous stalk; *b*, basidiospore; *s*, sterigma. Fixed in 10 per cent formalin and stained in iron-alum haematoxylin. ×800.

be 10 to 15 μ thick; to show details of nuclei, 3 μ is thick enough (Fig. 64).

In *Hydnum* and *Polyporus*, cut out pieces about 3 or 4 spines or 3 or 4 pores in width and about 1 cm. long. A rectangular piece which will allow the transverse sections of the spines or pores to be about 4 mm. wide and 1 cm. long cuts better than a piece which will give square sections.

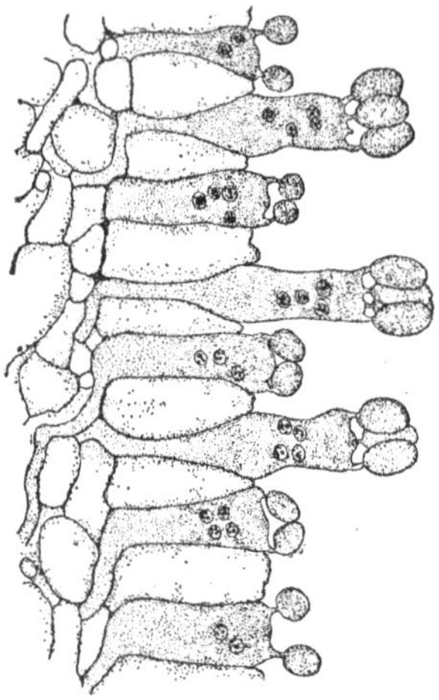

Fig. 64.—*Coprinus:* young basidia with four nuclei which, later, pass into the spores; fixed in chromo-acetic acid and stained in safranin, gentian-violet, orange. ×780.

In *Boletus*, simply strip off the hymenium and cut into pieces which will give transverse sections of the tubes.

In *Lycoperdon*, *Bovista*, *Geaster*, and *Scleroderma*, longitudinal sections of the entire fructification can be cut in paraffin as long as the fresh material is easily sliced with a Gillette blade.

Young stages of *Cyathus*, *Crucibulum*, and *Nidularia* cut easily in paraffin; somewhat older stages can be cut in celloidin, but mature stages fail to cut by any of our present methods.

It is often desirable to secure differential staining of the fungus and its host. Some of the methods previously mentioned secure this result and give excellent detail, but do not make the mycelium stand out sharply in contrast with the tissues of the host. A special method by B. T. Dickson generally gives a good differentiation. He uses Magdala red and light green, using the Magdala red in a 2 per cent solution in 85 per cent alcohol, and the light green in a 2 per cent solution in clove oil to which has been added a few drops of absolute alcohol. His schedule is as follows:

1. Dissolve paraffin in xylol and wash in absolute alcohol.
2. Wash in 95 and 85 per cent alcohol.
3. Stain with Magdala red 5 to 10 minutes.
4. Remove surplus stain and wash in 95 per cent alcohol.

5. Stain in light green in clove oil for 1 to 3 minutes.
6. Wash in absolute alcohol, or in carbol turpentine.
7. Clear in xylol and mount in balsam.

The time factors will vary slightly with different material. If the light green overstains slightly it may not interfere much with the differentiation. Mycelium, spores, amoebae, and bacterioidal tissue stain red and the host tissues green. If tissues do not stain readily, mordant in a 1 per cent solution of potassium permanganate in water for 2 to 5 minutes, wash in water, pass through the alcohols to 85 per cent alcohol, and stain. The mordant does not keep.

This combination has given good results with the following material: *Plasmodiophora brassicae, legume tubercles, Albugo candida, Phytophora infestans, Uromyces caryophyllinus, Puccinia graminis,* and many others.

CHAPTER XIX

BRYOPHYTES

The Bryophytes, comprising the two groups of Liverworts (*Hepaticae*) and Mosses (*Musci*), present a great diversity of structure, some being so delicate that good preparations are very uncertain, while others are so hard that it is difficult to get satisfactory sections. Between these extremes, however, there are many forms which readily yield beautiful and instructive preparations.

If but one fixing agent should be suggested for the entire group, it would be chromo-acetic acid with 1 g. chromic acid and 2 c.c. acetic acid to 100 c.c. of water. It should be allowed to act for at least 24 hours. For morphological study, Dr. Land uses a formalin alcohol solution (6 c.c. commercial formalin to 100 c.c. of 50 per cent alcohol), which he has tested in extensive collections in various tropical regions, where it has been impracticable to use the chromic series, with its tedious washing and changing of alcohols. Material may be left in the formalin alcohol solution until needed for use, a convenience which will hardly be appreciated by those who are always within reach of a laboratory.

For general study, the small, delicate forms may be mounted whole in Venetian turpentine.

Instead of treating forms in a taxonomic sequence, we shall consider first the gametophyte structures under the headings *thallus*, *antheridia*, and *archegonia*, and shall then turn our attention to the *sporophyte*.

HEPATICAE

Some of the liverworts are floating aquatics, but most of them grow on logs or rocks or upon damp ground. They are found at their best in damp, shady places. Many of them may be kept indefinitely in the greenhouse. *Riccia*, *Marchantia*, *Conocephalus*, *Asterella*, and many others vegetate luxuriously, and often fruit if kept on moist soil in a shady part of the greenhouse, and they do fairly well in the ordinary laboratory if covered with glass and protected from too intense light. *Riccia natans* is a valuable type for illustrative purposes. It floats freely on the surfaces of ponds and ditches. Early in the spring (during April in the Chicago region) it produces antheridia; then, for a short time (about the first of May) both antheridia and archegonia, and still later only archegonia. Sporophytes then appear as black dots along the grooves. After the spores are shed, the thallus remains sterile for the rest of the season. *Marchantia* and similar forms are not difficult to establish out of doors. A rather damp, shady spot close to the north side of a building is best. Scrapings from a board which has been nearly burned up make the best fertilizer to scatter on the soil, if one is to cultivate *Marchantia.* Such freezing as *Marchantia* receives in the vicinity of Chicago does not prevent it from appearing again the next spring. If it is desirable to have material throughout the year, the out-of-door culture may be made in a box which can be brought into the laboratory or greenhouse in the winter. A box 3 feet long, 2 feet wide, and 1 foot deep will be convenient. It should have a glass cover; an old window will do. There should be about 6 inches of dirt in the box. A mixture of sand, loam, and charred scrapings will make a good substratum for *Marchantia.* If one is to raise liverworts in the laboratory, it is absolutely necessary to note carefully the conditions under which they grow in the field.

The living plants are very desirable, since they not only furnish the best possible material for habit work and the coarser microscopic study, but they also enable one to secure complete series in the development of the various organs.

If even a small room in a greenhouse is available, liverworts can be grown in great variety and abundance. On one side of the room, have a pile of rocks. Half of this space should be occupied by limestone rocks, held in place with as little mortar as possible. There should be some shale and some porous red brick. The whole should be arranged so that water may trickle down from above. A pipe with holes $\frac{1}{16}$ inch in diameter will furnish enough water. The other three sides may be built up of various rocks, and some clay, so as to form a table about 1 m. high. A small fountain, with a bowl a couple of feet in diameter, built of rocks, will add to the efficiency. If a few well-supported cement tanks be placed above the principal pile of rocks, *Isoetes* and all the water ferns may be grown there, besides *Elodea, Myriophyllum, Chara,* and other forms constantly needed in laboratory work.

The Thallus.—In many cases it will not be necessary to make a special preparation for the study of the thallus, since preparations of antheridia, archegonia, or sporophytes may include good sections of vegetative portions. This is particularly true of forms like *Riccia*, where the various organs are not raised above the thallus. In forms like *Marchantia*, where the antheridia, archegonia, and sporophytes are borne upon stalked receptacles, it is better to make separate preparations to show the structure of the mature thallus. Sections intended to show the structure of the mature thallus should be 15 to 25 μ in thickness, but sections to show the growing point and development of the thallus should not be thicker than 10 μ. The apical region of the Jungermanniaceae (Figs. 65, 66) affords an excellent opportunity for studying the development of the plant body from a single apical cell. If mixtures containing osmic acid are used for fixing, there may be difficulty in the staining, even after using peroxide of hydrogen.

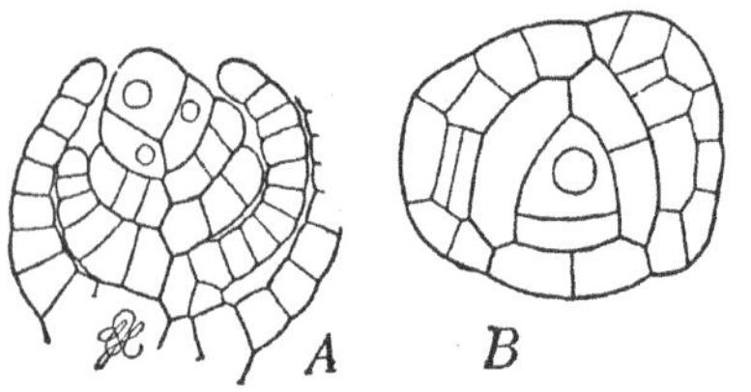

FIG. 65.—*Ptilidium ciliare*: *A*, longitudinal section, and *B*, transverse section of the apical region of the leafy gametophyte. Fixed in chromo-acetic acid and stained in Delafield's haematoxylin. ×420.

Chromo-acetic mixtures, without osmic acid, are better for the apical region. Chromo-acetic acid, followed by Delafield's haematoxylin, is good for the apical cells and developing regions, but a light counter-stain with erythrosin improves preparations of the mature thallus. Safranin, with light green, or safranin with crystal-violet, will give clear views of the growing region. The latter combination is particularly good for forms like *Pellia*, where even the apical cell is more or less vacuolated, since it not only brings out the cell walls, but stains plastids and other cell contents (Fig. 66). The chloroplasts and leucoplasts are well differentiated by this stain. After corrosive sublimate-acetic, a vigorous staining in a mixture of acid fuchsin and iodine green often brings out the walls very sharply. After corrosive sublimate-acetic the material may be stained in bulk with alum cochineal or alum carmine, thus giving fairly good preparations and saving considerable labor.

Antheridia.—It is not difficult to get good preparations showing the development of antheridia. In forms like *Conocephalus*, *Asterella*, *Pellia*, etc., cut out small portions of the thallus bearing the antheridia. The piece should not be more than 1 cm. long and 5 mm. wide, preferably smaller. For the development of the antheridia of *Marchantia*, select young antheridiophores which still lie close to the thallus. With the antheridiophore, cut out a small piece of the thallus, about 5 mm. in length. For general development, cut 10 μ, but for details of spermatogenesis, sections should not be thicker

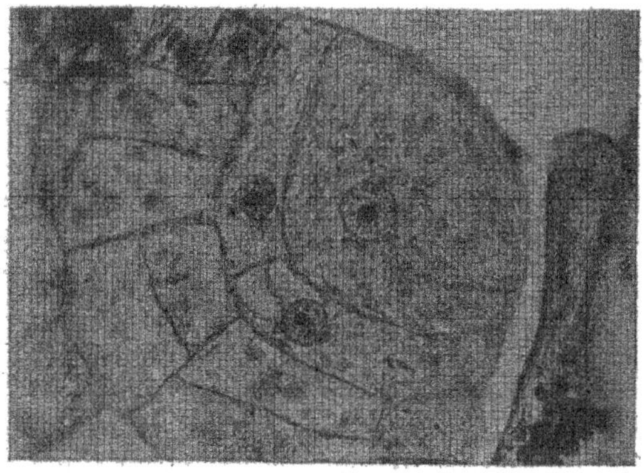

FIG. 66.—*Pellia epiphylla:* photomicrograph of apex of gametophyte showing apical cell and segments; fixed in chromo-acetic-osmic acid and stained in safranin, gentian-violet, orange. The negative was made by Dr. Köhler at the Zeiss factory in Jena, Germany.

than 5 μ (Fig. 67). Sections should be stained in iron-alum haematoxylin. The cells are very small and the contents very dense; consequently, the staining must be very critical to show the blepharoplasts and chromosomes or, in later stages, the transformation of spermatids into sperms. If the material is perfectly infiltrated and imbedded, a constant cooling with ice should make it possible to secure smooth sections as thin as 3 μ, or even 2 μ.

If sperms are found escaping, transfer them to a small drop of water on a clean slide, invert the drop over a 1 per cent solution of osmic acid for 2 or 3 minutes, allow the drop to dry up, pass the slide through the flame 2 or 3 times, as in mounting bacteria, and then stain sharply in acid fuchsin. This should show the general form of the antherozoid, and will usually bring out the cilia.

The Archegonia.—The methods for archegonia are practically the same as for antheridia. Too much stress cannot be laid upon the importance of carefully selecting the material. Use very small pieces, and, before placing them in the fixing agent, trim them to such a shape that the position of the archegonia will be known accurately even after the pieces are imbedded in paraffin. Since air is likely to be caught between the perigynium and the archegonium,

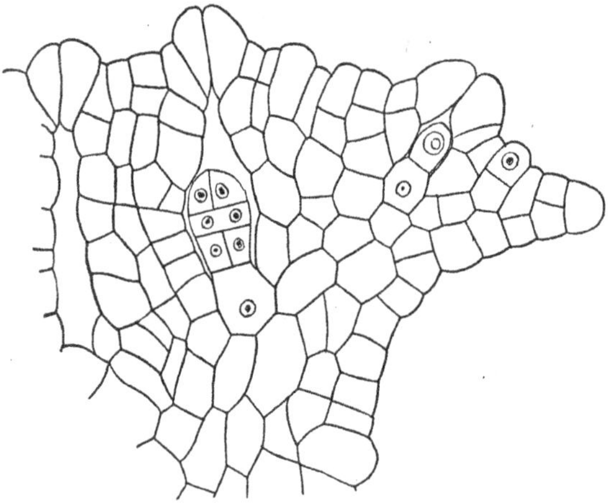

FIG. 67.—*Marchantia polymorpha:* early stages in the development of antheridia; from an unpublished drawing by Dr. W. J. G. Land. ×600.

it is worth while to use an air-pump as soon as the material is put into a chromo-acetic mixture. With the formalin alcohol solution, material is likely to sink promptly and the pump is not necessary.

For stages like those shown in Figure 68*A* and *B*, safranin with anilin blue or light green is a good stain and 7 to 10 μ is about the right thickness.

For stages like *C*, in such forms as *Marchantia*, where the necks are long and often somewhat curved, it is better for general purposes to use sections about 15 μ in thickness. If it is desired to obtain preparations showing the cutting off of the ventral canal cell, the development of the oösphere, and the process of fertilization, the sections should be from 3 to 6 μ in thickness.

For archegonia containing young embryos, like that shown in *D*, Delafield's haematoxylin without any counter-stain gives beautiful preparations when the staining is well done. It is easier for the beginner to get good preparations with the safranin, gentian-violet, orange combination.

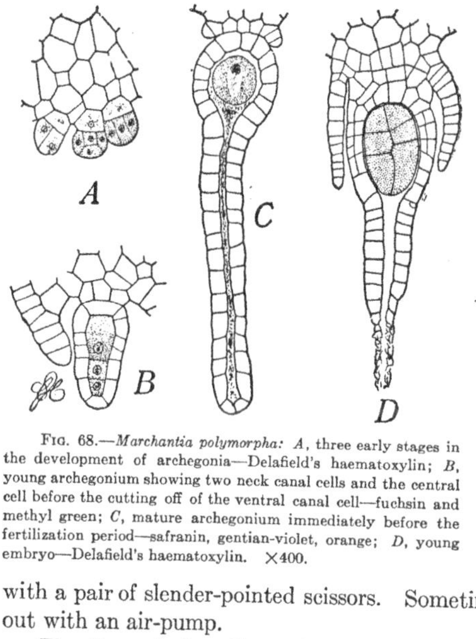

FIG. 68.—*Marchantia polymorpha: A*, three early stages in the development of archegonia—Delafield's haematoxylin; *B*, young archegonium showing two neck canal cells and the central cell before the cutting off of the ventral canal cell—fuchsin and methyl green; *C*, mature archegonium immediately before the fertilization period—safranin, gentian-violet, orange; *D*, young embryo—Delafield's haematoxylin. ×400.

In *Riccia natans* (*Ricciocarpus natans*) the direction of the axis of the archegonium at every stage in the development must be known; otherwise, there will be few good longitudinal sections.

In forms like *Porella* and *Scapania*, the involucre covering the archegonia is likely to hold a bubble of air, which will delay or even prevent fixing. The best plan is to cut off the offending leaf with a pair of slender-pointed scissors. Sometimes the air can be got out with an air-pump.

The Sporophyte.—Sporophytes in early stages of development often yield good preparations without very much trouble, but in later stages they are frequently difficult to cut on account of the secondary thickening of the capsule wall and the stubborn exine of the mature spores. Great care must be taken to get *Riccia natans* into paraffin without shrinking, and the same thing may be said of other forms which have such loose tissue with large air cavities. Formerly, we resorted to celloidin for stages like that shown in Figure 69. The gradual processes already described have obviated the difficulty, so that the student should be able to get thin paraffin sections as free from distortion as were the old celloidin sections. But even with well-fixed material great care must be taken not to let the paraffin get too hot. Remember that in most paraffin ovens the temperature is different in different parts of the oven. Do not let the temperature of the paraffin go above 53° C., and, preferably, not above 52° C. In *Riccia natans* it is even more difficult to get median longitudinal sections of the sporophyte than of the archegonium. Sections perpendicular to the groove, whether longitudinal or transverse, are almost sure not to give median longitudinal sections of the sporophyte, and these are the sections the beginner is sure to cut. Examine the material and note very exactly the orientation of the sporophyte; then, for fixing, cut out sections about 2 mm. thick, taking these sections in such a plane that paraffin sections parallel to the thick section will give the desired median longitudinal sections of the sporophyte.

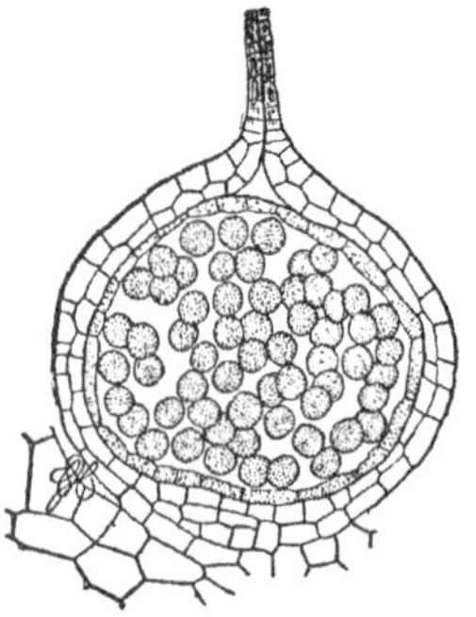

FIG. 69.—*Riccia natans:* young sporophyte inclosed in the archegonium; spore mother-cell stage—Delafield's haematoxylin. All the cells of the sporophyte, except a single peripheral layer (dotted in the figure), produce spores. Celloidin section 30 μ in thickness. ×104.

In forms like *Pellia* and *Aneura*, it is desirable to show the sporophyte still inclosed in the calyptra (Fig. 70). For such sections, we should recommend fixing in formalin alcohol. Aqueous fixing agents are likely to cause trouble on account of air bubbles. For cytological studies, the calyptra must be removed and a thin slab should be cut from opposite sides of the capsule to facilitate fixing and infiltration. Chromo-acetic acid, with the addition of a little osmic acid, is best for fixing. In *Pellia* and *Conocephalus* the spores are very large and have a rather thin wall. Both these genera show a peculiar intrasporal development of the gametophyte, i.e., the gametophyte develops to a considerable extent before it ruptures the spore wall and before it is shed from the capsule (Fig. 71). Mitotic figures during the first three divisions in these spores are exceptionally beautiful and are very easy to stain with the safranin, gentian-violet, orange combination, the chromosomes taking a very brilliant red, while the asters take the violet. Achromatic structures are very prominent during these three divisions, but become less and less conspicuous as division progresses and before the intrasporal stage is over, the radiations are scarcely demonstrable.

For the older sporophytes of *Marchantia,* it is better not to cut the whole receptacle. Remove the radiating branches. The sporophytes are in radiating rows, alternating with the branches. A piece 2 mm. wide can be cut so as to include two of the radiating rows, one on each side of the stalk, and such a piece will include early stages in other rows. By taking such care, you can get median longitudinal sections of nearly all the sporophytes. For class work, 5 to 10 μ is a good thickness, but for figures, especially the reduction mitoses in the spore mother-cells, the sections should not be thicker than 2 or 3 μ.

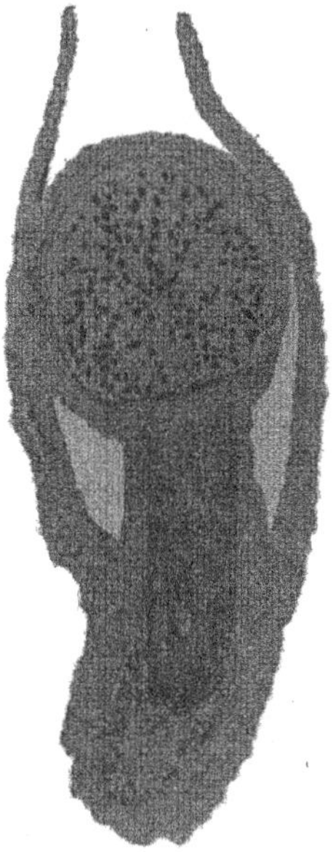

FIG. 70.—*Pellia epiphylla:* photomicrograph of longitudinal section of sporophyte still inclosed in the calyptra. The deeply lobed spore mother-cells are dividing. Eastman Commercial Ortho film, Wratten E filter (orange); arc light; J. Swift and Son 1-inch lens; exposure, 1 second. Negative by Dr. P. J. Sedgwick. ×30.

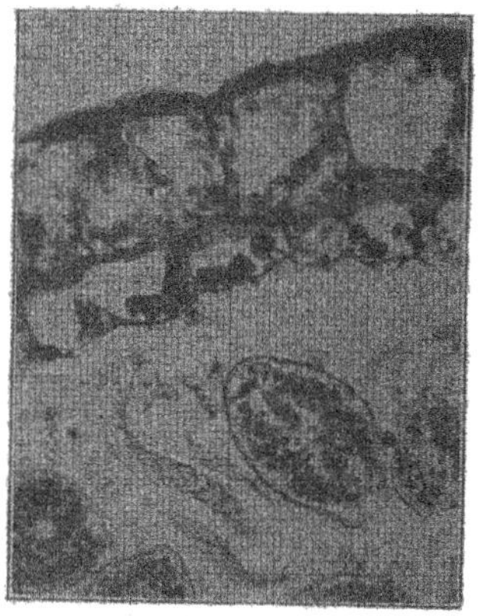

FIG. 71.—*Pellia epiphylla:* photomicrograph of spore germinating while still within the capsule. Fixed in chromo-acetic–osmic acid, and stained in safranin, gentian-violet, orange. Negative by Miss Ethel Thomas. ×276.

Among the Bryophytes no form affords a better opportunity for studying the development of spores than *Anthoceros*, since a single longitudinal section of the sporophyte may show all stages, from earliest archesporium to mature spores (Fig. 72). The sporophyte is even more difficult to orient than that of *Riccia natans*. Cut a slice 1 or 2 mm. thick, so as to orient the visible portion of the sporophyte, and trust to luck for the orientation of the foot. For studies like *A* and *B*, chromo-acetic material cut 10 μ thick

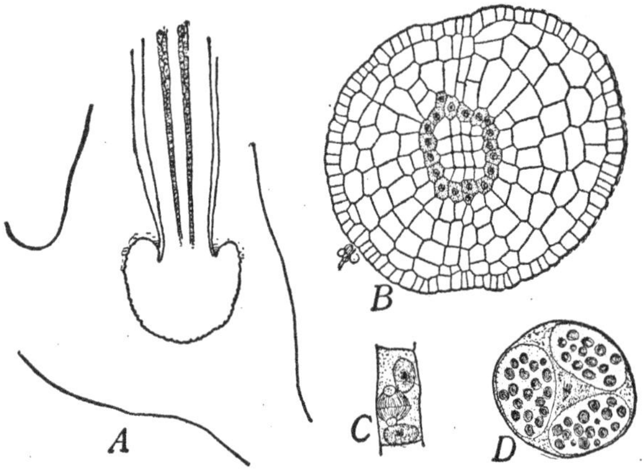

FIG. 72.—*Anthoceros laevis:* *A*, longitudinal section of lower portion of sporophyte imbedded in the gametophyte; ×45; *B*, transverse section of lower portion of sporophyte; ×200; *C*, vegetative cell from lower portion of the sporophyte; ×560; *D*, spore mother-cell showing three of the four chloroplasts with numerous starch grains; the nucleus in the metaphase of the first division; ×560.

and stained in Delafield's haematoxylin is very good. The starch grains in the chloroplasts take a beautiful violet color with the safranin, gentian-violet, orange combination. With so many stages in a single section, it will be impossible to stain all of them well. A stain which will show the mother-cells and their divisions will be too deep for the mature spores, and a stain which shows the spores well will be too faint for the mother-cells. It is better to stain some preparations for one feature and others for another. It is not worth while to steer a median course.

CHAPTER XX

BRYOPHYTES

MUSCI

In general, the mosses are more conspicuous than the liverworts and easier to collect. Many of the most desirable forms fruit only in the spring, but something can be found throughout the summer and autumn and some, like species of *Sphagnum*, pass the winter with the antheridia and archegonia in advanced stages of development.

Material is more troublesome to fix than in the liverworts because small bubbles of air hinder penetration of the fixing fluid. Use an air-pump. Older archegonia and capsules which have turned brownish add to the difficulties of the technic.

The special chromo-acetic–osmic-acid solution, or this solution without the osmic acid, fixes well. If an air-pump is not available at the time of fixing, Land's formalin alcohol solution (6 c.c. commercial formalin to 100 c.c. of 70 per cent alcohol) will be more satisfactory.

Protonema.—Protonema of some moss can always be found at any season. Look for greenish patches resembling *Vaucheria*. Such mats show the developing protonema and young leafy plants. Very young mats of moss will also show good protonema, but are not likely to show young buds. The brownish bulbils, which are quite common in mosses, can be seen with a good pocket lens. The little *Webera*, almost always found on the pots in the fernery or on the benches in greenhouses, quite frequently shows this mode of reproduction. Protonema is easily grown from spores.

Permanent mounts are very easily made. Simply wash away the dirt with water and put the material into 50 per cent glycerin, and let the glycerin concentrate. Mount in glycerin or glycerin jelly for permanent mounts. Seal thoroughly. Such mounts, with no fixing or staining, may retain the green color for many years.

If you do not insist upon keeping the green color, much clearer mounts can be made by fixing in formalin acetic acid, about 10 c.c. of formalin and 5 c.c. of acetic acid to 100 c.c. of water, and staining in eosin, or in Magdala red and anilin blue. Mount in Venetian turpentine.

Antheridia.—It is easy to find material for a study of antheridia, because, in so many cases, the antheridial plants can be detected at once without even a pocket lens. *Funaria,* with its bunch of antheridia as large as a pinhead, is extremely common everywhere. Spring is the best time to collect it, but it is found fruiting in the autumn and sometimes in summer; besides, it is easily kept in the greenhouse, where it may fruit at any time. *Bryum roseum* has a large cluster of antheridia surrounded by radiating leaves, making it easy to recognize. Other species of *Bryum* and species of *Mnium,* like *M. cuspidatum,* make good sections. *Polytrichum* has a large cluster of antheridia surrounded by reddish leaves, so that the whole is sometimes called the moss "flower." In fixing this or the closely related *Atrichum* (*Catharinea*), cut a small slab from two sides, so as to leave a flat piece to cut for longitudinal sections. This trimming will greatly facilitate fixing and infiltration. A single antheridial plant of *Polytrichum* often furnishes a fairly complete series of stages in the development

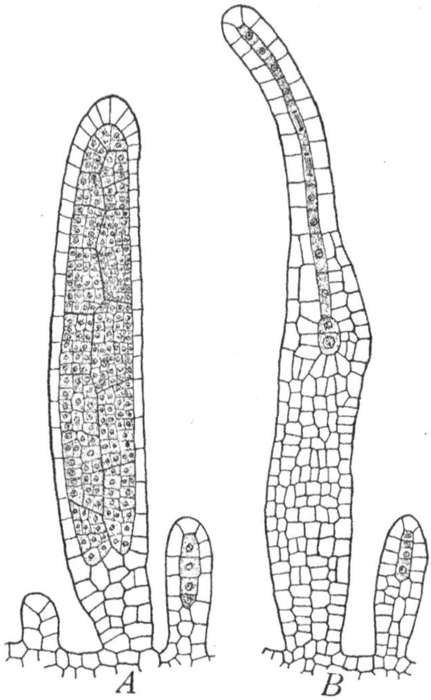

FIG. 73.—*Mnium cuspidatum: A*, nearly mature antheridium in the center, with a young archegonium at the right, and at the left a still younger stage which may develop into either an antheridium or an archegonium; *B*, a nearly mature archegonium with a young archegonium at the right. Fixed in formatin, acetic alcohol (formalin 10 c.c., acetic acid 5 c.c., 50 per cent alcohol, 100 c.c.) and stained in safranin, gentian-violet, orange. ×240.

of antheridia. Transverse sections show not only the antheridia, but also good views of the peculiar leaf of this genus. In all cases the stem should be cut off close up to the antheridia, for many of the moss stems, after they have begun to change color, cut like wire.

Sections to show the development of the antheridium should be 5 to 10 μ in thickness. The safranin, gentian-violet, orange is a good

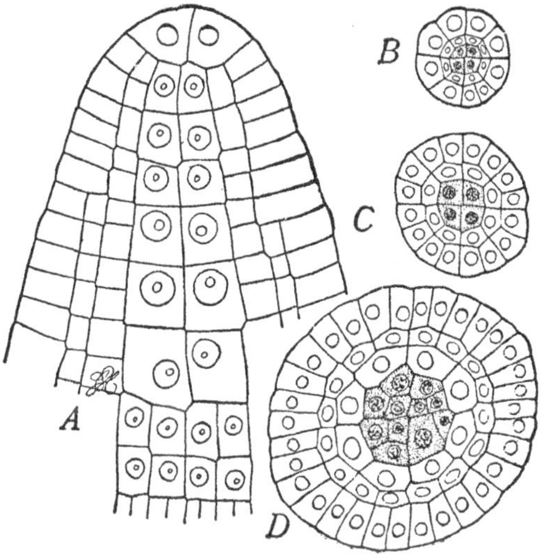

FIG. 74.—*Funaria hygrometrica:* *A*, apex of young sporophyte showing endothecium and amphithecium—chromo-acetic acid and Delafield's haematoxylin; 10 μ; ×420; *B*, *C*, and *D*, transverse sections of a sporophyte of the same age as *A*, taken at different levels; ×255.

combination (Fig. 73*A*). For details of spermatogenesis, sections should not be thicker than 3 μ. Iron-haematoxylin is a better stain for the chromatin and blepharoplasts.

Although sections 20 to 50 μ in thickness can be cut to show topography, it is far better to study such stages in the fresh material. When a particularly fine view is secured in this way, a permanent preparation may be made by putting the piece into 10 per cent glycerin without any fixing or staining, and allowing the glycerin to concentrate. Then mount in glycerin jelly.

Archegonia.—Since the necks of the archegonia are usually long and more or less curved, it is necessary, for habit work, to cut sections as thick as 20 or 30 μ in order to get a view of an archegonium in a single section (Fig. 73*B*). Mayer's albumen fixative is not likely to hold such sections to the slide. Use Land's fixative. Here, as in case of antheridia, it is better to use fresh material,

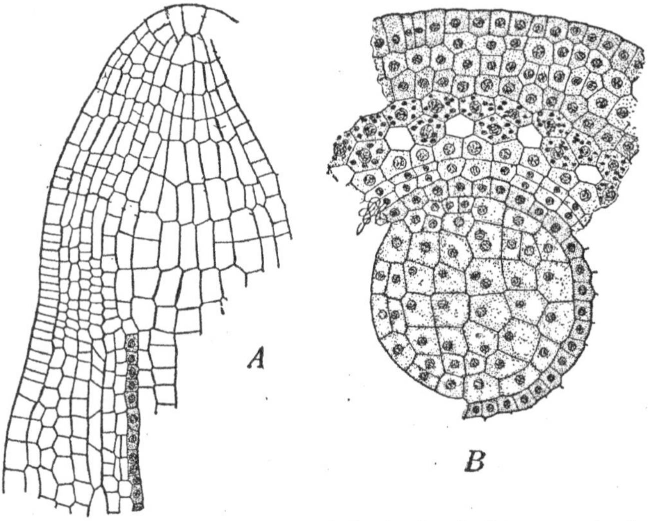

FIG. 75.—*Funaria hygrometrica:* *A*, longitudinal section of capsule; *B*, transverse section of capsule of about the same age as *A*—Delafield's haematoxylin and erythrosin; 10 μ. The columella, archesporium, outer spore case, two layers of chlorophyll-bearing cells, and the beginning of the air spaces can be distinguished at this stage. ×420.

putting particularly good pieces into 10 per cent glycerin for glycerin jelly mounts.

For the development of the archegonium, trim away the leaves which usually cover the cluster. Fix in chromo-acetic acid with a little osmic acid and cut 5 to 10 μ thick. For a study of the ventral-canal cell and fertilization, sections should not be thicker than 5 μ.

There is a general impression that the antheridia and archegonia of *Sphagnum* are rare and hard to find. Dr. George Bryan, who made an extensive study of *Spagnum subsecundum,* found that antheridia appear in August and archegonia in September. In examining acres of this species, he did not find a sterile plant.[1]

Sporophyte.—It is often difficult to get good mounts of sporophytes. In the younger stages the calyptras are likely to interfere

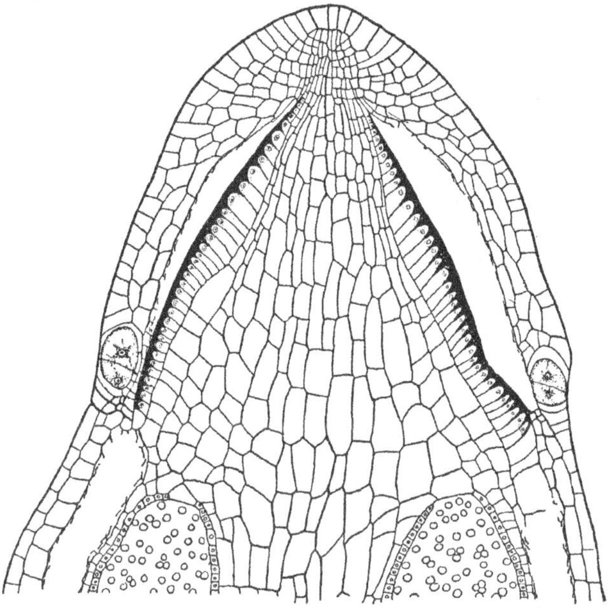

FIG. 76.—*Mnium:* upper part of capsule showing two teeth of the peristome cut longitudinally; *a*, annulus. Fixed in formalin alcohol and stained in safranin and light green. ×120.

with cutting, while in the older stages the peristome, or hard wall of the capsule, occasions the trouble. If an attempt is made to remove the calyptra in young stages, like *A* of Figure 74, the apex of the sporophyte usually comes with it. While picro-acetic acid material cuts more easily, chromo-acetic acid followed by Delafield's haematoxylin gives so much sharper differentiation in stages

[1] *Botanical Gazette,* **59**:40–56, 1915.

like those shown in Figure 75 that it is better to use ice or Land's cooler and get preparations from chromic material.

Stages like that shown in Figure 75 are cut with comparative ease, for the calyptra is easily removed, and the capsule wall is not yet hard enough to occasion any difficulty. Safranin, gentian-violet, orange is a good stain. The cell walls stain so sharply that they are not obscured by a stain which will bring out the cell contents.

Later stages, after the peristome has begun to differentiate, are likely to occasion difficulty in cutting. *Bryum* cuts as easily as any (Fig. 76). For the development of the peristome, fix in formalin alcohol and stain in safranin and anilin blue, or in safranin and light green. Safranin and Delafield's haematoxylin is also an excellent stain for the older stages in the differentiation of the capsule.

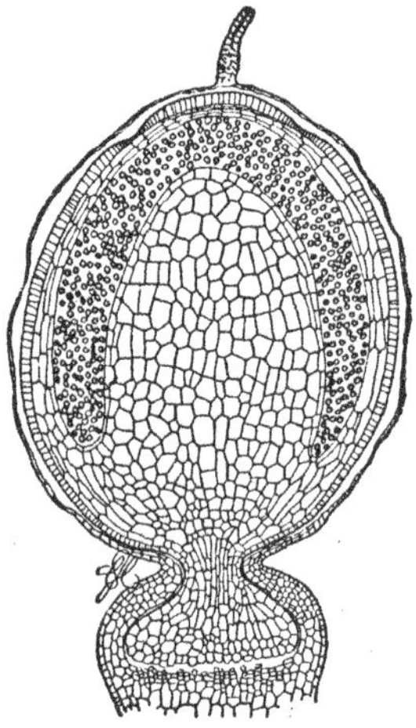

FIG. 77.—*Sphagnum:* longitudinal section of sporophyte showing also the upper portion of the pseudopodium and the calyptra—Delafield's haematoxylin. ×24.

Beautiful mounts of the peristome are easily and quickly made. Take a capsule at the stage when the operculum is just ready to fall off, or has just fallen off; with a sharp razor cut off the end of the capsule just below the line of the annulus; put it into absolute alcohol for 10 minutes, clear in clove oil, transfer to xylol, and mount in balsam. If several are placed on a slide, some one side up and some the other, with some complete and some teased a little, there will be good views of the entire peristome and also good views of teeth and cilia.

The mature sporophytes of *Sphagnum* are exceptionally hard to cut. It will be worth while to prick the capsule with a needle when the material is collected. This will allow the fixing agent to penetrate readily, and will also facilitate the infiltration of paraffin or celloidin. The puncture causes only a slight damage, and need not reach the really valuable portion which is to furnish the median longitudinal sections.

The younger stages in the sporophyte of *Sphagnum* like that shown in Figure 77, and also the antheridia, archegonia, and the peculiar development of the leaves cut easily in paraffin.

CHAPTER XXI

PTERIDOPHYTES

This group includes the Lycopodiales, Sphenophyllales, Psilotales, Pseudoborniales, Equisetales, and Filicales. The Sphenophyllales and Pseudoborniales occur only as fossils and the Psilotales are confined to tropical and subtropical regions. The Lycopodiales are commonly called club mosses or ground pines, the Equisetales are called horsetail rushes or scouring rushes, and the Filicales are the common ferns. The Ophioglossaceae, a family of the Filicales, is often treated as an order. Two of its genera, *Ophioglossum* and *Botrychium*, are widely distributed and well known. Material, except in the Psilotales, is abundant and so easily recognized that anyone who pays a little attention to collecting can, in a single season, get a fine supply for a study of the group. Some desirable forms may not be present in all localities, but these will be few, and can be secured at a reasonable price from those who make a business of collecting.

The technic for Sphenophyllales will be found under "Special Methods" (chap. xii). Nothing but impressions has yet been found in Pseudoborniales. Gametophytes of Psilotales have been found only recently. Their young sporangia cut easily, but older stages are very refractory and should receive extreme care in dehydrating, clearing, and infiltration. No further directions will be given for these rather inaccessible orders.

LYCOPODIALES

Lycopodium.—The genus is evergreen, and consequently some stage in development can be secured at any season. In general, the tropical species are easier to cut than the temperate. Without any regard to taxonomic sequence, we shall consider the vegetative structure, the strobili, and the prothallia.

Vegetative structure.—Formalin alcohol, with or without the addition of acetic acid, is an excellent fixing agent and, quite contrary to prevalent notions, the staining capacity of material seems to improve with several months' immersion. We prefer the following formula: formalin 10 c.c., acetic acid 5 c.c., 50 per cent alcohol 100 c.c.

The growing points of stems and roots cut easily in paraffin, and when the material becomes too hard to cut in paraffin it can be cut without any imbedding. It is easier to get good sections of *L. lucidulum* and *L. inundatum* than of drier species, like *L. obscurum* and *L. clavatum*. Safranin and Delafield's haematoxylin is a reliable stain. Safranin with anilin blue or light green is also good, and the light green gives particularly clear views of the phloem.

If young sporelings are available they afford a beautiful example of a very primitive type of stele, in transverse section showing an exarch protostele with 4 or 5 radiating arms of metaxylem, each tipped with a comparatively large group of protoxylem cells. In most species, this simple radial stele of the sporeling passes into a complicated, banded stele in the adult plant. Even in the adult plant the protoxylem and metaxylem are easily distinguished in sections near the growing point of the stem or root (Fig. 78). Not only in *Lycopodium*, but in any vascular plant, sections at this age are useful in pointing out the protoxylem.

Sections of the stem and root of *Lycopodium complanatum*, mounted on the same slide, show an interesting parallelism of structures. Transverse sections of the stem of *Lycopodium pithyoides*, a Mexican species, show not only the stem structures but excellent transverse sections of roots which grow down through the cortex.

The strobilus.—For longitudinal sections, cut a slab from each side of the strobilus to insure fixing and infiltration. If a strobilus, or similar organ, is simply halved, both pieces are likely to curve. Among north temperature species, *Lycopodium inundatum* is the most easily cut. A young strobilus 1 cm. in length may show all stages from the archesporium to the spore mother-cell. Iron-haematoxylin is the best stain for differentiating the archesporial cells. The divisions in the spore mother-cell stain intensely, so that care must be taken not to overstain.

Strobili of *Lycopodium dendroideum* or *L. obscurum* 6 or 7 mm. in length show a beautiful series in the development of the sporangium from the earliest stages up to tetrads. These young stages fix well in chromo-acetic acid, with or without a little osmic acid.

The gametophyte.—In most species the gametophyte, or prothallium, is subterranean, tuberous, and has no chlorophyll; in other species the prothallium is partly subterranean and partly aerial, the aerial portion being green and bearing the archegonia and antheridia.

In the third edition of this book, published in 1915, the statement was made that no one had yet discovered prothallia of *Lycopodium* in the United States. Two years later, a teacher in the high school at Marquette, Michigan, Dr. E. A. Spessard, announced the discovery

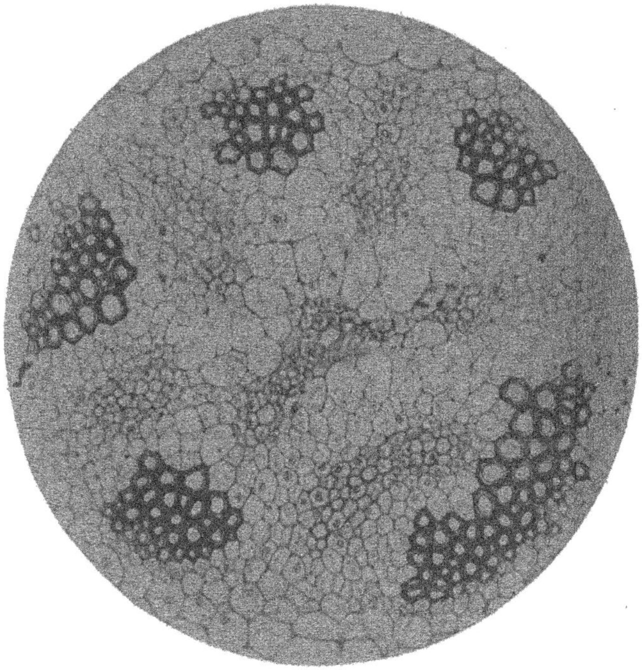

FIG. 78.—*Lycopodium lucidulum:* photomicrograph of transverse section of stem near the apex. The five deeply staining groups of thick-walled cells are the protoxylem and alternating with them are groups of phloem cells; the large, thin-walled cells are metaxylem. Eastman Commercial Ortho film, Wratten E filter (orange); arc light; Spencer 4-mm. objective. N.A. 0.65; ocular ×6; exposure, 6½ seconds. Slide by Dr. J. J. Turner, negative by Dr. P. J. Sedgwick. ×360.

of the prothallia of four species (*Botanical Gazette*, **63**:51–65, 1917) and gave such clear directions for finding prothallia that already two papers have appeared, announcing the discovery of prothallia. One of these papers, by Dr. Alma Stokey and Dr. Anna Starr describes seven stations for prothallia in Massachusetts; and the other, by Mr. O. Degener, adds four more stations in Massachusetts and mentions the finding of prothallia near the crater of Kilauea, Hawaii. Both papers appeared in the *Botanical Gazette* of March, 1924. With these papers describing the finding of prothallia, it is probable that prothallia of Lycopodium, especially the older prothallia with sporelings attached, will become familiar objects in botanical laboratories.

Once found, the technic is easy. Fix in the special chromo-acetic-osmic-acid solution, or this with the acetic reduced to 2 or even to 1 c.c. Stain in iron-alum haematoxylin and orange for the development of archegonia and antheridia. Use safranin, gentian-violet, orange for the development of the embryo. For the endophytic fungus which stains well with the last-named combination, try the method of differentiation of pathogen and host, described at the close of the chapter on fungi.

The prothallia of *Lycopodium inundatum,* the only species in the United States known to have aerial green prothallia, have not yet been discovered in this country, although they have been found in Europe.

It would seem natural to get the prothallia by germinating the spores, but here again no one has had any notable success, except Bruchmann. In some species, the spores do not germinate for several years, but when the prothallia are once developed they continue to bear archegonia and antheridia for several years. The spores of *L. Selago* germinate in 3 to 5 years after shedding; those of *L. clavatum* and *L. annotinum* in 6 to 7 years. In *L. clavatum* and *L. annotinum* archegonia and antheridia develop in 12 to 15 years after the spores are shed. *L. inundatum* germinates more promptly—in 10 days to 6 months—but no one has succeeded in keeping a culture up to the archegonium stage.

Botanists in *Lycopodium* localities should look for prothallia. Since the subterranean forms are perennial and as large as a grain of wheat, some reaching a length of 1.5 cm., it would seem as if they should be found wherever *Lycopodium* grows.

Selaginella.—Material of *Selaginella,* in all phases of the life-history, is easy to secure, but not so easy to handle after it is obtained. As many as 340 species, mostly tropical, have been described, only 3 of which are common in the range of Gray's *Manual.* Of these 3, *Selaginella apus* is best for sections. It is found in moist or wet situations on the borders of ponds, along ditches, or on moist meadows. While the plant is very small, it has large spores. Several of the tropical species are common in greenhouses, and they fruit abundantly.

Vegetative structure.—Growing points and root-tips are easily cut in paraffin. In most species, the older parts of the stem are too hard and brittle to cut in paraffin and are too small to cut well free-hand. It might be worth while to try hydrofluoric acid or the cellulose acetate method when that method becomes developed. At present, patience and a sharp knife seem to be the only reliance. Some of the tropical species, like *Selaginella Wildenovii,* have stems nearly as large as a lead pencil, with polystelic structure, and are not hard to cut. The vascular cylinder is an exarch protostele or, when poly-stelic, each bundle is an exarch protostele. It is exceptionally easy to get a brilliant, differentiated stain when once the sections are cut.

The strobilus.—Very young strobili cut easily in paraffin, but after the megaspore coats begin to harden, there are few objects which make more trouble than the strobili of *Selaginella.* For stages up to the young megaspores, fix in chromo-acetic acid, with or without the addition of a little osmic acid; but for later stages use hot corrosive sublimate acetic acid in 50 per cent alcohol. If this is not available, use formalin, acetic acohol. Sometimes sections of the difficult later stages stay on the slide with Mayer's albumen fixative. Even stages like that shown in Figure 79 should always stay on with Land's fixative.

The strobili of most species are square in transverse section. To get longitudinal sections showing the relations of sporangia, sporo-phylls and axis, cut diagonally, from corner to corner, never parallel to the flat side. For archesporial cells, use iron-haematoxylin; for young megaspores and the development of spore coats, use safranin, gentian-violet, orange; for later stages, use safranin and light green.

The gametophytes.—In most cases the spores germinate while still within the sporangium and, in some cases, like *Selaginella apus,* the female gametophyte develops up to the archegonium initial stage before shedding. If strobili of this species at the stage shown in Figure 79 be broken off and laid down on moist ground so as to keep the sporangium moist, dehiscence may not be vigorous enough to discharge the megaspores; but development of archegonia and antheridia will continue, fertilization will take place, and an embryo will be developed while the megaspore is still inclosed within the sporangium. Such a structure satisfies the definition of a seed. Ordinarily, to get the later stages, shake the spore out from the older strobili into a Petri dish with the bottom covered by several thicknesses of wet filter paper. There is enough nutritive material in the spores to carry them up to the sperm and egg stage. The female gametophytes within the old spore coats generally orient themselves in the

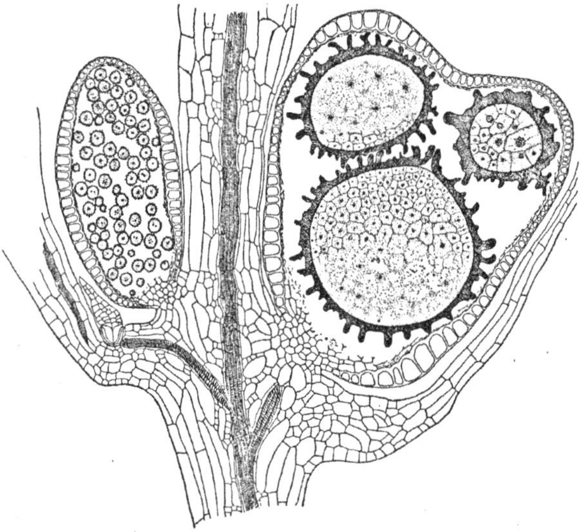

FIG. 79.—*Selaginella apus:* longitudinal section of strobilus showing a microsporangium with germinating microspores on the left; on the right, three of the four megaspores with gametophytes near the archegonium initial stage; fixed in formalin alcohol, cut in paraffin, and stained in safranin and light green; from a preparation by Dr. W. J. G. Land. ×80.

paraffin, the base of the spore being down and the archegonium end of the gametophyte being up. A little of some nutrient solution, added to the water, will carry the development up to the cotyledon stage. The young embryo, with its two cotyledons, its root, and the megaspore still attached makes an instructive preparation when mounted whole in Venetian turpentine.

The various stages of the female gametophyte and embryo are not hard to stain; but the walls throughout the development of the male gametophyte are very thin and extremely hard to stain. Safranin and light green is a good combination. Light green in clove oil may prove more satisfactory than the alcoholic solution.

Isoetes.—This peculiar genus is widely distributed and 16 of its 64 species occur within the United States. It looks so much like a sedge that it is easily overlooked, even when rather abundant. As a genus, it is hydrophytic, growing in wet places or even under water. A recent monograph by Dr. Norma Pfeiffer not only gives an ecological, morphological, and taxonomic account (with keys in English), but gives hundreds of stations, a feature which will enable many to find material.

Vegetative structure.—The short, thick stem, even in old plants, cuts easily in paraffin. Fix in formalin alcohol and stain in safranin and light green. Sporelings with stems about 2 mm. in diameter and young plants with stems up to 5 mm. in diameter are best for a study of the peculiar vascular system of this plant. These young stages fix well in chromo-acetic acid and are not hard to cut.

Sporangia.—All the sporangia of the plant may be said to constitute a single strobilus of the *Selago* type. Both longitudinal and transverse sections should be cut. The stem is so short that, in a plant of medium size, a longitudinal section may include the stem, the sporangium, and the sporophyll, up to the top of the ligule. Such sections, 10 to 15 μ, or even 20 μ in thickness, are best for demonstration. Transverse sections through the whole cluster of sporophylls show the arrangement of megasporophylls and microsporophylls and also the relations of the sporangia to sporophylls.

The gametophytes.—The spores are shed in the uninucleate stage, and consequently it is not so easy to find the germination as in the case of *Selaginella*. When the large megasporangium begins to decay, let the megaspores dry naturally. They retain their power of germination for a year at least. Simply wet them with tap water and the earlier stages are easily secured, quite clean and ready for cutting. There must be soil in the dish for later stages. Try a similar method for microspores. Also, look at the top of the stem of old plants for stages developing naturally. The cell walls of the male gametophyte, as in the case of *Selaginella*, are rather hard to differentiate. Use anilin blue or light green.

CHAPTER XXII

PTERIDOPHYTES

EQUISETALES

This order was large and prominent in the Carboniferous age, but now only a single family, the Equisetaceae, survives. Its only genus, *Equisetum*, contains 24 pieces, 10 of which occur within the Gray's *Manual* range. *Equisetum* is often called the "scouring rush," because the rough stems have been used for scouring kettles. The roughness is due to silica. Species, like *E. hiemale*, which contain much silica, must be treated with hydrofluoric acid before the older parts can be cut in paraffin.

Vegetative Structure.—The roots are very small, but have large cells and easily yield good preparations. If a handful of *Equisetum fluviatile* or *E. hiemale* growing in water be pulled up, scores of root-tips may be secured in a few minutes. Fix in chromo-acetic acid with a little osmic acid. In case of such small objects it is a good plan to add a few drops of eosin to the alcohol during the process of dehydrating, in order that the material may be seen more easily. The slight staining does no damage, even if more critical stains are to be used after the sections are cut. Longitudinal sections of the roots may also be obtained by cutting transverse sections of the nodes.

The growing points of stems may be cut with ease in paraffin. *E. arvense* is particularly favorable on account of the numerous apical cells which may be found in a single preparation.

The "fertile" stem of *Equisetum arvense* is so free from silica that it can be cut in paraffin without any difficulty. The adult vegetative stem of *E. arvense*, and all stems which contain so much silica, must be treated with hydrofluoric acid before imbedding in paraffin. However, nearly all of these stems can be cut freehand, before fixing, without removing the silica. Fix freehand sections in 95 per cent alcohol. Material for paraffin sections should be fixed in formalin, alcohol, acetic acid; or in hot alcohol, corrosive sublimate acetic. Safranin and anilin blue, with or without a little orange, is a good combination.

The Strobilus.—*E. arvense* affords the most favorable material for a study of the development of sporangia, since the strobilus contains almost no silica and, even in its latest stages, is easily cut in paraffin. In this species, the young strobili, in the Chicago region, can be distinguished from vegetative buds in July; sporogenous tissue is well advanced by the middle of August; and the reduction divisions occur late in August or early in September (Fig. 80). The

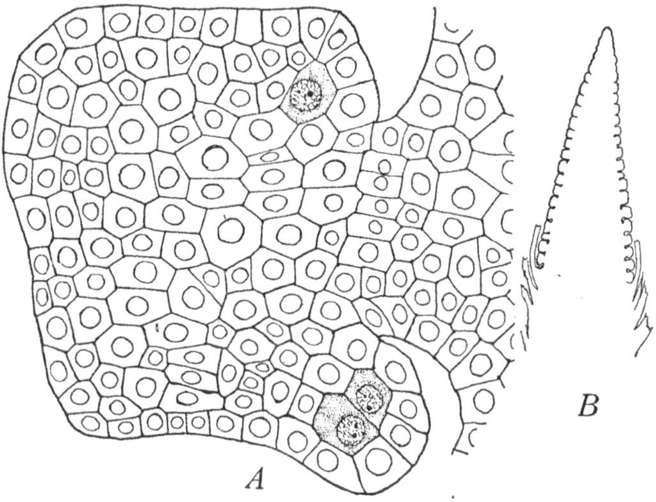

FIG. 80.—*Equisetum arvense:* *A*, section of a sporangiophore showing beginning of sporogenous tissue, early August condition; *B*, topography of the strobilus at this stage. *A*, ×580; *B*, ×8.

spores are not shed until the following April. If you know a patch of this species which "fruits" every year, dig up the horizontal underground stem in July. The tip of the main axis is almost sure to be a strobilus. Dissect away the scale leaves and fix the strobilus in chromo-acetic acid with a little osmic acid. August and September stages are easy to recognize. If strobili are brought into the laboratory in December or January, they shed their spores within a week.

The spores of *Equisetum* are excellent for illustrating hygroscopic movements. Shake out the spores from the strobili and let them dry thoroughly. They can be kept dry for years. When wanted for demonstration, put a few on a slide, moisten a little, and watch the movements under the microscope. Strobili of other species, like *E. fluviatile and E. hiemale,* contain a large amount of silica and, consequently, only the younger stages cut well in paraffin. Hydrofluoric acid damages the cell contents more or less. In species like these, all stages in the development are found in a single season.

The Gametophytes.—The spores of *Equisetum* germinate as soon as they are shed, but, like all spores with a considerable amount of chlorophyll, they do not long retain the power of germination. A comparatively small percentage will germinate a week after shedding, and after a month, there may be no germination at all. There is no difficulty in growing prothallia to maturity and securing stages in the embryo, if fungi or blue-green algae do not appear and ruin the cultures. Use Costello's method for fern prothallia, as described on page 263. Consult also a paper by Dr. Elda Walker.[1] This paper corrects some previous misconceptions in regard to gametophytes of *Equisetum,* gives a full account of the gametophyte of *E. laevigatum,* and also directions for finding gametophytes as they occur in nature.

The prothallia fix well in chromo-acetic acid. The younger stages may be stained in iron-alum, haematoxylin and mounted in Venetian turpentine. The older stages, even of *E. arvense,* are too large for such mounts. *E. laevigatum* has prothallia a centimeter in diameter.

For the development of antheridia, the blepharoplast, and the development of the sperm, fix in Flemming's weaker solution and stain in iron-haematoxylin. The sperm of *Equisetum* is the largest in Pteridophytes.

[1] *Botanical Gazette,* 71:378–391, 1921.

CHAPTER XXIII

PTERIDOPHYTES

FILICALES

This order includes the common ferns and also the Ophioglossaceae, which, in the previous edition of this book, was treated as an order, Ophioglossales, co-ordinate with Filicales. Some of the ferns are sure to be available in almost any locality, and all stages in the life-history are easily secured, except early stages in the Ophioglossaceae.

Vegetative Structure.—From a technical standpoint, the vegetative structures of Filicales present a wide range of conditions, some being so soft that the greatest care must be taken to get them into paraffin, while others are so hard that it is almost impossible to cut them at all.

The stem.—Growing points, even of the largest ferns, can be cut in paraffin. If the growing point is covered with dense hairs or ramentum, either remove the covering entirely or, in case of rather fleshy ramentum, remove only the scales which are beginning to turn brownish. The white scales will fix and cut. Use chromo-acetic acid (1 g. chromic acid and 1 c.c. acetic acid to 100 c.c. water). Unless mitotic figures are particularly desirable, it is just as well not to add any osmic acid. For illustrating the development of the stem from the apical cell, sections 10, 15, or even 20 μ are not too thick.

Older portions of the stem, or rhizome, in most ferns are easily cut while fresh, the sections being transferred to 95 per cent alcohol after cutting. But even fairly well-developed rhizomes, after the xylem has become lignified sufficiently to stain sharply in safranin, can be cut in paraffin, and much finer sections can be obtained than by cutting without imbedding (Fig. 81). In digging up rhizomes, do not merely dig down until the rhizome can be grasped and then pull it up, for such material is sure to show the pericycle of the bundles torn away from the parenchyma. Dig carefully around the rhizome and then with a very sharp knife cut off pieces which are perfectly free. The pieces can be wrapped in wet paper and taken to the laboratory. Then, if they are to be cut without imbedding, cut into pieces about 3 cm. long; but material which is to be imbedded should be in pieces not longer than 1 cm.

Pteris aquilina is a good form to practice with. For freehand sections, cut as thin as possible, allow 15 to 30 minutes in 95 per cent

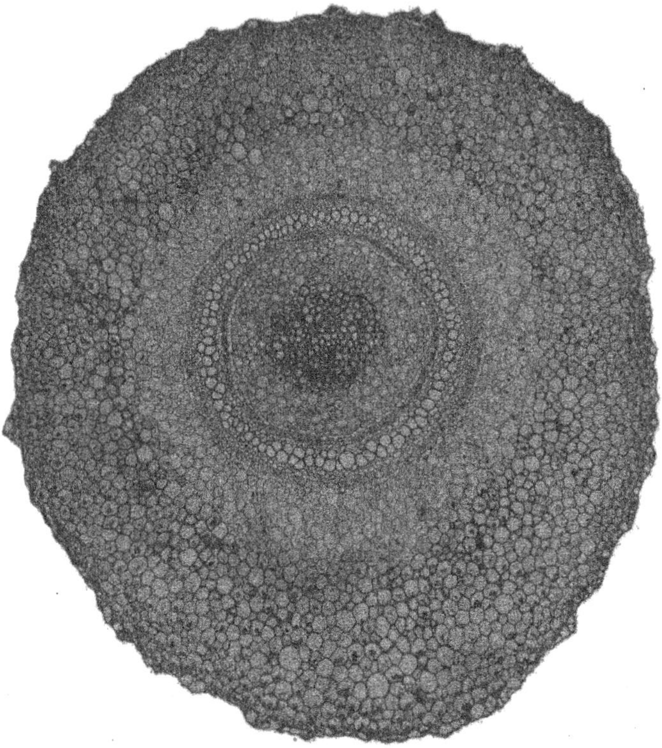

Fig. 81.—*Dicksonia punctilobula:* photomicrograph of transverse section of rhizome. Fixed in formalin-acetic alcohol, and stained in safranin and light green. Paraffin section 10 μ thick. Eastman Commercial Ortho film, Wratten E filter (orange); arc light; J. Swift and Son 1-inch lens; exposure, 1 second. Negative by Dr. P. J. Sedgwick. ×33.

alcohol, transfer to 50 per cent alcohol and stain over night in safranin. Rinse in 50 per cent alcohol and stain in light green. Paraffin sections 2 or 3 cm. back of the apex are very instructive, for only the protoxylem will be sufficiently lignified to stain red with safranin, the metaxylem still being thin walled and staining with the light green. This rhizome affords an excellent illustration of a mesarch polystele.

Dicksonia punctilobula has a small rhizome, often on the surface of the soil or rock, so that it is easy to get good clean pieces. About 2 or 3 cm. back of the growing point, the xylem is well lignified, but the material still cuts well in paraffin. One could hardly find a better illustration of a mesarch amphiphloic siphonostele (Fig. 81).

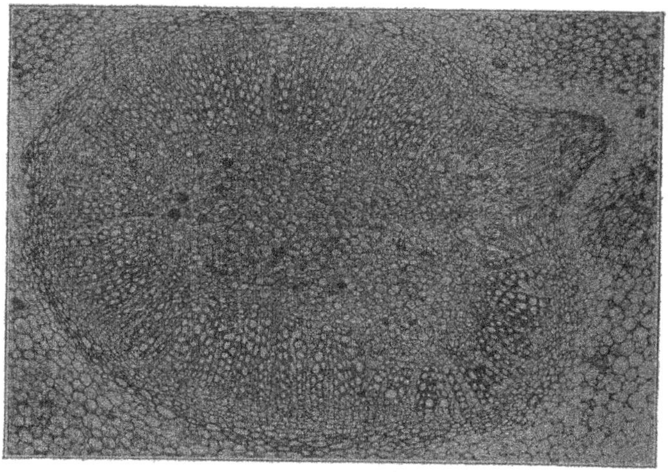

FIG. 82.—*Botrychium obliquum:* photomicrograph of transverse section of rhizome. Fixed in formalin-acetic acid-alcohol, and stained in safranin, gentian-violet, orange. Paraffin section, 10 μ. Eastman Commercial Ortho film, Wratten E filter (orange); arc light; J. Swift and Son 1-inch lens; exposure, 1 second. Negative by Dr. P. J. Sedgwick. ×45.

Botrychium is widely distributed but individual plants are not abundant. The stem is erect, subterranean, and has an endarch siphonostele with secondary wood. Trim away the roots, which are very thick and fleshy in *B. obliquum,* fix in formalin-alcohol-acetic acid, and imbed in paraffin. Even the older parts of old stems can be cut in paraffin if you are sufficiently careful (Fig. 82.) Transverse sections from the base of the bud down to the secondary wood will give a beautiful series in the development of the stele.

The bud at the top of this rhizome is an interesting object. The leaf is in its fourth year when it appears above ground, and, consequently, the bud contains young leaves of three successive seasons. Two of the three show a differentiation into sterile and fertile portions.

In *Osmunda*, and in many other ferns of similar habit, the rhizome is surrounded by the very hard leaf bases. Good sections of the central cylinder can be secured only by dissecting away these hard leaf bases and any hard portions of the cortex before attempting to cut sections. A short distance back of the growing point will be found a region which will show practically all the structures of the mature stem, which will be easy to cut. Even in this region the leaf bases should be dissected away. From the apical cell back to the region where the sclerenchyma is beginning to turn brown, the material is easily cut in paraffin. Older portions should be cut freehand. *Osmunda* affords an excellent illustration of the mesarch siphonostele (Fig. 83).

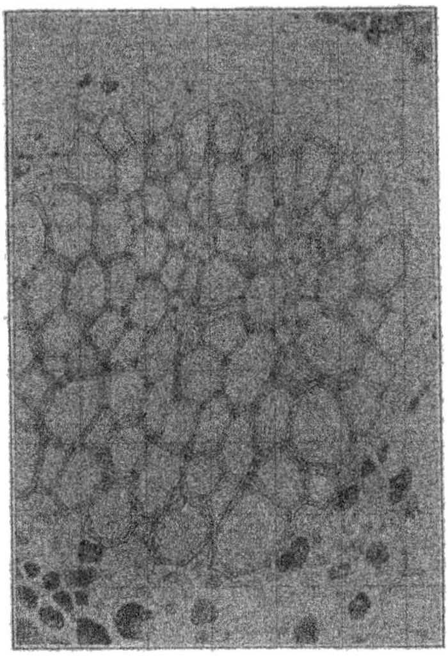

Fig. 83.—*Osmunda cinnamomea:* photomicrograph of single bundle of the mesarch siphonostele. Stained in safranin and anilin blue. Eastman Commercial Ortho film, Wratten C (blue-violet) filter; Bausch and Lomb 4-mm. objective, N.A. .65; exposure, 4 seconds. Negative by Dr. P. J. Sedgwick. ×108.

The rhizome of *Adiantum* affords a good illustration of leaf gap and leaf trace. The vascular cylinder is a mesarch siphonostele; but there are few sections like Figure 81, because the clyinder is so interrupted by leaf gaps. This rhizome cuts well without imbedding. The petiole of *Botrychium*, in transverse sections below the fertile spike, shows the interesting leaf-trace situation which proves that the fertile spike consists of a pair of pinnae fused together.

The ferns of the Gray's *Manual* range afford no very satisfactory material for illustrating the protostele, although protosteles occur in *Lygodium* and *Trichomanes*. The most satisfactory material is *Gleichenia*, a very common and very beautiful fern in tropical and subtropical regions, but almost never seen in greenhouses nor even in botanical gardens. Formalin alcohol material is easily cut without imbedding and is easy to stain.

The stems of tree ferns require special treatment. With the large leaf bases partly cut away with a sharp razor, transverse sections are easily cut for a considerable distance below the apex. Material fixed in formalin alcohol cuts very well. If fresh material is to be cut, the softer portions should be flooded with alcohol after each section. Farther down, there will be a region where sections can be cut without any flooding, and still farther down, it will be difficult or impossible to cut sections across the whole stem. Sections 1 or 2 cm. thick, cut smooth on the ends, may be kept in 95 per cent alcohol or in glycerin in large glass dishes of the Petri dish pattern. Better still, clear such sections in xylol and preserve in cedar oil.

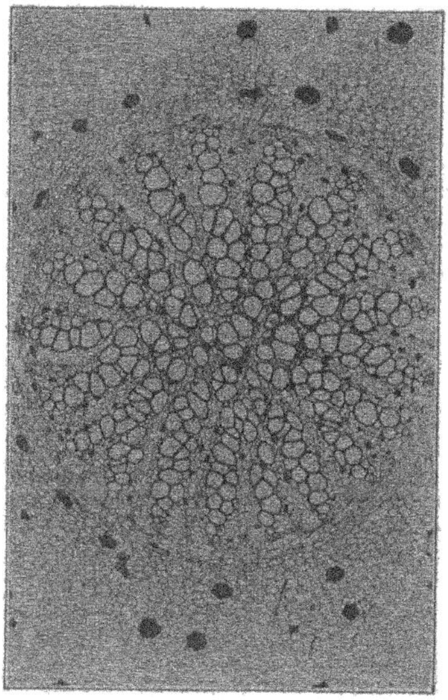

Fig. 84.—*Angiopteris evecta:* photomicrograph of a transverse section of a root, showing the polyarch, exarch siphonostele. Eastman Commercial Ortho film, Wratten E (orange) filter; arc light; J. Swift and Son 1-inch lens; exposure, ¾ second. Negative by Dr. P. J. Sedgwick. ×25.

The root.—The roots of Filicales develop from a strong apical cell. For mitotic figures and the development of the root from the apical cell, fix the tip in chromo-acetic acid with a little osmic acid. If the development of the root is the principal object, stain in safranin and light green, or in the safranin, gentian-violet, orange combination; if mitotic figures are to be studied, stain in iron-haematoxylin with a very light counter-stain in orange. The comparatively large root-tips of *Botrychium* are excellent for the apical cell and its segments. *Dicksonia punctilobula* can also be recommended; but even the very small root-tips of most of our ferns will yield good preparations.

Roots of tree ferns are sometimes available in greenhouses. In some species the stem is covered by a dense felt of small roots, some of which will be white and soft at the tip. These roots are likely to have about the diameter of onion root-tips, and the beauty of preparations made from them could hardly be excelled. In the tropics, where the plants are often in the spray of cataracts and the lower part of the trunk is often washed by mountain streams, a thousand tips might be secured from a single specimen.

The older roots of *Botrychium*, especially the large fleshy roots of *B. obliquum*, cut very easily and show a simple exarch protostele with 4 or 5 protoxylem points.

The roots of *Angiopteris* and *Marattia*, which become as large as a lead pencil, may be secured in some greenhouses. They cut easily after fixing in formalin alcohol and furnish a fine example of the exarch protostele, common to all roots (Figs. 84 and 85).

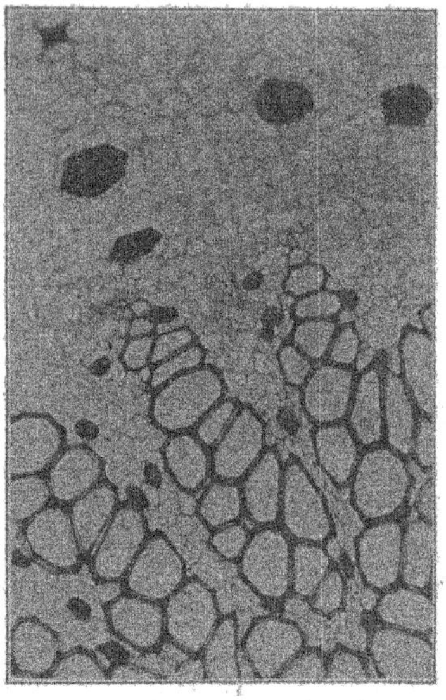

FIG. 85.—*Angiopteris evecta:* photomicrograph showing a detail of the stele shown in Figure 84. Stained in safranin. Eastman Commercial Ortho film, Wratten E (orange) filter; Bausch and Lomb 8-mm. objective, N.A. .50; exposure, 1½ seconds. Negative by Dr. P. J. Sedgwick. ×60.

The structure of the leaf will appear in sections cut to show the sporangia.

The Sporangia.—To illustrate the character of the annulus, select sporangia which are just beginning to turn brown. Fix in formalin alcohol and dehydrate as if for paraffin sections; after the absolute alcohol, transfer to 10 per cent Venetian turpentine. Staining is neither necessary nor desirable.

The various relations of sorus and indusium are best illustrated by rather thick sections (10 to 20 μ) of material in which the oldest sporangia have barely reached the spore stage. Fix in formalin alcohol and stain in safranin and anilin blue.

For the development of sporangia use hot corrosive sublimate–acetic acid–alcohol, or chromo-acetic acid (1 g. chromic acid and 2 c.c. acetic acid to 100 c.c. of water). A larger proportion of figures will be secured by adding 5 or 6 drops of 1 per cent osmic acid to this solution. Sections should not be more than 10 μ in thickness and, for mitotic figures, 5 μ is thick enough.

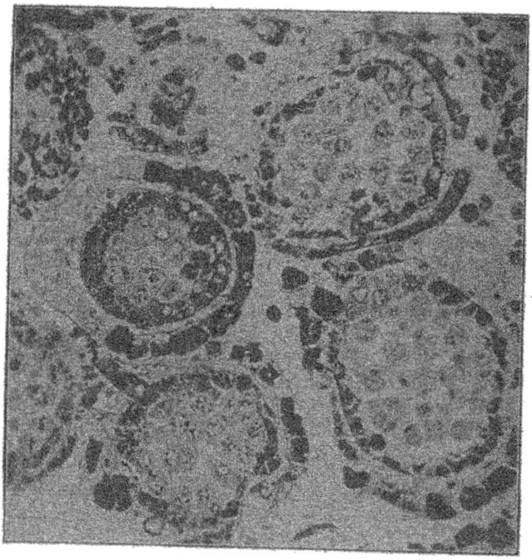

Fig. 86.—*Osmunda cinnamomea:* photomicrograph of sporangia with spore mother-cells in various stages of division; fixed in Flemming's weaker solution and stained in iron-alum haematoxylin; from a preparation by Dr. S. Yamanouchi. Negative by Miss Ethel Thomas. ×114.

For the reduction of chromosomes, the sections should not be thicker than 5 μ. *Osmunda* is particularly good for this purpose because the number of chromosomes is comparatively small. The young sporangia of *Osmunda cinnamomea* and *O. Claytoniana* show the mother-cell stage in the autumn, but the division into spores does not occur until the following spring, in the vicinity of Chicago, the mitotic figures being found during the latter part of April (Fig. 86). *O. regalis* does not reach the mother-cell stage in the autumn. Material for mitosis should be collected during the first two weeks in May.

Various species of *Pteris* are common in greenhouses and are very good for development of sporangia. Any fern of the *Aspidium* type will yield a good series, and some, like *Cyrtomium*, may show a fine series in a single sorus. *Marattia*, which is likely to be found in botanical gardens, will illustrate the "synangium" type; *Angiopteris* has a sporangium which forms an easy transition to that of the Cycadales.

For the development of a typical sporangium of the eusporangiate type, nothing is better than *Botrychium*. Buds of *B. virginianum* taken in September or October show sporangia with well-marked sporogenous tissue. For a study of the development of sporangia, cut off the fertile portion and fix it separately, using chromo-acetic acid with 5 or 6 drops of 1 per cent osmic acid to 100 c.c. of the solution. Stain in iron-alum haematoxylin. The reduction divisions in the spore mother-cell take place after the leaf arrives above the surface.

The Prothallia.— Prothallia can usually be found on the pots in the ferneries of greenhouses. Ripe spores of some fern or other can be obtained at any greenhouse at any time in the year, and spores of most of our native ferns germinate well and produce good prothallia, even if the sowing is not made for several months after the spores have been gathered.

Fine prothallia of *Pteris aquilina* have been grown two years after the spores were gathered. Some, however, must be sown at once, or they will not germinate at all. Spores which are large and contain enough chlorophyll to make them appear greenish should be sown at once. The spores of the common *Osmunda regalis*, and of the other members of the genus, must be sown as soon as ripe, or they fail to germinate. The prothallia of *O. regalis*, if carefully covered with glass, may be kept for a long time, and they become quite large. Prothallia of this fern in the writer's laboratory produced ribbon-like outgrowths 5 mm. wide and more than 5 cm. in length. These prothallia continued to produce archegonia, antheridia, and ribbon-like outgrowths for more than a year, when they suddenly "damped off." Lang watered prothallia with a weak solution of permanganate of potash, which kills the fungi but does not injure the prothallia. He does not state the strength of the solution, but 4 or 5 crystals to 1 liter of water seems to be effective.

The prothallia of most ferns will grow for a long time under such conditions. *Pteris aquilina* and many other ferns often furnish a good supply of antheridia 3 weeks after sowing, and the archegonia appear soon after, but it is well to make sowings 6 weeks before material is needed for use. In *P. aquilina* and in many others, if the spores are sown too thickly only antheridial plants will be ob-

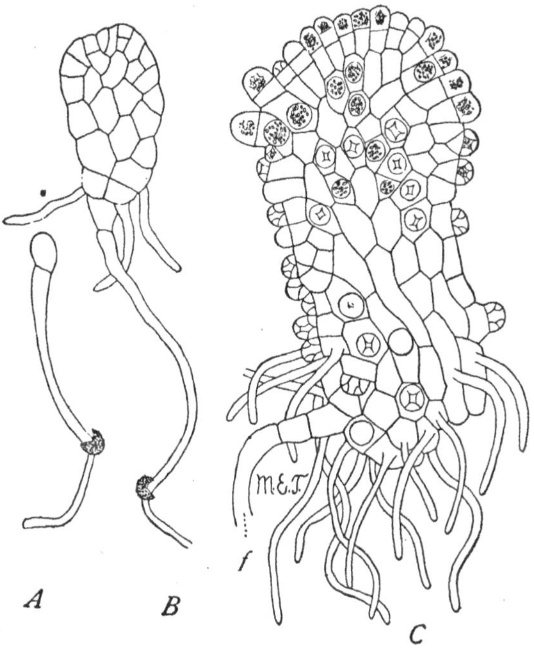

FIG. 87.—*Pteris aquilina:* *A*, filamentous stage; *B*, the apical cell has been established and several segments have been cut off; the figure shows the initial rhizoid and also three rhizoids coming from the main body of the prothallium; *C* an older prothallium covered with antheridia in various stages of development; from a drawing by Miss M. E. Tarrant.

tained (Fig. 87). If prothallia are to produce archegonia, they must have sufficient room and nutrition.

The best method we have ever seen for growing fern prothallia was devised by Mr. M. Costello, head gardener at the University of Chicago. The diagrammatic Figure 88 will make the method clear. Select a clean flower-pot, as porous as possible, and pack it full of wet *Sphagnum*. Wet the outside of the pot and invert it in a pan of water. Sow the spores on the surface of the pot and cover with a bell jar. No further wetting is necessary, except to take care that the water in the pan does not dry up. With *Pteris longifolia*, there may be antheridia in 2 weeks; archegonia in 3 or 4 weeks; and in 5 or 6 weeks, abundant sporophytes in various stages. Prothallia grown by Costello's method are entirely free from soil and, consequently, very convenient for cutting or for mounting a hole.

While there should always be a study from living material, it is worth while to make permanent mounts, even for habit study. For such study, the prothallia should be mounted whole. Fix in the special chromo-acetic–osmic-acid solution. If the material shows any tendency to break up, use 2 c.c. of acetic acid instead of 3 c.c. In cities where water is treated with copper or other substances, the difficulty may sometimes be due to the water rather than to any excess of acetic acid. Formalin–acetic acid (10 c.c. formalin, 5 c.c. acetic acid to 100 c.c. water) is good for material which is to be mounted whole. Stain some in iron-alum haematoxylin, and some in Magdala red and anilin blue. Mount in Venetian turpentine, using material from each stain for each mount. Select stages so that each preparation will show the filamentous stage, the apical cell stage, the group of initials stage, and also antheridia and archegonia.

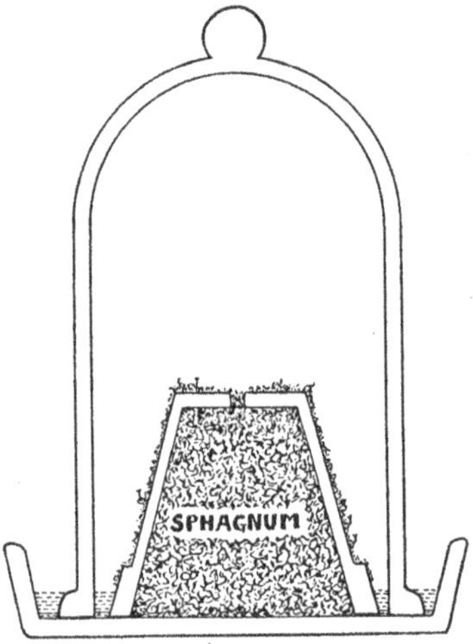

FIG. 88.—Costello's method for growing fern prothallia.

For sections, the chromic series is better than formalin or corrosive sublimate. If the gradual processes of dehydrating, clearing, and infiltrating have been carefully observed, about 15 or 20 minutes in the bath should be sufficient. About 10 μ is a good thickness for such views as are shown in Figures 89, 90, 91. Safranin, gentian-violet, orange is a good stain.

For the development of the antheridium and sperm, and especially for the blepharoplasts and their transformation, cut about 3 μ in thickness and use iron-haematoxylin; for the development of archegonia, cut at 5 μ and stain in the safranin, gentian-violet, orange combination.

The gametophyte of *Botrychium* is subterranean and tuberous. It sometimes reaches a length of 7 to 12 mm. and a thickness of 4 to 5 mm. Usually, it is not more than 5 or 6 mm. long and 2 or 3 mm. thick. Gametophytes showing the development of antheridia and

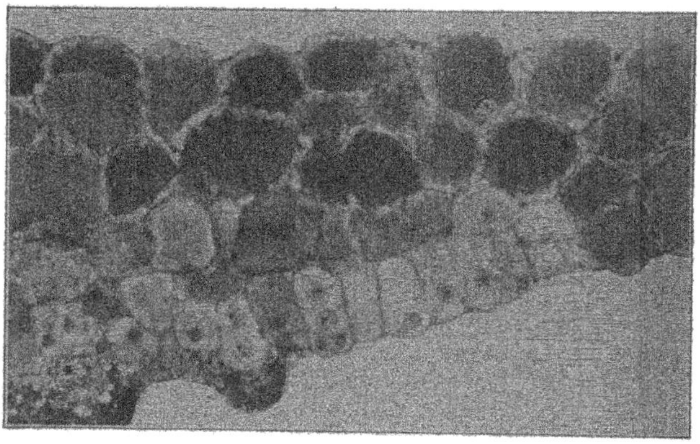

FIG. 89.—*Osmunda cinnamomea:* photomicrograph of vertical section from the notch toward the base of the prothallium showing four stages in the development of the archegonium—chromo-acetic acid; safranin, gentian-violet; from a preparation by Dr. W. J. G. Land. Negative by Miss Ethel Thomas. $\times 425$.

archegonia are not likely to be more than 2 or 3 mm. long and 1 or 2 mm. thick. Near large plants, look for small sporelings, not more than 1 or 2 cm. in height. Dig very carefully and you may find the gametophytes attached. The soil should be examined for smaller specimens. Most of the gametophytes will be found at a depth of 1 to 3 cm. Fix in chromo-acetic acid.

No one has yet succeeded in raising the prothallia from the spores. The prothallia always contain an endophytic fungus, but even when this is present no one has succeeded in raising prothallia.

The prothallium of *Ophioglossum* is harder to find or, perhaps it would be better to say, harder to recognize, for it is also subterranean and tuberous and, besides, looks so much like the root that you may not recognize it, even when you have it in your hand.

The Embryo.—Instructive mounts of the whole embryo, with the prothallium still attached can be made by the Venetian turpentine method. Iron-alum haematoxylin, with the stain not too deep,

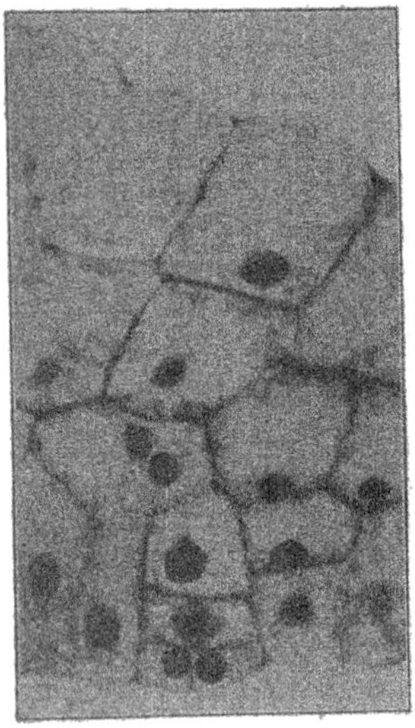

FIG. 90.—*Osmunda cinnamomea:* photomicrograph of vertical section of prothallium with an early stage in the development of the archegonium, showing the basal cell, two neck cells, and, between them, the cell which is to give rise to the neck canal cell, the ventral canal cell, and the egg—chromo-acetic acid; safranin, gentian-violet; from a preparation by Dr. W. J. G. Land. Negative by Miss Ethel Thomas. ×425.

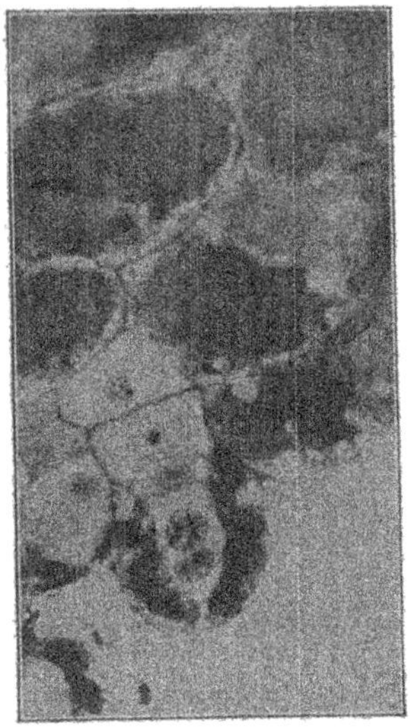

FIG. 91.—*Osmunda cinnamomea:* photomicrograph of a vertical section with a young archegonium, showing the neck canal cell with two nuclei, the ventral canal cell, the egg, and the basal cell—chromo-acetic acid; safranin, gentian-violet; from a preparation by Dr. W. J. G. Land. Negative by Miss Ethel Thomas. ×293.

is good; Magdala red and anilin blue are more transparent and will show the structure of the root, if the two stains are well balanced.

For sections, cut longitudinally, perpendicular to the prothallium. *Pteris longifolia* may show the young embryo within 3 or 4 weeks; *Osmunda*, somewhat later.

Nephrodium is generally parthenogenetic. The embryos, developed while the prothallia are still very small, come from tissue cells of the prothallium. *Pteris cretica* is frequently parthenogenetic.

The Heterosporous Filicales.—The four genera, *Pilularia, Marsilia, Salvinia,* and *Azolla* are aquatic, the first two growing rooted but more or less submerged, and the other two floating freely on the water. *Marsilia* is the most available and convenient laboratory type of this group. It is easily grown in a pond or in an aquarium in the greenhouse. In setting it out in a pond, select a place with a gently sloping bank, so that part of the material may be under water and part may creep up the bank. In the greenhouse, a rectangular aquarium may be tilted to secure the same conditions. The portions which are not under water will continue to fruit during the summer and autumn. The whole sporocarp cuts easily in paraffin during the development of sporangia, the division of the spore mother-cells, and even during the earlier stages in the formation of spores. Except in the case of the youngest sporocarps, it is better to cut off a small portion at the top and at the bottom to facilitate fixing and infiltration. The mother-cell stage and the young spores will be found in sporocarps which are just beginning to turn brown. In nature, no further nuclear divisions take place within the sporangium until the next spring, but the wall of the sporocarp becomes extremely hard. Sporocarps for germinating should not be collected until they are so hard that it is impossible to crush them between the thumb and finger. They can be kept in a box until needed for use. When you find them in good condition, make a big collection, for they retain their power of germination almost indefinitely, sporocarps from poisoned herbarium specimens 50 years old germinating readily. Sporocarps which have been kept in 95 per cent alcohol for years germinate almost as quickly as those which have been kept in a dry box.

To germinate sporocarps, cut away a portion of the hard wall along the front edge and place the sporocarp in a dish of water. The gelatinous ring with its sori will sometimes come out in a few minutes. In less than 24 hours, sometimes within 10 or 12 hours, microspores, starting from the one-cell stage, will produce the mature sperms; and the development of the female gametophyte is equally rapid. Starting with the uninucleate megaspore, the stage found when the gelatinous ring comes out, the archegonium may be developed and fertilization may occur within 12 hours; and within 36 hours, stages like that shown in Figure 92 may be reached. At the end of a week, there may be green sporophyte more than a centimeter in length.

It is obvious that material should be fixed at frequent intervals if one is to secure a series of stages in the development of the gametophytes and embryo.

For sections showing the development of the antheridium and sperms, it is better to remove the megaspores from the sorus, since they occasion considerable difficulty in cutting. Fix in the special chromo-acetic-osmic-acid solution, cut 5 μ thick, and stain in iron-alum haematoxylin.

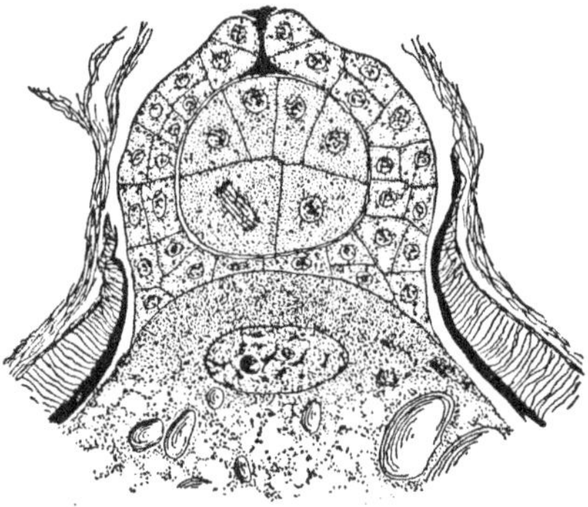

FIG. 92.—*Marsilia quadrifolia:* upper portion of megaspore with an archegonium containing a young embryo. ×212.

The older megaspores, with the archegonium or embryos at the apex, are very hard to cut. Fix in hot alcoholic corrosive sublimate–acetic, or in alcohol-formalin–acetic acid. It will facilitate infiltration and cutting it you prick each megaspore with a sharp needle before fixing. Since the archegonium is at the apex of the megaspore, the pricking need do no damage. If sections come loose from the slide, use Land's fixative. Cut 5 to 10 μ thick and stain in safranin, gentian-violet, orange.

The sperm, which in *Marsilia* has an unusually large number of turns in the spiral, is easily mounted whole. When the sperms have become numerous, put several megaspores upon a slide and heat gently until dry. Then wet the preparation in any alcohol and stain sharply in acid fuchsin. Dehydrate in absolute alcohol for at least 10 minutes, clear in clove oil, and mount in balsam. Such a preparation will often show a score of sperms in the gelatinous funnel leading down to the neck of the archegonium.

Azolla is not difficult to obtain, and it is easy to get a series of stages in the development of the micro- and megasporangia; but it is not at all easy to find the gametophytes, since the spores germinate only after they have been set free by the decay of the plant. *Azolla* does not fix well in any of the chromic-acid series, because it catches so much air that it will not sink. Alcohol-formalin-acetic, or hot alcoholic corrosive sublimate–acetic acid, formalin (4 g. corrosive sublimate, 5 c.c. formalin; 5 c.c. acetic acid, 100 c.c. of 50 per cent alcohol) can be recommended for both *Azolla* and *Salvinia*. Both of these forms grow well in the greenhouse, floating on the water in tanks or aquaria; but *Azolla* seldom fruits under such conditions. *Salvinia* sometimes produces microsporangia in the greenhouse, but megasporangia are comparatively rare.

CHAPTER XXIV

SPERMATOPHYTES

Material in this group ranges from structures so delicate that they require more skill and patience than the coenocytic algae to structures so hard that the method for rock-sections is the best way to get preparations. We cannot hope to give even approximately complete directions for making preparations, but must be content to give a few hints which may prove helpful in collecting material and in securing mounts of the more important structures. We shall consider the gymnosperms and the angiosperms separately, although in many respects the technic is the same for both.

GYMNOSPERMS—CYCADALES

Cycas revoluta, the Sago Palm, can be found in almost any large greenhouse which keeps decorative plants. The large conservatories of city parks may keep, in addition, some species of *Ceratozamia* or *Encephalartos*. Only one cycad, *Zamia*, occurs in the United States, and it is confined to Florida. In *Encephalartos* and *Ceratozamia* the development of the ovule, and even the development of the female gametophyte up to the fertilization period, takes place quite naturally in the greenhouse, where pollination is not likely to occur; but in other genera, the female cones, or at least their ovules, nearly always abort unless fertilization takes place. The vegetative structures are natural enough, but, with the exception of leaves and small roots, are not so available, since material of the stem would mean damage to the plant.

The Vegetative Structures.—All the vegetative structures cut rather easily.

The stem.—*Zamia*,[1] which grows in various parts of Florida, is the most available material. Directions for handling the stem are given on page 132.

Stems of the larger cycads are not likely to be obtained, except in the field, and they are confined to tropical and subtropical regions.

[1] Material of *Zamia* can be obtained at $1.00 a plant (express, collect) by addressing the Plant Pathologist, Agricultural Experiment Station, University of Florida, Gainesville, Fla. This is not a commercial matter, but an accommodation to botanists.

They cut better while fresh; consequently, if one can get material, it is a good plan to send it to the laboratory and have it cut before fixing. Even transverse sections are not difficult to cut while fresh (Fig. 93). A piece of cycad trunk 15 to 30 cm. in diameter and 20 cm. in length will survive a journey of 6 weeks or even 2 months, if care be taken to coat the exposed ends with a mixture of melted paraffin and moth balls, using 3 or 4 moth balls as large as marbles to half a kilo of paraffin. If material is to be fixed before cutting, use 6 to 10 per cent formalin in water.

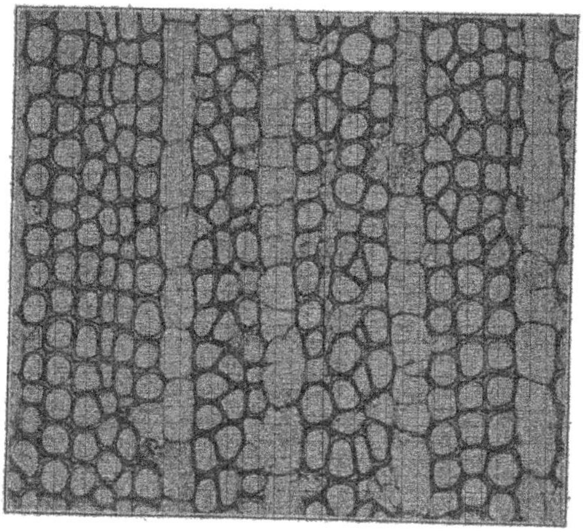

Fig. 93.—*Dioon spinulosum:* photomicrograph of transverse section of wood of *Dioon spinulosum*, cut from fresh material. ×105.

Dr. La Dema Mary Langdon succeeded in cutting paraffin sections of the adult wood of *Dioon spinulosum.*[1] She fixed 1 to 2 cm. cubes of adult wood in formalin-acetic acid-alcohol (6 c.c. formalin, 3 c.c. acetic acid, 100 c.c. of 50 per cent alcohol). After thorough washing, 24 to 48 hours in running water, the blocks were softened for 3 to 6 weeks in 50 per cent hydrofluoric acid in water. With dry material, the cubes were boiled and cooled repeatedly to remove air. The usual gradual processes of dehydrating, clearing, and infiltrating

[1] *Botanical Gazette,* **70**:82–84, 1920.

were then followed and sections were cut with a sliding microtome, with the knife placed obliquely, as in cutting celloidin.

The course of the vascular bundles, as they pass to the cones, is quite peculiar. Instructive preparations may be made by cutting longitudinal sections, about 3 mm. thick, through the apex of the stem and, without staining, clearing thoroughly and mounting in balsam. In this way we have mounted sections 5 cm. long, 15 cm. wide, and 3 mm. thick.

The course of the bundles in the xylem zone and in the cortex may be traced by clearing the cubes in xylol and then transferring to equal parts of xylol and carbon disulphide. Placed on a glass plate with an electric light bulb beneath, the bundles are quite distinct.

The root.—Small roots, up to a centimeter in diameter, are easily cut freehand. The tender root-tips and also the peculiar "root-tubercles" should be fixed in chromo-acetic acid and imbedded in paraffin.

The leaves.—The young tender leaves should be fixed in formalin alcohol and imbedded in paraffin. The adult leaves are rigid and cut well freehand. It is a good plan to tie several leaflets together with a string and then cut across, about a centimeter beyond the tied portion, so that the whole will be like a leaflet 5 or 6 mm. thick. Dip the whole thing in paraffin two or three times. Of course, there is no infiltration but the paraffin holds the leaflets in place. Cut in a sliding microtome with an oblique stroke. The sections fall out from the paraffin, which is easily skimmed away. Fix the sections for an hour in 95 per cent alcohol, stain in safranin and light green, clear in clove oil, transfer to xylol, and mount in balsam.

Spermatogenesis.—Except in the earliest stages, the staminate cones are too large to be cut whole. The individual sporophylls, with their sporangia, cut easily up to the formation of microspores; then the sporangium wall hardens rapidly and cutting becomes difficult. Up to the young microspore stage, fix in chromo-acetic–osmic-acid solution (1 g. chromic acid, 1 c.c. acetic acid, 1 c.c. 1 per cent osmic acid to 100 c.c. water). With a larger proportion of acetic acid, our results have not been satisfactory. Fix later stages in formalin–acetic acid–alcohol. Transverse sections are more instructive and are more easily cut, since the peripheral end of the sporophyll can be cut only in younger stages. In all the genera of cycads, the microspore germinates while still within the sporangium, the pollen grain at the time of shedding consisting of a prothallial cell, a generative cell, and a tube cell. For preparations at the shedding stage, shake the cone over a piece of paper and pour the pollen into water. After 15 or 20 minutes, put it into the fixing agent. In wind-pollinated plants, the pollen would look shriveled if you put it into a fixing agent before it regained its turgidity.

The pollen, mounted whole, makes beautiful preparations. Fix in formalin-acetic acid (formalin 10 c.c., acetic acid 5 c.c., water

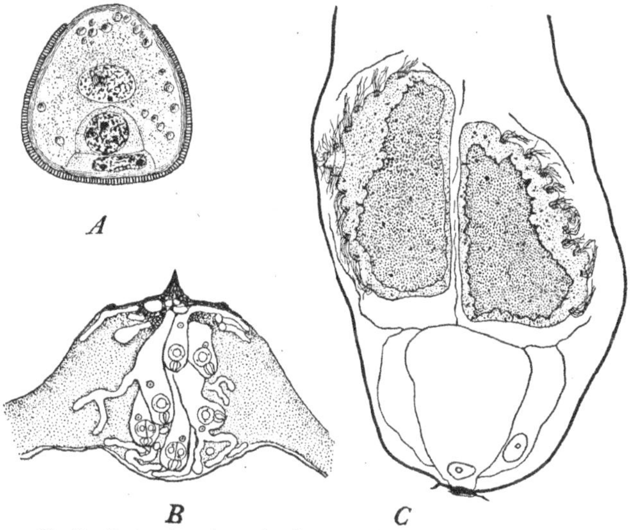

FIG. 94.—*Ceratozamia mexicana: A*, pollen grain which has been in a sugar solution for two days; ×876; *B*, nucellus with numerous pollen tubes; ×17; *C*, basal end of pollen tube showing the persistent prothallial cell; outside it the stalk cell; and, above, the two sperms still inclosed in the sperm mother-cells; ×156.

100 c.c.). Stain in iron-alum haematoxylin and follow the Venetian turpentine method. Carmalum, as described in the Power's methods on page 178, is also satisfactory. If some material is put into a 5 or 10 per cent sugar solution for two or three days, early stages in the formation of the pollen tube may be added. In a week or two the tubes may reach several times the length of the pollen grain; but, as far as we know, the generative cell has never divided in culture solutions (Fig. 94*A*).

The development of the pollen tube and its structures must be studied in sections of the nucellus. As soon as the integument is removed the nucellus is exposed and the position of the pollen tubes is easily determined, since the haustorial portions of the tubes form brownish lines radiating from the nucellar beak. Having learned the location of the pollen tubes, it is better not to remove the integument, but to remove the female gametophyte; then cut from the underside of the nucellus against the hard, stony layer of the integument so as to remove a small piece of the nucellus 5 to 7 mm. square, according to the species. Fix in chromo-acetic-osmic acid (1 g. chromic acid, 2 c.c. acetic acid, 1 c.c. 1 per cent osmic acid to 100 c.c. water). Nothing surpasses iron-alum haematoxylin for all the stages in the development of the male gametophyte (Fig. 94*A–C*). By using a 2 per cent iron-alum for 5 or 6 hours and staining over night, or even 24 hours, the stain can be drawn so precisely that the portion of the cilia between the blepharoplast and the surface can be differentiated from the free portion (Fig. 95).

The pollen tubes, with their sperms, make instructive preparations when mounted whole. Fix the nucellus, with its pollen tubes, as if for paraffin sections. About 6 per cent formalin in water has proved successful. Wash in water for half an hour and stain in aqeous safranin, 1 or 2 hours. Extract the stain until it is satisfactory, and then transfer to 10 per cent glycerin and follow the Venetian turpentine method. When the turpentine becomes thick enough for mounting, tease the pollen tubes from the nucellus and mount with pieces of cover-glass under the cover to prevent crushing. Iron-alum haematoxylin, which is so satisfactory for staining sections of sperms, is the most unsatisfactory stain we have tried for staining the cycad pollen-tube structures whole.

The immense sperms of the cycads, more than 100 μ in diameter, can be seen with the naked eye. If the tip of the nucellus with the pollen chamber be removed when the sperms are mature, the large pollen tubes, 300 to 500 μ in diameter, are very conspicuous. Even with the naked eye, the movements of the sperms can be seen; and with a pocket lens one can see some details. Place the piece of nucellus on a slide in a drop of water and examine with a 16 mm. objective. The amoeboid movements of the sperm and also a quick movement reminding one of the sudden jerky movement of *Vorticella*, are easily seen; and, by closing the diaphragm a little, some movement of the cilia should be distinguishable. *Zamia* is the only genus available in the United States; but even its rather small female cones, as they near the fertilization stage, will keep 4 or 5 weeks after they have been removed from the plant.

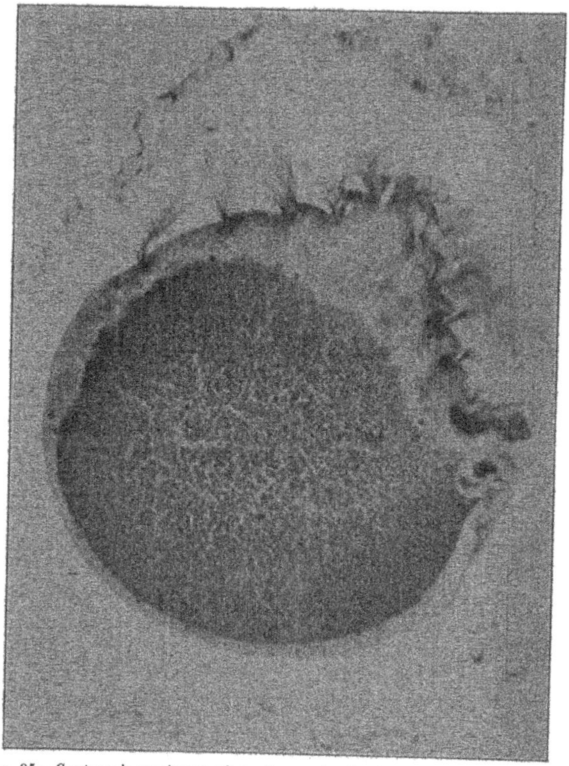

FIG. 95.—*Ceratozamia mexicana:* photomicrograph of a longitudinal section of a sperm, showing the very large nucleus, thin sheath of protoplasm which is thicker at the apical end, and blepharoplast cut across in various places and bearing numerous cilia. At this stage, the sperm is swimming in the pollen tube. Cramer, contrast plate. Negative by Miss Ethel Thomas. ×506.

In the region of Miami, Florida, pollination takes place late in December or early in January; blepharoplasts appear in March, and swimming sperms are found during the first 2 weeks of June. In northern Florida, all stages are 2 or 3 weeks later. If one is so fortunate as to get cones of *Ceratozamia*, the sperms should be found from the last week in June through the second week in July.

Oögenesis.— The ovules of the Cycads and *Ginkgo* are very large, and, when mature, thin sections cannot be cut by any method yet discovered. In younger stages it is not difficult to get good sections of the entire ovule. Slabs should be cut from two sides of the ovule to facilitate fixing and infiltration. During free nuclear stages in the endosperm, and even during earlier stages in the formation of walls, care must be taken that the slabs may not cut into the endosperm, or even too near to it; for the endosperm is so turgid that it may even break out, and, at least, will be distorted. Even after the ovule approaches its full size, it can be cut entire, until the stony layer begins to harden. Paraffin sections of the entire ovule, cut 15 to 20 μ thick, and stained rather lightly in safranin, gentian-violet, orange, make very instructive preparations. When the fresh ovule can no longer be cut easily with a razor, it is not worth while to try to cut it in paraffin. Interesting preparations may be made by cutting from the median longitudinal portion of the ovule a slab about 5 mm. thick. The slab should be fixed, washed, dehydrated, and cleared in xylol. It should then be kept in a flat-sided bottle. Such a preparation shows the integument, micropyle, nucellus with its beak, pollen tubes, the stony and fleshy layers, general course of vascular bundles, and the female gametophyte with its archegonia.

For thin sections of the archegonia, a cubical piece with an edge of 6 or 8 mm. should be cut from the top of the endosperm with a very sharp, thin blade. The slightest pressure upon the archegonia will ruin the preparations.

Chromo-acetic-osmic acid (1 g. chromic acid, 2 c.c acetic acid, and 1 c.c. of 1 per cent osmic acid to 100 c.c. of water) is a good fixing agent for all stages in oögenesis. During the free nuclear stage and early wall stage in the female gametophyte, some plasmolysis is to be anticipated. Hot alcoholic corrosive sublimate–acetic acid sometimes fixes these stages with less distortion.

Sporophyte.—During the period of simultaneous free nuclear division, which follows the fertilization of the egg, the mitotic figures are quite striking and are easily stained (Fig. 96).

After the embryos begin to grow down into the endosperm, oblong pieces containing the embryos should be cut out.

After the cotyledons appear, useful preparations may be made by dissecting out the entire embryos, which may be fixed in chromo-acetic acid, washed, stained in eosin or in Delafield's haematoxylin, placed in 10 per cent glycerin, and mounted by the Venetian turpentine method. Since the suspensors become long and irregular, each embryo should be placed in a separate dish, lest the suspensors become entangled and broken.

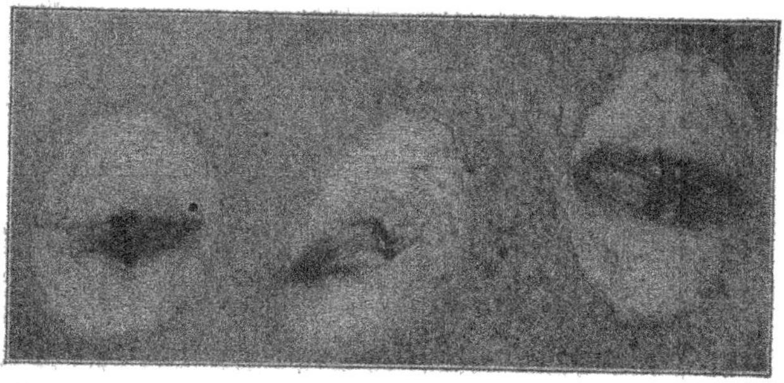

FIG. 96.—*Zamia floridana:* photomicrograph of a small portion of a proembryo showing simultaneous free nuclear division. Fixed in chromo-acetic-osmic acid and stained in safranin, gentian-violet, orange. Cramer contrast plate; Spencer 4-mm. objective, N.A. .65; ocular ×4; yellowish-green filter; camera bellows, 50 cm.; arc light; exposure, 6 seconds. Negative by Miss Ethel Thomas. ×413.

After the stony layer becomes hard, it is better to use a small fret saw for opening the ovule. Before the embryo has pushed down into the endosperm, the ovule should be sawed in two transversely. The endosperm and nucellus can then be picked out and treated as desired. After the tip of the embryo reaches the middle of the endosperm, the ovule should be sawed open longitudinally.

GYMNOSPERMS—GINKGOALES

From the standpoint of technic, the Ginkgoales, now represented only by *Ginkgo biloba*, are less difficult than the *Cycadales*, but the difficulties are somewhat similar.

The Vegetative Structures.—The adult stem is harder to cut than *Pinus*, but good sections should be secured by boiling in water and soaking for a few days in glycerin. Transverse sections of the "spur" shoots are easily cut. They have a comparatively large pith and narrow zone of wood, thus contrasting sharply with a long shoot of the same diameter, which has a small pith and wide zone of wood.

The petiole of the leaf and the peduncle of the ovule look alike; but transverse sections show two bundles in the petiole and four in the peduncle. Both cut easily in paraffin.

Spermatogenesis.—The entire staminate cone, even at the time of shedding pollen, can be cut in paraffin. For the latest stages, however, it is better to remove the sporophylls and cut them separately, since the sections must not be thicker than 5 μ, if they are to show the internal structures of the pollen grain.

The young staminate cones become recognizable in June; by September, they have nearly or quite reached the spore mother-cell stage, but the division of the spore mother-cell does not take place until the following April. In these early stages the bud scales should be carefully dissected away before fixing. Pollen is shed early in May. Fix in chromo-acetic acid, with or without a little osmic acid, cut 5 μ thick, and stain in iron-alum haematoxylin. There are four cells in the pollen grain at the time of shedding (Fig. 97).

Pollen tubes and their structures must be studied in sections of the nucellus. Fertilization, in the Chicago region, occurs about the middle of September.

Oögenesis.—Young ovules about 0.25 mm. in length are found about the middle of April; the megaspore mother-cell and its division into four megaspores are found about the first of May; the free nuclear stage in the development of the female gametophyte extends from the first week in May to the first week in July; during July, walls appear; then the archegonium initials and the growth of the archegonium, the ventral canal cell being cut off the second week in September; fertilization, free nuclear division in the sporophyte, and the beginning of walls may all be found before the end of September; cotyledon stages belong to October, and when the seeds fall in November the embryo extends throughout nearly the entire length of the endosperm. This is the winter resting stage, but, planted in the greenhouse, the seeds germinate without any resting period, as in the case of cycads.

For all stages in oögenesis and development of the embryo, use chromo-acetic acid. The free nuclear stages in both gametophyte and sporophyte are almost sure to plasmolyze. The special chromo-acetic-osmic-acid solution is better than one with less acetic acid. Hot alcoholic corrosive sublimate–acetic acid will cause very little shrinking if the pieces are not too large; but figures are not as satisfactory as in chromic material.

For sections of the entire ovule, use safranin, gentian-violet, orange; for free nuclear stages in both gametophyte and sporophyte,

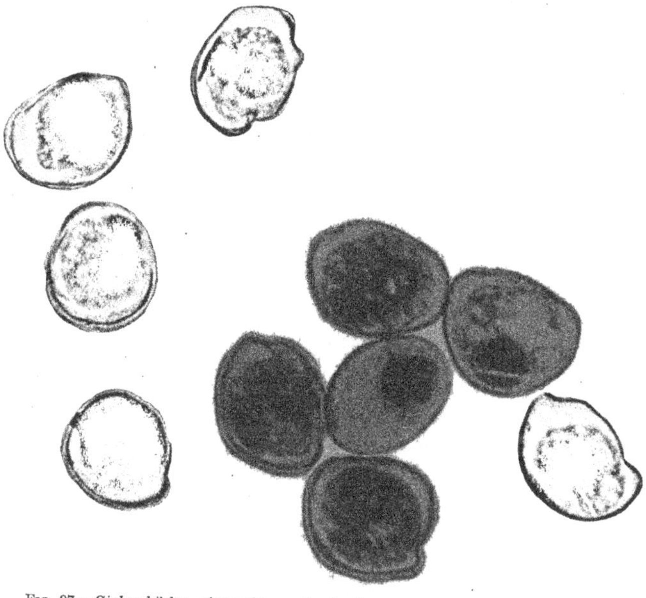

FIG. 97.—*Ginkgo biloba:* photomicrograph of microspores at shedding stage, showing two prothallial cells, generative cell, and tube cell. Stained in iron-alum haematoxylin. Eastman Commercial Ortho film, Wratten B filter (green); Spencer 4-mm. objective, N.A. .66; ocular ×6, arc light; exposure, 3½ seconds. Negative by Dr. P. J. Sedgwick. ×654.

use iron-haematoxylin with a touch of orange; for the megaspore membrane, safranin seems to be the best stain.

GYMNOSPERMS—CONIFERALES

Since *Pinus* is an available laboratory type, we shall describe methods for demonstrating various phases in the life-history of this genus, hoping that the directions will enable the student to experiment intelligently with similar forms. The dates are for *Pinus Laricio* in the vicinity of Chicago, but dates will be different for different species and even for the same species in different regions; *P. Laricio*, at Chicago, sheds pollen about the middle of June, but *P. maritima* at Auckland, New Zealand, sheds its pollen about the first of October. After a year's collecting in any region, there should be no difficulty, since the dates do not vary much from year to year.

The Vegetative Structures.—The stem, root, and leaf will be treated separately.

The stem.—The vascular cylinder is an endarch siphonostele, a type which, with few exceptions, is found throughout the living gymnosperms.

The young stem in its first year's growth is green and soft and is easily cut in paraffin. The best time to collect material is soon after the young shoot has emerged from the bud scales in the spring. Resinous material, like these young stems, do not fix well in aqueous media. With a thin safety-razor blade, cut the stem transversely into pieces about 5 mm. in length; fix in formalin alcohol, imbed in paraffin, and stain in safranin and light green. Longitudinal sections of the buds in winter or early spring condition are instructive for comparison with longitudinal sections of the ovulate cone. Trim away most of the bud scales and cut a slab from opposite sides, leaving a piece 2 or 3 mm. thick to be imbedded. The bud, and also selected pieces of the young stem, will show the structure of the young leaf. Later in the season, even the first year's shoot should be cut without imbedding. The two- and three-year shoots and all older material can be cut freehand, without imbedding, and, preferably, before fixing. Such sections are transferred directly from the knife to 95 per cent alcohol.

For the structure of the adult stem, select a clear board and, for transverse sections, cut out pieces about 2 cm. long and 6 to 10 mm. square; for longitudinal sections, use pieces about 2 cm. long, with 5 and 10 mm. for the other two faces. Cut from the face which will give sections 5×10 mm. Orient carefully, so that the longitudinal radial sections shall be exactly parallel with the rays, and the longitudinal tangential sections exactly tangential to the rays. Leave the sections in 95 per cent alcohol for 15 or 20 minutes before staining. Stain for at least 24 hours in safranin, extract the stain until only a faint red color is left in the cellulose walls, and then stain in Delafield's haematoxylin. Stain some of the sections in safranin and crystal-violet and some in safranin and light green. Of course, every preparation should contain transverse, longitudinal radial, and longitudinal tangential sections.

The safranin and Delafield's haematoxylin should show the bars of Sanio very clearly, as should also the safranin and crystal-violet. If a preparation shows the bars of Sanio and differentiates the torus of the bordered pit, the technic is good (Fig. 98).

Jeffrey's maceration method will isolate the tracheids and other cells, which can then be stained and mounted in balsam. Such preparations show features which might be overlooked in sections.

The root.—The primary root should be studied in the embryo while it is still contained in the seed. Collect material in September, October, or at any later date. If material is collected in winter, the seeds should be soaked in water for a day or two before fixing. In any case, remove the testa and cut a thin slab from opposite sides of the endosperm to facilitate fixing and infiltration. For secondary roots and also for the structure of the stele in the primary root, germinate the seeds and fix material after the hypocotyl has reached a length of 3 or 4 cm. The seeds of *Pinus edulis*, commonly called Piñon, or edible pine, can be obtained in most cities. They are particularly good for a study of the mature embryo and the seedling.

The older roots are treated like the stems. The structure of roots 2 or 3 mm. in diameter is wonderfully regular (Fig. 99).

The leaves.—The leaves of our common gymnosperms cut readily in paraffin while they are young and tender, but as they approach maturity it is a fuitless task to attempt paraffin sections.

Good sections may be obtained in great quantities with little trouble by the following method: Make a bunch of the needles as large as one's little finger, wrap them firmly together with a string, allowing about ¼ inch of the bunch to project above the wrapping; dip two or three times into melted paraffin; and then dip into cold water to harden the paraffin. While there is no infiltration, the paraffin holds the needles in place for cutting. Fasten in a sliding microtome and cut with the knife placed obliquely. Place the sections in water as they are cut and the paraffin can be easily skimmed away. Then fix the sections in 95 per cent alcohol for half an hour; transfer to 70 per cent alcohol, where they should remain for about 5 minutes; then to 50 per cent alcohol, where they should be kept until the green color, due to chlorophyll, disappears. Stain in safranin and light green.

Spermatogenesis.—In October the clusters of staminate cones which are to shed their pollen in the coming spring are already quite conspicuous. The cones should be picked off separately, and the

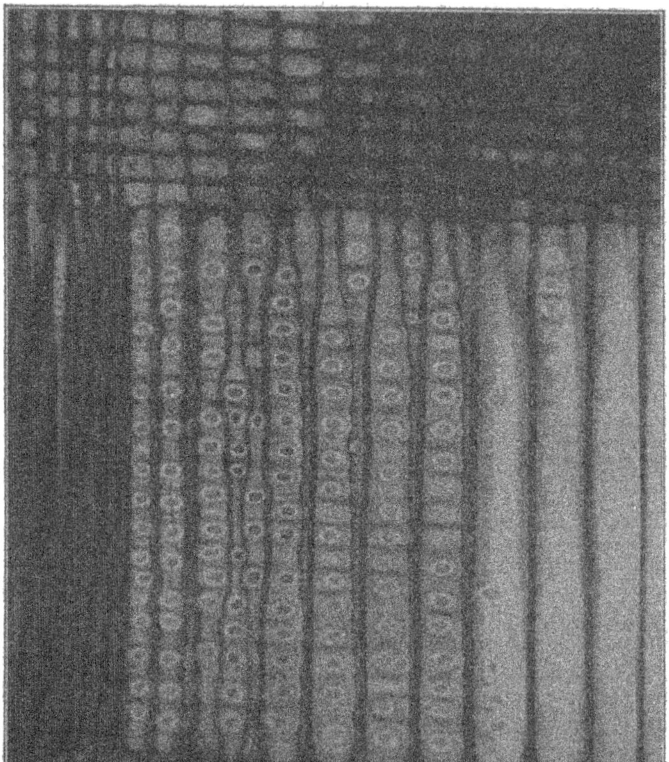

FIG. 98.—*Pinus Strobus:* photomicrograph of longitudinal radial section of mature wood of stem, showing bordered pits and bars of Sanio. Safranin and crystal-violet. Eastman Commercial Ortho film, Wratten B filter (green); Spencer 4-mm. objective, N.A. .65; arc light; exposure, ½ second. Negative by Dr. P. J. Sedgwick. ×166.

scales should be carefully removed so as to expose the delicate greenish cone within. At this time the sporogenous cells are easily distinguished. Material collected in January, or at any time before growth is resumed in the spring, shows about the same stage of development. If it is desired to secure a series of stages with the least possible delay, a branch bearing numerous clusters of cones may be brought into the laboratory and placed in a jar of water. Growth is more satisfactory in case of branches broken off in the

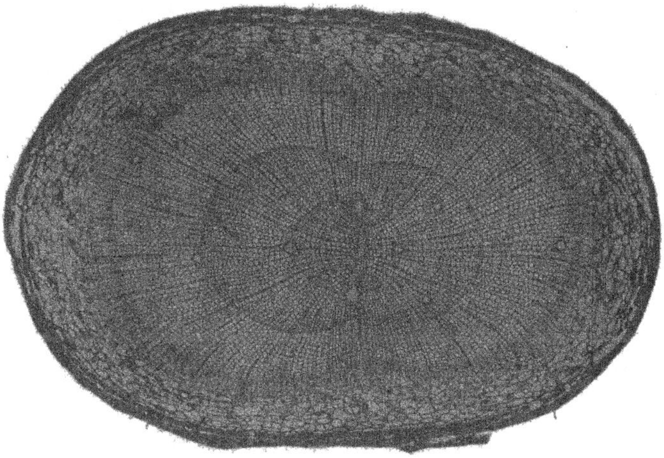

FIG. 99.—*Picea nigra:* photomicrograph of transverse section of root. Fixed in alcohol and stained in safranin and Delafield's haematoxylin. Eastman Commercial Ortho film, Wratten B filter (green); J. Swift and Son 1-inch objective; arc light; exposure, ¾ second. Negative by Dr. P. J. Sedgwick. ×28.

winter than in those brought in before there has been any period of rest. The material can be examined from time to time, and a complete series is easily secured. The mitotic figures in the pollen mother-cells furnish exceptionally instructive preparations. The two mitoses take place during the last week in April and the first week in May. Staminate cones which will yield mitotic figures can be selected with certainty by examining the fresh material. Crush a microsporangium from the top of the cone and one from the bottom, add a small drop of water and a cover to each, and examine. If there are pollen tetrads at the bottom, but only undivided spore mother-cells at the top, it is very probable that longitudinal sections of the cone will yield the figures. If a drop of methyl green be allowed to run under the cover, it will enable one to see whether figures are present or not. When desirable cones are found, slabs should be cut from two sides, in order that the fixing agent may penetrate more rapidly and that infiltration with paraffin may be more thorough.

The later stages, showing the germination of the microspores, furnish better sections if the cones are cut transversely into small pieces about 5 mm. thick. It is very easy to get excellent mounts of the pollen just at the time of shedding, which, in *Pinus Laricio* in the vicinity of Chicago, occurs near the middle of June. Shake a large number of cones over a piece of paper, thus securing an abundance of material; slide the pollen off from the paper into a bottle half-full of water and shake a little to wet the pollen grains, because pollen of wind-pollinated plants is likely to be more or less shriveled at the time of shedding. A few minutes in water will make the pollen turgid. Fix in formalin-alcohol-acetic acid. If material is so abundant that you can lose much of it and still have plenty left, fix in chromo-acetic-osmic acid. Staining, especially the staining of mitotic figures, is likely to be more brilliant after fixing in the chromic series. However, most of the mitoses take place before the pollen is shed or after it reaches the nucellus. Infiltration in the bath will not require more than 30 minutes. When the infiltration is complete, pour out into a paper tray or an imbedding dish, just warm enough to allow the pollen to settle to the bottom. A layer of pollen 3 mm. thick, with enough paraffin to make the cake about 5 mm. thick, will be satisfactory for cutting. Or, the paraffin may be poured out upon a piece of cold glass. Still another method is to leave the paraffin in a shell or vial during infiltration in the bath, and then let it cool in the bottle. After the paraffin is hard, break the bottle. Break the bottle carefully, cut off the lower portion of the paraffin containing the pollen, mount it on a block in the usual manner, and trim away some of the paraffin so that two parallel surfaces will make the sections ribbon well. Sections should not be thicker than 5 μ. Material in this stage shows a large tube nucleus, a somewhat lenticular (generative) cell with a more deeply staining nucleus, and, lastly, two small prothallial cells quite close to the spore wall. The prothallial cells cannot always be detected at this stage, and there may be some doubt as to whether two such cells are always present. The division of the lenticular cell into "stalk cell" and "body cell," and also the division of the body cell into the two male cells, must be looked for in sections of the nucellus of the ovule.

Abies balsamea is a better type for illustrating spermatogenesis, since the pollen mother-cells and the pollen grains are much larger

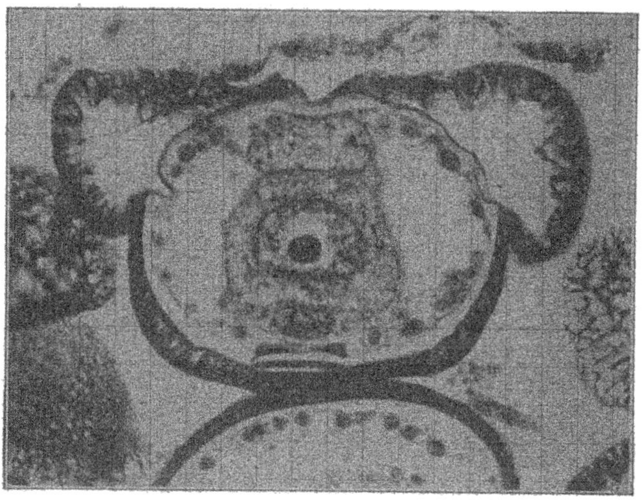

FIG. 100.—*Abies balsamea:* photomicrograph of pollen; one complete section showing two prothallial cells, the stalk cell, generation cell, and tube cell with starch grains. Fixed and cut while still within the microsporangium. Stained in safranin and gentian-violet. Eastman Commercial Ortho film, Wratten B filter (green); Spencer 4-mm. objective, N.A. .66; ocular ×6; arc light; exposure, 6 seconds. Preparation by Dr. A. H. Hutchinson, negative by Dr. P. J. Sedgwick. ×784.

and the division of the generative cell into the "stalk" and "body" cells takes place before the pollen is shed (Fig. 100).

Araucaria and *Agathis* are the best forms for illustrating numerous prothallial cells. *Podocarpus* and *Taxodium* are also good. *Thuja* or *Juniperus* may be used to illustrate the entire absence of prothallial cells, a very advanced condition; while both these genera have highly developed sperms, like those of the cycads and *Ginkgo,* except that they lack cilia. This is a good illustration of the fact that one structure may advance while another remains primitive.

Oögenesis.—In *Pinus Laricio* the rudiment of the ovulate strobilus, which is to be pollinated in June, can be detected in the preceding October. The collection of this stage is very uncertain, because there seems to be no mark distinguishing buds containing ovules from buds which are only vegetative. By collecting numerous buds from the tops of vigorous trees which are known to produce an abundance of strobili, a few buds containing the desired stages may be obtained. In May, after the strobili break through the bud scales, material is easily collected. Up to the time of pollination the entire ovulate strobilus cuts easily in paraffin. Longitudinal sections of the cone at this time give good views of the bract and ovuliferous scale bearing the ovules. The integument is very well marked, and in the nucellus one or more sporogenous cells can usually be distinguished. As soon as the scales close up after pollination, the cone begins to harden and soon makes trouble in cutting. Even before the scales close up, it is better to cut a slab from opposite sides of the cone; after the scales close, it is almost a necessity. For sections of the whole cone, fix in formalin alcohol. Dr. Hannah Aase succeeded in cutting complete series of paraffin sections from cones of *Pinus Banksiana* more than 2 cm. in length. She fixed them in formalin alcohol, and used prolonged periods in dehydrating, clearing, and infiltrating. Land's bichromate of potash and glue fixative was used in fixing the sections to the slide. Such series of sections of large cones were necessary for an investigation of the vascular anatomy.

For a study of the ovule and the structures within it, better preparations will be obtained by carefully cutting off the pair of ovules from the scale. In free nuclear stages of the female gametophyte, which begin in the autumn, are interrupted by winter, and are completed in May, plasmolysis is likely to occur. After walls appear there is less danger. From the middle of May to the first of July collections should be made at intervals of two or three days, since during these six weeks the gametophyte completes the free nuclear stage and develops cell walls, the archegonium completes its entire development, the egg is fertilized, and the sporophyte may reach the suspensor stage.

At the stages shown in Figure 101*B–D*, it is a good plan to remove the female gametophyte with its proembryos from the ovule; but at the stages shown in Figures 101*A* and 102 the pollen tubes with their contents are rapidly working their way through the nucellus toward the archegonia, and consequently, in some of the material, it is better to retain enough of the tissues of the ovule to keep the nucellus in place. In later stages, after fertilization has taken place, the developing testa should be removed with great care, for a very slight pressure is sufficient to injure the delicate parts within.

With any fixing agent of the chromic-acid series, the free nuclear stages of the female gametophyte and, later, the archegonia and proembryo stages like those shown in Figure 101, are likely to show

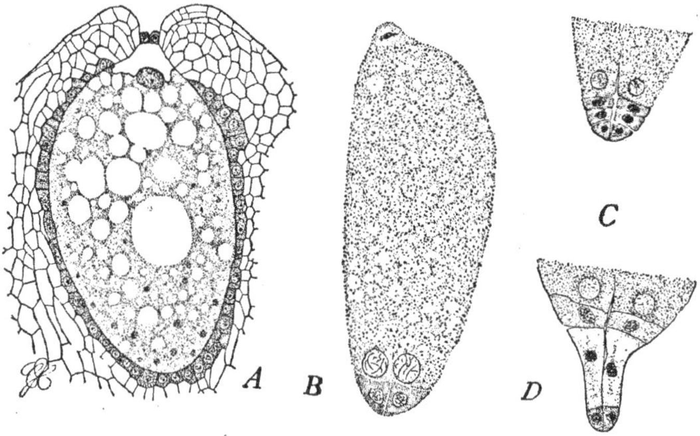

FIG. 101.—*Pinus Laricio:* *A*, top of prothallium with an archegonium just before the cutting off of the ventral canal cell; fixed in Flemming's weaker solution and stained in Haidenhain's iron-alum haematoxylin; collected June 18, 1897; *B*, *C*, and *D*, early stages in the development of the embryo; fixed in chromo-acetic acid and stained in safranin, gentian-violet, orange; collected July 2, 1897. ×104.

plasmolysis. Stages like those shown in Figure 101*B–D*, without any shrinking, were secured by Miss Ethel Thomas by using hot alcoholic corrosive sublimate–acetic acid with formalin (4 g. corrosive sublimate, 5 c.c. acetic acid, 5 c.c. formalin, 100 c.c. 70 per cent alcohol). Figures like that shown in Figure 101 are better in chromo-acetic–osmic acid, but, in general, we should recommend the corrosive sublimate solution.

The period at which the various stages may be found varies with the species, the locality, and the season. In *Pinus Laricio* the megaspore mother-cells appear as soon as the young strobili break through the bud scales. At Chicago, in the season of 1897, material collected May 27 did not yet show archegonia; the ventral-canal cell was cut off about June 21 (see Fig. 102), the fusion of the egg

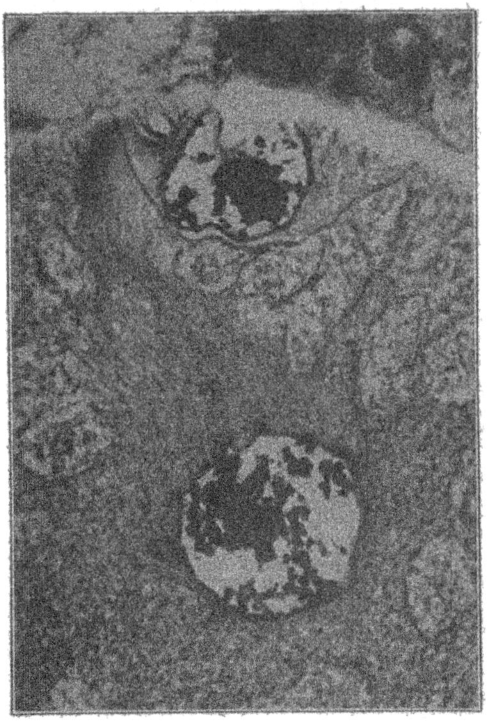

FIG. 102.—*Pinus Laricio:* photomicrograph showing the formation of the ventral canal cell; usually this cell is not so large in proportion to the egg; fixed in Flemming's weaker solution and stained in safranin, gentian-violet, orange; the preparation was made in 1897, the photomicrograph in 1915; Cramer contrast plate; 4-mm. objective; ocular ×4; Abbé condenser; yellowish-green filter and also a strong filter used in outdoor photography; camera bellows, 75 cm.; arc light; exposure, 7 minutes. Negative by Miss Ethel Thomas. ×587.

and sperm nuclei occurred about a week later, and stages like Figure 101*B*, *C*, and *D*, were common in material collected July 2. In the season 1896 all the stages appeared about 2 weeks earlier. In *Pinus sylvestris* the stages appeared a little earlier than in *Pinus Laricio*.

After the stage shown in Figure 101*A* has appeared, it is necessary to collect every day until the stage shown in Figure 101*D* is reached. If collections are made at intervals of 3 or 4 days, the most interesting stages, like the cutting off of the ventral-canal cell, fertilization, and the first divisions of the nucleus of the oöspore, may be missed altogether. It should be mentioned that all the ovules of a cone will be in very nearly the same stage of development; consequently, it is worth while to keep the ovules from each cone separate. Stages like that shown in Figure 102 are rare in miscellaneous collections, but if ovules from each cone are kept separate and this figure is found, the rest of the ovules from that cone will be likely to show some phase of this interesting mitosis.

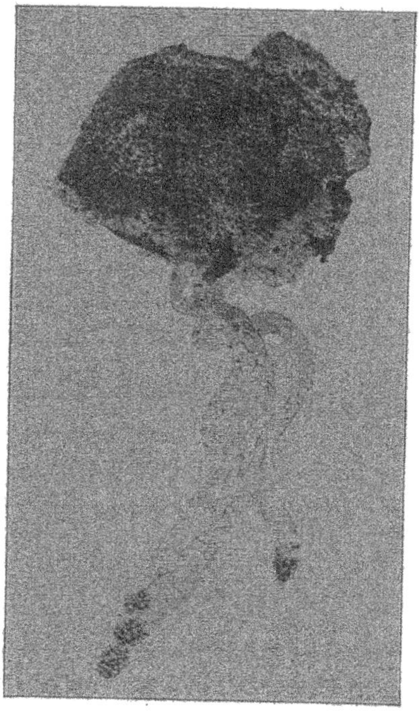

FIG. 103.—*Pinus Banksiana:* photomicrograph of young embryos teased out by the method described in the text; from a preparation by Dr. J. T. Buchholz; Cramer contrast plate; 16-mm. objective; no ocular or Abbé condenser; camera bellows, 75 cm.; safranin filter; arc light; exposure, 17 seconds. Negative by Miss Ethel Thomas. ×54.

Thuja and *Juniperus* are good types to illustrate the archegonium complex and the large, highly organized male cells. In *Thuja* a series from the appearance of archegonium initials to young embryos may be collected between June 10 and June 20. In *Juniperus* pollination occurs late in May and fertilization takes place 12½ months later. The megaspores are formed late in April and the development of the female gametophyte occupies about 6 weeks.

The Embryo.—The early stages of the sporophyte, usually designated as the proembryo, have been mentioned already.

From the time when the suspensors begin to elongate up to the appearance of cotyledons, instructive preparations can be made by mounting the embryo whole. Dr. J. T. Buchholz has developed a method for handling these small objects. Remove the testa and then, under water, hold the endosperm gently with forceps and press the neck and upper part of the archegonium with a needle, pressing, and at the same time drawing the needle away, so as to pull the young embryo out. Some of the embryos will be broken, but by careful manipulation more than half should be entirely uninjured. Fix in formalin (5 per cent in water), stain in Delafield's haematoxylin, transfer to 10 per cent glycerin, and continue with the Venetian turpentine method. A preparation made in this way is shown in the photomicrogaph (Fig. 103).

These stages, and all subsequent stages, are easily cut in paraffin without removing the embryo from the endosperm. Cut a thin slab from opposite sides of the endosperm, fix in chromo-acetic acid, with or without a little osmic acid, imbed in paraffin, and stain in safranin and gentian-violet. This will give a good view of the abundant starch and other food stuff, and at the same time will bring out sharply the cell walls of the embryo.

GYMNOSPERMS—GNETALES

Of the three peculiar genera belonging to this order only one, *Ephedra*, occurs in the United States. *Welwitschia* is found only in Damaraland, South Africa, and *Gnetum* is tropical and subtropical. *Gnetum* thrives in the greenhouse, but the other two have not been cultivated successfully. All show vessels in the secondary wood, an angiosperm character. The strobili can be fixed in formalin-alcohol-acetic acid, but in *Ephedra* the dry chaffy scales must be dissected away before completing the dehydration and infiltration with paraffin. If you can secure material of *Ephedra*, Dr. Land's researches[1] present a very complete life-history, with dates of various stages and suggestions for fixing and staining.

[1] *Botanical Gazette*, **38**:1–18, 1904, and **44**:273–292, 1907.

CHAPTER XXV

SPERMATOPHYTES

ANGIOSPERMS

If one should master the technic for all features of this group, he would be prepared to deal with all the rest; for embryo sacs are so delicate that they are as difficult as the free nuclear stage in the female gametophyte of a gymnosperm, while the peach stone needs a petrotome rather than a microtome. Between these extremes there is everything imaginable in structure, chemical composition, and consistency.

Some hints will be given, but the student will gradually learn what should be cut freehand and what should be imbedded; what stages in floral development, what stages in the development of the embryo sac, or what stages in spermatogenesis are likely to be correlated with easily recognized field characters, and what fixing agents are likely to give the best results with various kinds of material.

The Vegetative Structures.—In stems, roots, and leaves the more delicate structures should be imbedded in paraffin and the more rigid structures should be cut without imbedding at all; but it should be remembered that the range of structures which can be imbedded in paraffin can be increased by the use of hydrofluoric acid; and also that hydrofluoric acid or cellulose acetate will extend almost indefinitely the range of structures which can be cut without imbedding.

The stem.—The vascular cylinder of the angiosperms is either an endarch siphonostele, or a polystele derived from it. For a study of the development of the stem, the common geranium (*Pelargonium*) may be recommended. Near the base of a fresh stem, about 1 cm. in diameter, cut freehand sections and fix them in 95 per cent alcohol for 10 to 20 minutes; transfer to 70 per cent alcohol to extract the chlorophyll, and then stain in safranin and light green. Such sections will show both primary and secondary structures in the stele and in the cortex. Higher up, there will be secondary structures only in the stele; and still higher up will be found the origin of interfascicular cambium. All these can be cut without imbedding, but the earlier stages showing the differentiation of protoxylem, metaxylem, and the origin of secondary xylem are too soft for successful freehand sections. Fix in alcohol-formalin–acetic acid (10 c.c. formalin, 5 c.c. acetic acid, and 100 c.c. of 50 per cent alcohol) and imbed in paraffin.

For a study of woody stems, *Tilia americana* (basswood) is good, and shoots from 5 to 10 mm. in diameter are easy to cut. Very hard stems like *Hicoria* (hickory) and *Quercus* (oak) must be boiled and treated with hydrofluoric acid, if you expect to cut shoots more than 5 to 7 mm. in diameter. However, with a good sharp knife and a rigid microtome much larger sections can be cut without resorting to hydrofluoric acid. Of course, veneer machines cut very large and fairly thin, smooth sections from the most refractory woods.

While a random selection of stems would furnish material for practice in technic, we suggest that the stem of *Clintonia* shows a good siphonostele in a monocotyl; the rhizome of *Acorus calamus* is a good type for the amphivasal bundle and, although a monocotyl still shows a differentiation into stele and cortex; *Zea Mais*, universally used but not characteristic of monocotyls, shows scattered bundles, but not the amphivasal condition; *Aloe*, *Dracaena*, or *Yucca* will illustrate secondary wood in monocotyls. *Iris* has a highly developed endodermis in the rhizome; and *Nymphea* or *Nuphar* will show scattered bundles in a dicotyl.

Lenticels and *tyloses* are abundant and typical in *Menispermum*, and very thin sections can be cut without imbedding; but both these structures are well developed while the stem can still be cut in paraffin without previous treatment in hydrofluoric acid.

The sieve tubes of the phloem are easily demonstrated in *Cucurbita Pepo*, the common pumpkin; other members of the family furnish good material. Take pieces of stem about 1 cm. long and not too hard to cut in paraffin, fix in formalin alcohol, and stain in safranin, gentian-violet, orange. The tropical *Tetracera*, one of the Dilleniaceae, has sieve plates so large that they are easily seen with a pocket lens. The phloem area is so large in the larger stems that it can be cut out for imbedding in paraffin long after the entire stem has become too hard for paraffin sections. Safranin and gentian-violet is a good stain for sieve tubes. It was once thought that these large sieve tubes afforded an obvious illustration of the continuity of protoplasm; but, as a matter of fact, the actual protoplasmic connections are scanty and hard to demonstrate. Iron-alum haematoxylin and orange will differentiate the strands if you are careful.

Roots.—It has long been known that the root-tip furnishes constantly available material for a study of mitosis (Fig. 104). An onion thrown into a pan of water will soon send out numerous roots. Soak beans in water for several hours and then plant them in loose, moist sawdust. In a greenhouse, with "bottom heat," the primary root will be long enough in 2 or 3 days. The large, flat beans, especially *Vicia Faba,* are very favorable. The root-tips of *Trillium grandiflorum, Tradescantia virginica, Podophyllum peltatum, Arisaema triphyllum,* and *Cypripedium pubescens* may be mentioned as known to be favorable; but it is very possible that the best root-tip has not yet been tried.

Cell division does not proceed with equal rapidity at all hours of the day. Kellicott has shown that in the root-tips of *Allium* there are in each 24 hours two periods at which cell division is at the maximum, and two at which it is at the minimum. The maximum periods are shortly before midnight (11:00 P.M.), and shortly after noon (1:00 P.M.). The minima, when cell division is at the lowest ebb, occur about 7:00 A.M. and 3:00 P.M. When cell division is most vigorous, there is little elongation, and when cell division is at the minimum, cell elongation is at the maximum. Consequently, root-tips of *Allium* should be collected about, 1:00 P.M. or 11:00 P.M. Lutman, later, made observations upon periodicity of mitosis in the desmid, *Closterium;* and in 1915, Karsten made a comparatively extended study of periodicity in various stems and roots, together with notes on algae.

It is safe to say that the maximum number of mitoses in root-tips will be found shortly after noon (1:00 P.M.) and shortly before midnight (11:00 P.M.) (Fig. 104). It is certain, however, that abundant mitoses may be found at other times—even at 3:00 P.M.—in sporangia of ferns, in anthers of angiosperms, in endosperm, and in free nuclear stages of the embryo of gymnosperms.

Mitotic figures play such an important part in the development of the plant and in modern theories of heredity that it is worth while to acquire a critical technic in fixing and staining these structures. Use the various fixing agents—Flemming's weaker solution, chromo-acetic acid with or without a little osmic, Benda's fluid, Bouin's fluid, corrosive sublimate with acetic acid, and any others. Make

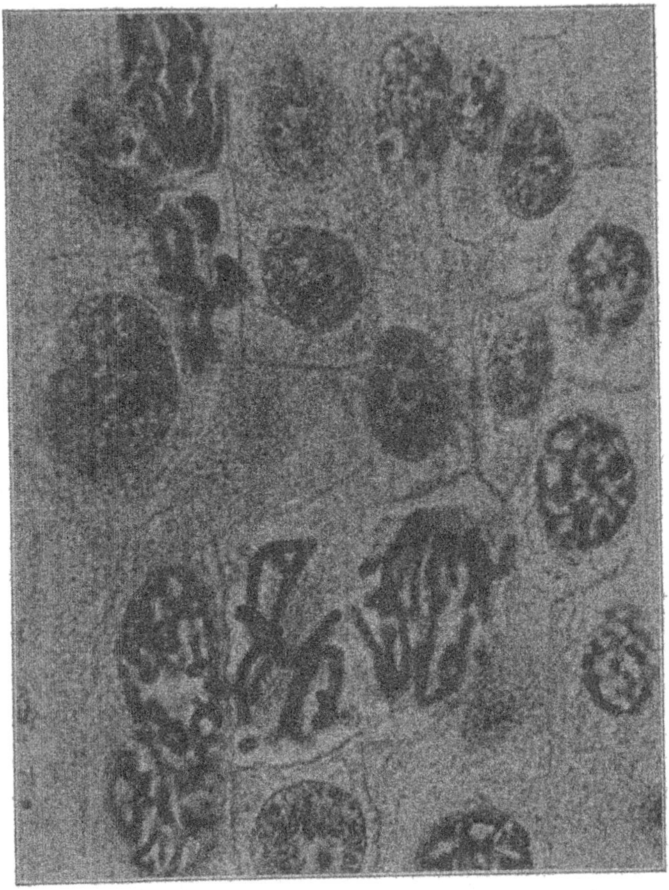

FIG. 104.—*Trillium grandiflorum:* photomicrograph of longitudinal section of root-tip, fixed in chromo-acetic-osmic acid at 1:00 P.M., and stained in iron-alum haematoxylin. Eastman Commercial Ortho film, Wratten H filter (blue); Spencer 2-mm. fluorite objective; ocular ×6; arc light; exposure, 12 seconds. Negative by Dr. P. J. Sedgwick. ×1,174.

yourself master of Haidenhain's iron-alum haematoxylin; then add the safranin, gentian-violet, orange combination; then safranin and anilin blue; and then experiment for yourself, but remember that the triumphs of modern cytology have been won with iron-haematoxylin and that you cannot read intelligently the literature of the past twenty-five years until you have gained at least an approximate mastery of this stain. Of course, dehydration, clearing, and infiltration must be very gradual. The schedules by Yamanouchi and by Sharp, on pages 45 and 46, will repay careful study.

In staining with safranin, gentian-violet, orange, allow the alcoholic safranin to act for 16 to 24 hours; then extract it with 50 per cent alcohol, slightly acidulated with hydrochloric acid, if necessary, until the stain has almost disappeared from the spindle; then pass through 70, 85, 95, and 100 per cent alcohol; stain in gentian-violet dissolved in clove oil, or in a mixture of clove oil and absolute alcohol, for 5 to 20 minutes; follow with orange dissolved in clove oil, remembering that this will weaken the safranin and sometimes the gentian-violet; finally use pure clove oil to differentiate the gentian-violet. Leave the slide in xylol for 2 to 5 minutes to remove the clove oil and to hasten the hardening of the balsam.

If you use aqueous gentian-violet or crystal-violet, after the safranin is satisfactory, transfer to water and then to the violet. After staining in violet, dip in water to remove the excess of stain and then dehydrate rapidly in 95 per cent and absolute alcohol, differentiate in clove oil, and then transfer to xylol.

The structure and development of the young root will be shown, to some extent, in preparations made for mitotic figures. The origin of dermatogen, periblem, plerome, and also of protoxylem, is well shown in *Zea Mays*. An ear of sweet corn, as young and tender as you can find on the market, will furnish material. Cut out from the grain a rectangular piece about 2×3 mm. and 4 or 5 mm. long; if you want to show also the structure of the entire grain, take a section the entire length of the grain, perpendicular to the flat side of the grain, and about 2 mm. wide. Cut the latter longitudinally; the rectangular pieces are sufficient for transverse sections. Fix in chromo-acetic acid. The roots of *Hordeum vulgare* (barley) might also be suggested.

The roots of *Ranunculus repens* and *Sambucus nigra* furnish good illustrations of the radial arrangement of xylem and phloem.

Smilax shows the radial arrangement, with a large number of poles and a very highly differentiated endodermis. The origin of secondary xylem and phloem is well shown in *Sambucus nigra*. *Vicia Faba* shows very clearly the origin of secondary roots. *Pistia stratiotes*, although not so generally available, is splendid for showing the origin of secondary roots. The arrangement of cells in the young

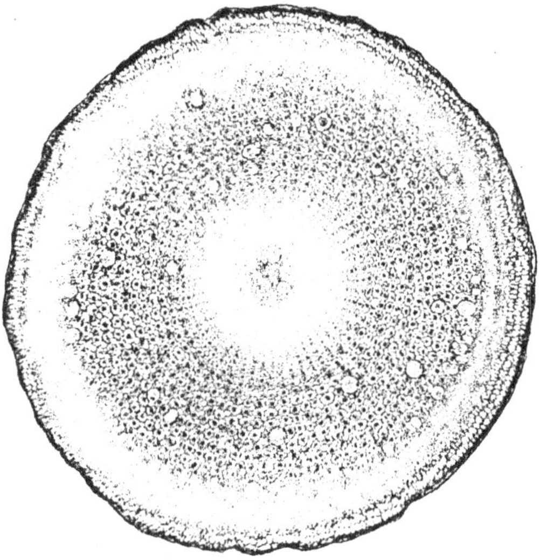

FIG. 105.—*Sparganium eurycarpum:* photomicrograph of transverse section of young root; fixed in chromo-acetic acid and stained in Bismarck brown; Cramer contrast plate; 16-mm. objective; ocular ×4; no Abbé condenser; yellowish-green filter; camera bellows, 1 meter; exposure, 8 seconds. ×90.

roots of aquatic or semi-aquatic plants is often extremely regular (Fig. 105).

The leaf.—Young and tender leaves should be fixed in formalin alcohol and cut in paraffin. Cut sections freehand whenever there is sufficient rigidity. Resort to pith only when necessary. In cutting sections of a leaf like that of *Lilium*, lay one leaf on another until you have a bundle of them which will be nearly square in transverse section. Wrap the bundle with string for about 15 mm.; cut the bundle transversely so that about 5 mm. of the bundle will project beyond the tied portion. Dip in melted paraffin, as already suggested for *Pinus*, fasten the tied portion in the sliding microtome, and cut with the knife placed obliquely. About 15 to 20 μ is a good thickness for general leaf structure. In case of large leaves, cut 1 cm. wide and 3 cm. long and tie them together to make a good bundle for cutting.

Of course, for the finest preparations, imbed in paraffin. The common lilac, *Syringa*, has a good leaf to illustrate palisade and spongy parenchyma; the privet, *Ligustrum*, is also excellent.

Buds will furnish beautiful preparations of young leaves and, at the same time, will show the vernation. Cut the bud transversely, a little above the middle; remove the bud scales, if they promise to cause trouble; retain only enough tissue at the base of the bud to hold the parts in place. Fix in formalin alcohol and stain in safranin and light green.

Epidermis stripped from the leaf, fixed in 10 per cent formalin in water for a day or two, and then stained in safranin and anilin blue, will give excellent views of stomata. The development of *stomata* is particularly well shown in *Sedum purpurascens*, even in leaves which have reached the adult size. The epidermis is very easily stripped from a leaf of *Sedum*. If the big *Sedum maximum* is available, pieces of epidermis 6 or 7 cm. long and 2 or 3 cm. wide are easily stripped off, almost free from any underlying tissue. The epidermis of *Lilium* and *Tradescantia* show fine, large stomata, but it is not so easy to strip off. In these two genera the stomata, as in nearly all leaves, show only the adult structure.

Floral Development.—For a study of floral development very young buds are necessary, and it is best to select those forms which have rather dense clusters of flowers, in order that a complete series may be obtained with as little trouble as possible.

The usual order of appearance of floral parts is (1) calyx, (2) corolla, (3) stamens, and (4) carpels; but if any of these organs is reduced or metamorphosed, their order of appearance may be affected.

Floral development is easily studied in the common *Capsella bursa-pastoris*. The best time to collect material is late in March or early in April. Dig up the plant, carefully remove the leaves, and in the center of the rosette a tiny white axis will be found. A series of these axes from 3 to 9 mm. in length, and from 1.5 to 3.5 mm.

in diameter will give a very complete series of stages in the development of the floral organs. Preparations from the apex of the shoot taken after the inflorescence appears above ground are not to be compared with those taken early in the season, because the pedicels begin to diverge so early that median longitudinal sections of the flowers are comparatively rare. Fix in chromo-acetic acid and stain in Delafield's haematoxylin. The sections should be longitudinal and about 5 μ thick. *Capsella* shows the hypogynous type of development. The order of appearance of floral parts is (1) calyx, (2) stamens, (3) carpels, and (4) petals. The ovary is compound (syncarpous).

Ranunculus, which is also hypogynous, will illustrate the development of the simple (apocarpous) ovary. The ovules appear quite early, so that the archesporial cell, or even the megaspores, may be seen while the carpel is still as open as in any gymnosperm. The whole structure is a simple strobilus.

Rumex crispus (Yellow Dock) is also a good hypogynous type, and the densely clustered flowers afford a fine series of stages. Besides, in transverse sections, the early stages in spermatogenesis are very clear.

In the willows, *Salix*, the bud scales must be removed and the copious hairs should be trimmed off as much as possible with scissors, after which the catkin should be slabbed a little on opposite sides to facilitate penetration. This is a fine illustration of a compound strobilus.

The cat-tail, *Typha*, presents a simple type of floral development. The leaves should be dissected away long before the flowers can be seen from the outside. The cylindrical clusters, varying in diameter from 2 or 3 mm. up to the size of one's finger, will afford a complete series of stages. Until the spadix reaches the diameter of a lead pencil, transverse sections are easily cut. For later stages, the outer part of the spadix should be sliced off so that only enough spadix is retained to hold the florets in place.

Prunus and many other members of the Rosaceae furnish examples of the perigynous type of development. In many of them the floral parts do not occur in the usual succession.

The epigynous type is well shown in the Compositae. The order of appearance is (1) corolla, (2) stamens, (3) carpels, and (4) calyx (pappus).

The common dandelion, *Taraxacum officinale*, affords an excellent series with little labor. Examine vigorous plants which have, as yet, no flowers or buds in sight. Dig up the plant and dissect away the leaves. If there is a white cluster of flower buds, the largest not more than 4 mm. in diameter, cut out the cluster, leaving only enough tissue at the base to hold the buds in place. Larger heads should be cut separately.

Our most common thistle, *Cirsium lanceolatum*, shows the floral development with unusual clearness, but the preparation of the material is somewhat tedious. The involucre, which is too hard to cut, must be carefully dissected away. Retain only enough of the receptacle to hold the developing florets in place. A series of sizes with disks varying from 3 to 10 mm. in diameter will show the development from the undifferentiated papilla up to the appearance of the archesporial cell in the nucellus of the ovule. The Canada thistle, *Cirsium arvense*, is equally good, but it is more difficult to dissect out the desirable parts. In the common sunflower, *Helianthus annuus*, the young floral parts, like the mature head, are so very large that a satisfactory study may be made with a low-power objective. As in the case of the thistle, the involucre must be trimmed away and only enough of the receptacle retained to hold the florets together.

Erigeron (we have cut *E. philadelphicus* and *E. annuus*) furnishes a beautiful example of epigynous floral development, and the heads are so densely clustered that, in a single section, one may find various stages from heads with undifferentiated disk up to heads with florets showing pappus, corolla, stamens, and carpels.

Spermatogenesis.—The earlier stages in spermatogenesis will be found in the preparations of floral development. The origin of the archesporium, the origin of sporogenous tissue, and the formation of the tapetum are beautifully shown in longitudinal and in transverse sections of the anthers of *Taraxacum* and many other Compositae. Transverse sections of the head of *Taraxacum* or any similar head at the time when pollen mother-cells are rounding off in the center of the head, will show various stages from the mother-cells in the center to the tetrads of spores at the periphery. Transverse sections of the anther of *Polygala* give exceptionally well-defined views of the archesporial cells and sporogenous areas.

Lilium, Trillium, Galtonia, Iris, Tradescantia, Vicia, and *Podophyllum* can be recommended for demonstrating the nuclear changes involved in the formation of spores from the mother-cell (Fig. 106). Several species of *Lilium* are common in greenhouses, and these may be used where wild material is not available. In early stages, where the sporogenous cells have not yet begun to round off into spore mother-cells, it is sufficient to remove the perianth, retaining just enough of the receptacle to hold the stamens in place. Transverse sections show the six stamens and also the young ovary. After the

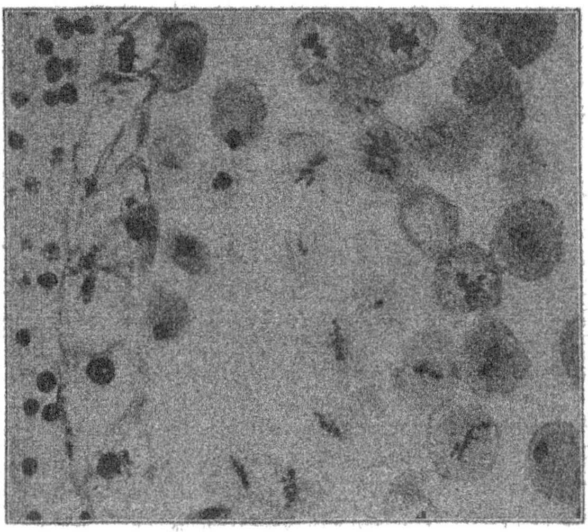

FIG. 106.—*Lilium* candidum: photomicrograph of mitosis in pollen mother-cells; in one of the pollen mother-cells the twelve chromosomes can be counted; from a preparation by F. L. Pickett. Negative by Miss Ethel Thomas. ×260.

spore mother-cells have begun to round off, each stamen should be removed so as to be cut separately. In securing the desirable stages showing the division of the mother-cell into microspores, much time and patience will be saved by determining the stage of development before fixing the material. Mitosis is more or less simultaneous throughout an anther. Long anthers are particularly favorable, since they may show a very closely graded series of the various phases of mitosis. An anther of *Lilium* may show mother-cells with nuclei in synapsis at the top, while the mother-cells at the bottom have reached the equatorial plate stage of the first division; or, the mother-cells at the top may show the first division, while those at the bottom show the second. Determine the stage by examining a few mother-cells before fixing.

From what has been said, it is evident that longitudinal sections should be cut to show mitosis. Transverse sections should be cut to show the general structure of the anther. It is not necessary to cut the stamens into pieces before fixing, since they are easily penetrated and infiltrated; in later stages the stamens *must* not be cut into pieces, since the pollen grains and even the pollen mother-cells are easily washed out.

The problem of fixing spore mother-cells has received much attention. In fixing mother-cells and the two mitoses by which spores are formed, investigators have used almost exclusively the chromo-osmo–acetic acid solutions of Flemming, some preferring the weaker solution and some the stronger. These solutions have been used in nearly all of the work of the Bonn (Germany) school. Osterhout experimented with forty fixing agents, and then concluded that Flemming's stronger solution was the best. Professor Grégoire and his students have made this their principal fixing agent.

In the third edition of this book we stated that, in spite of the weight of authority, we believed the value of solutions with such a large proportion of osmic acid had been overestimated and we suggested that half the amount of osmic recommended in Flemming's weaker solution would be sufficient. The weaker solution has 10 c.c. of 1 per cent osmic acid in 100 c.c. of the fluid. After an examination of thousands of figures in both reduction and vegetative divisions, we feel sure that 5 c.c. of the osmic is enough and, perhaps, 3 or 4 c.c. would be better. The formula for Flemming's weaker solution is often given as follows:

A	Chromic acid (1 per cent)	25 c.c.
	Glacial acetic acid (1 per cent)	10 c.c.
	Water	55 c.c.
B.	Osmic acid (1 per cent)	10 c.c.

Keep the mixture A made up, and add B as the reagent is needed for use, since the solution does not keep well. One seldom uses this reagent in large quantities. About 40 c.c. is as much as one is likely to need for any collection of anthers or root-tips. Take 36 c.c. of A and 4 c.c. of B. It will be worth while to try 36 c.c. of A and 2 c.c.

of B, or even 1 c.c. of B. If the regular formula is used, we should let it act for an hour, and then replace it by A, without any osmic acid. The osmic acid undoubtedly accelerates the killing of the protoplasm. This is seen more readily in animals. If *Cyclops* be brought into 30 c.c. of the solution A, the animals will swim for awhile; if 5 or 6 drops of 1 per cent osmic acid be added to the solution, the animals cease their movements almost instantly. Doubtless the osmic acid has the same effect upon plant protoplasm. Where fixing is slow, very few mitotic figures are found with the chromosomes midway between the equator and the poles. The addition of 10 drops of 1 per cent osmic acid to 50 c.c. of the solution just mentioned will secure as large a proportion of anaphases as solutions which are stronger in osmic acid, and there is no disagreeable blackening.

Farmer and Shove, in studying these mitoses and also vegetative mitoses in *Tradescantia*, secured better results with a mixture of 2 parts of absolute alcohol and 1 part glacial acetic acid. They allowed the fixing agent to act 15 to 20 minutes, then washed in absolute alcohol, and imbedded by the usual methods. This proportion of acetic acid seems entirely too large for any accurate work with chromatin, and we doubt whether the structure of the cytoplasm is normal when so much acetic acid is used.

The entire pollen mother-cell may be stained and mounted without sectioning. Two descriptions of technic appeared in 1912, one by Mann and the other by Pickett. Mann removes the pollen mother-cells before fixing and staining; Pickett fixes and stains the anther *in toto* and teases out the pollen mother-cells just before mounting.

In Mann's method, the anther is placed in a drop of water and the tip is cut off; a gentle tapping with a needle will then cause the pollen mother-cells to float out into the drop. Fix in Bouin's fluid, 4 to 8 hours, wash in 50 per cent alcohol until no color remains, and then stain in iron-haematoxylin. At this stage we should put the material into 10 per cent glycerin and follow the Venetian turpentine method.

Pickett fixed entire anthers in chromo-acetic acid for 30 hours, washed in water for 24 hours, and then passed up to 80 per cent alcohol. At this point, he stained in strong cochineal or Kleinenberg's haematoxylin for 5 days, then completed the dehydration, cleared in cedar oil, teased out the mother-cells, and mounted in balsam.

In dealing with the whole anther, it is necessary to select stains which will not overstain. Alum cochineal and Mayer's haem-alum might be suggested. It would be worth while to try a combination of the two methods. Fix the entire anther in chromo-acetic acid, wash in water, and then stain in iron-haematoxylin. When the last stage in staining is reached—the extraction of the stain in iron-alum—remove the pollen mother-cells and watch the differentiation; then wash in water and follow the Venetian turpentine method.

FIG. 107.—*Erythronium americanum:* photomicrograph of mature pollen grains; the one at the top, which is cut longitudinally, shows both the tube nucleus and the conspicuous generative cell; the other is cut transversely and shows the generative cell, but not the tube nucleus; stained in safranin and gentian-violet; from a preparation by Dr. Lula Pace; Cramer contrast plate; 4-mm. objective; ocular ×4; yellowish-green filter; bellows, 85 cm.; exposure, 3 minutes. ×615.

The pollen grain at the time of shedding generally consists of two cells, the tube cell and the generative cell, which afterward divides and forms two male cells or two male nuclei. *Lilium* and *Erythronium* furnish good illustrations of pollen shed in the two-cell stage (Fig. 107). In *Silphium, Sambucus,* and *Sagittaria* the generative nucleus divides before the pollen is shed. The division of the generative cell to form the two sperms takes place just before fertilization; consequently, in forms like *Silphium,* fertilization is likely to occur within less than 72 hours after the division of the generative cell.

Sections should not be more than 5 μ thick, if they are to show a clear differentiation of exine, intine, starch, and other structures. If sections have been stained in iron-haematoxylin, staining in safranin for from 3 to 7 minutes will give the exine a bright-red color and will not obscure the haematoxylin. A rather sharp stain in gentian-violet will stain the starch and also the intine. In *Asclepias* and many orchids, in which a common exine surrounds the entire mass of pollen grains, care must be taken not to overstain.

In many cases the pollen grains will put out their tubes in a 2 to 5 per cent solution of cane-sugar in water. Where the interval between pollination and fertilization is known (about 72 hours in *Lilium philadelphicum* and 96 to 100 hours in *L. canadense*), pieces of the stigma and style showing pollen tubes can be selected with some certainty.

Oögenesis.—As in spermatogenesis, the early stages will be found in preparations of floral development. The preparations of *Capsella* will show the origin and development of the nucellus (megasporangium) and also the megaspore mother-cell. The division of the megaspore mother-cell to form four megaspores takes place shortly before the bud begins to unfold. A massive megasporangium with several megaspore mother-cells may be found in *Ranunculus;* a megasporangium with only one megaspore mother-cell and only one layer of cells surrounding it may be found in any of the Compositae. *Senecio aureus* and *Erectites hieracifolium* are good and are particularly easy to cut. In *Trillium* and in *Cypripedium* the embryo sac is formed from two megaspores, which are not separated by walls. In *Lilium, Tulipa, Fritillaria, Erythronium,* and many others, the embryo sac is formed by all four megaspores, which are not separated by walls. In *Peperomia,* the Peneaceae, and some species of *Euphorbia,* the sac is formed by the four megaspores, not separated by walls, and the sac has 16 free nuclei. In *Plumbagella* the four megaspores, not separated by walls, constitute the mature sac, one of the megaspores functioning as the egg, two more fusing to form the endosperm nucleus, while the fourth megaspore aborts; so that the embryo sac, ready for fertilization, contains only two nuclei.

The reduction of chromosomes takes place during the two mitoses by which the mother-cell gives rise to four megaspores. The figures are much larger than in the corresponding mitoses in spermatogenesis but so much more tedious to secure that most studies in reduction have been based upon divisions in the pollen mother-cell. *Lilium* is quite favorable for a study of oögenesis, but it must be remembered that it is exceptional in having an embryo sac formed from four megaspores.

In very young stages, before the appearance of the integument, the ovary may be removed from the flower and placed directly in the fixing agent, but in later stages, such as are shown in Figure 113, strips should be cut off from the sides of the ovary in order to secure more rapid fixing and more perfect infiltration with paraffin. The dotted lines in Figure 111*C*, show about how much should be cut off. This is a much better plan than to secure rapid fixing and infiltration by cutting the ovary into short pieces, because the ovules will be in

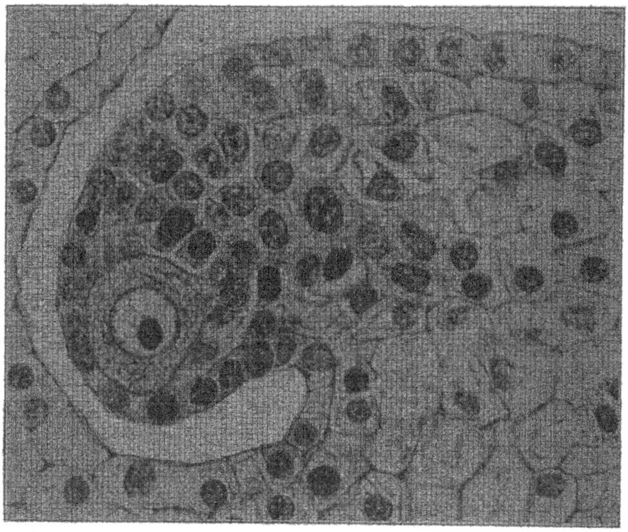

FIG. 108.—*Lilium canadense:* photomicrograph of part of transverse section of ovary, showing longitudinal section of ovule with megaspore mother-cell, just before the formation of integuments. Fixed in chromo-acetic acid and stained in iron-alum haematoxylin. Eastman Commercial Ortho film, Wratten H filter (blue); Spencer 4-mm. objective, N.A. .66; ocular ×6; arc light; exposure, 2 seconds. Negative by Dr. P. J. Sedgwick. ×400.

about the same stage cf development throughout the ovary, and when one finds desirable stages like those from which these photomicrographs were taken, it is gratifying to have these pieces as long as possible.

Chromo-acetic acid, with the addition of a little osmic acid, is good for fixing the entire series. Iron-haematoxylin, with a light touch of orange, is best for the chromatin. For general beauty and for the achromatic structures, the safranin, gentian-violet, orange combination has not been excelled. The photomicrographs (Figs. 108–110) illustrating the series from the archesporial cell (which, in this case, is also the primary sporogenous cell and the megaspore mother-cell) to the four megaspore nuclei will repay a careful study. One more mitosis produces the 8-nucleate embryo sac, but *Lilium* is not a good type for illustrative purposes, since the egg apparatus is not very definitely organized.

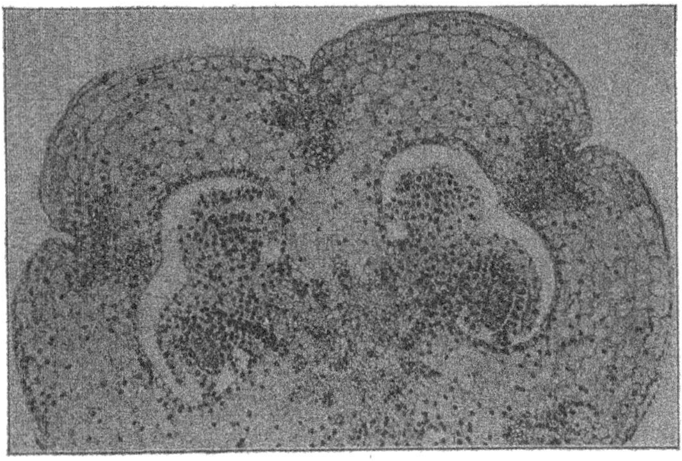

FIG. 109.—*Lilium philadelphicum:* photomicrograph of transverse section of ovary showing, in one of the ovules on the left, the first mitosis in the megaspore mother-cell; and, in one of the ovules on the right, the second mitosis which gives rise to the four megaspore nuclei—chromo-acetic acid; safranin, gentian-violet, orange. Cramer contrast plate; 16 mm.-objective; ocular ×4; yellowish-green filter and also a strong filter such as is used in outdoor work; camera bellows, 30 cm.; exposure, 2 minutes. Negative by MISS Ethel Thomas. ×64.

For the embryo sac at the fertilization stage, many of the Compositae are good. *Senecio aureus* is quite favorable, because it is easy to cut and the akenes do not spread. *Aster* gives an exceptional view of the antipodal region, but is rather hard to cut. Before fixing, trim the head as indicated in Figure 111. *Silphium,* especially *S. laciniatum,* furnishes an ideal view of the embryo sac. With thumbs and fingers grasp the two wings of the akene and carefully split it, exposing the single white ovule inside. This is rather tedious, but every ovule will yield a perfectly median longitudinal section of the embryo sac, and there is not the slightest difficulty in cutting. When the rays look their best, the embryo sac is ready for fertilization, or the pollen tubes may be entering; as the rays begin to wither, you will find fertilization or early stages in the embryo and endosperm. Sections should be about 10 μ thick.

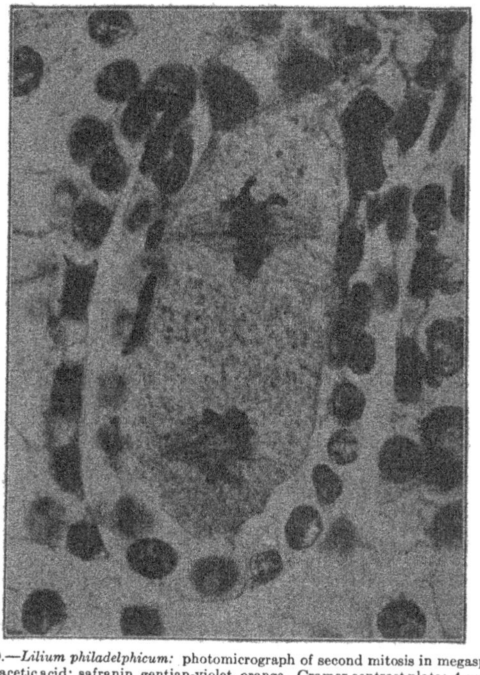

FIG. 110.—*Lilium philadelphicum:* photomicrograph of second mitosis in megaspore mother-cell—chromo-acetic acid; safranin, gentian-violet, orange. Cramer contrast plate; 4-mm. objective; ocular ×4; Abbé condenser camera bellows, 1 meter; yellowish-green filter and also a strong filter such as is used in outdoor work; camera bellows, 1 meter; exposure, 7 minutes. Negative by Miss Ethel Thomas. ×626.

The *Ranunculaceae*, especially *Anemone patens var. Wolfgangiana*, show a rather large, broad embryo sac, with highly organized egg apparatus and antipodals. Sections should be 10 to 20 μ thick.

For general views of the embryo sac, the safranin, gentian-violet, orange combination is recommended.

The peculiar embryo sac of *Plumbagella*, with only two nuclei, the egg nucleus and the endosperm nucleus, when ready for fertilization, is shown in Figure 112.

Fertilization.—The later stages cut to show the mature embryo sac will often show fertilization. The male and female nuclei almost invariably show a difference in staining capacity when the male nuclei are just discharged from the pollen tube. With cyanin and erythrosin, the male nucleus stains blue and the female red; hence the obsolete terms cyanophilous and erythrophilous. As the nuclei come into contact within the egg, they begin to stain alike, the male nucleus staining more and more like the female. In the final stages of fusion it is difficult, or impossible, to distinguish the two nuclei. The male nucleus which takes part in the "triple fusion" to form the endosperm nucleus behaves in the same way.

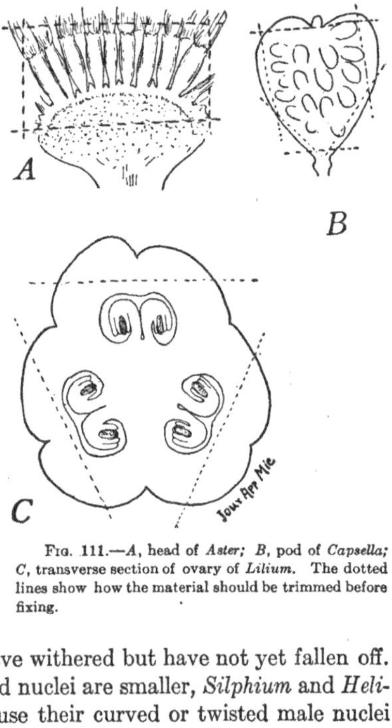

FIG. 111.—*A*, head of *Aster;* *B*, pod of *Capsella;* *C*, transverse section of ovary of *Lilium*. The dotted lines show how the material should be trimmed before fixing.

Lilium is a very good and always available type for illustrating fertilization (Fig. 113). Take ovaries from flowers whose petals have withered but have not yet fallen off. Though their embryo sacs and nuclei are smaller, *Silphium* and *Helianthus* are good types, because their curved or twisted male nuclei are easily distinguished from the spherical nuclei in the embryo sac. The embryo sacs of orchids are very small, but ovules are extremely numerous and the chances for securing the fusion of nuclei are correspondingly good. In *Cypripedium* the nuclei do not fuse in the resting condition, but the chromosomes of the two parents are perfectly distinct in the egg. The general statement that nuclei fuse in the "resting condition" is not correct, and probably the chromosomes of the two gametes never fuse.

The Endosperm.—Some of the preparations intended for fertilization will be likely to show early stages in the development of endosperm.

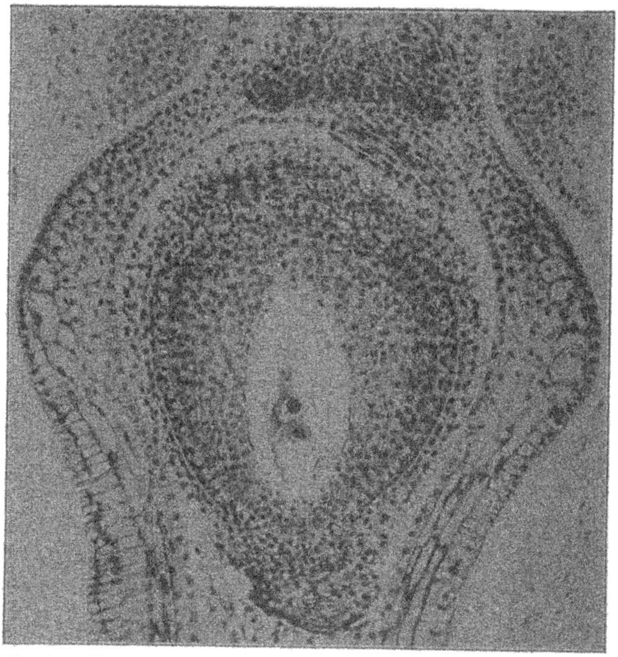

FIG. 112.—*Plumbagella micrantha:* longitudinal section of ovule showing the embryo sac, with the egg and endosperm nucleus ready for fertilization. Stained in iron-alum haematoxylin. Eastman Commercial Ortho film, Wratten E filter (orange); Bausch and Lomb 8-mm. objective N.A. .50; Spencer ocular ×6; arc light; exposure, 1 second. Preparation by Dr. K. von O. Dahlgren and negative by Dr. P. J. Sedgwick. ×208.

In rather long, narrow embryo sacs, a cell wall is likely to follow even the first division of the endosperm nucleus, so that the endosperm is cellular from the beginning. *Ceratophyllum, Monotropa,* and *Verbena* will furnish material of this type.

In large, broad embryo sacs, the formation of endosperm is almost sure to be initiated by a series of simultaneous free nuclear divisions. In large sacs walls then begin to appear at the periphery and wall formation gradually advances toward the center until the entire sac is filled with tissue. *Lilium, Peperomia,* and *Ranunnlus* furnish examples of this type.

An intermediate condition is seen in somewhat elongated embryo sacs of medium size, like those of Compositae. After a few free nuclear divisions, walls appear simultaneously throughout the entire sac. *Silphium laciniatum* is particularly good. Akenes from which the corolla has just fallen will furnish material.

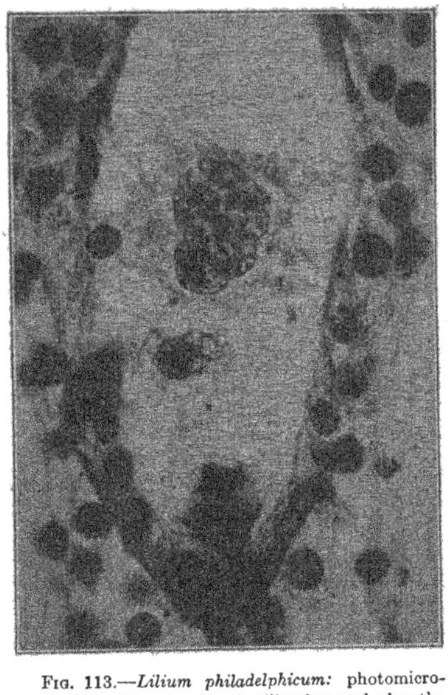

FIG. 113.—*Lilium philadelphicum:* photomicrograph of section showing fertilization and also the triple fusion; from a preparation and negative by Dr. W. J. G. Land. ×585.

The Embryo.—In most angiosperms the endosperm divides earlier than the fertilized egg and in some cases, like *Asclepias* and *Casuarina,* the free nuclear stage of the endosperm is completed and the cellular stage is well advanced before the first division of the egg. In some forms, like the aroids, the embryo is massive and differentiation into dermatogen, periblem, and plerome comes comparatively late; while in others, like the Cruciferae, the differentiation occurs very early. *Capsella* is a standard example of the latter type (Fig. 114). The stages shown in Figure 114*A–F*, will be found in pods about 3 mm. in length. These may be put directly into the fixing agent, but stages like *G* and *H*, which are found in pods about 5 mm. in length, should be trimmed as indicated in Figure 111*B*, before fixing. Formalin-alcohol-acetic acid is a good fixing agent; the special chromo-acetic-osmic acid, with 2 c.c. of acetic acid instead of 3 c.c., is also very good, and Delafield's haematoxylin stains better after the chromic series. Cut 5 to 10 μ thick and parallel to the flat face of the pod.

For a study of the monocotyl embryo, *Iris*, and especially *I. pseudacorus*, can be recommended. The embryo is straight, and cotyledon, stem-tip, and root are clearly differentiated before the endosperm

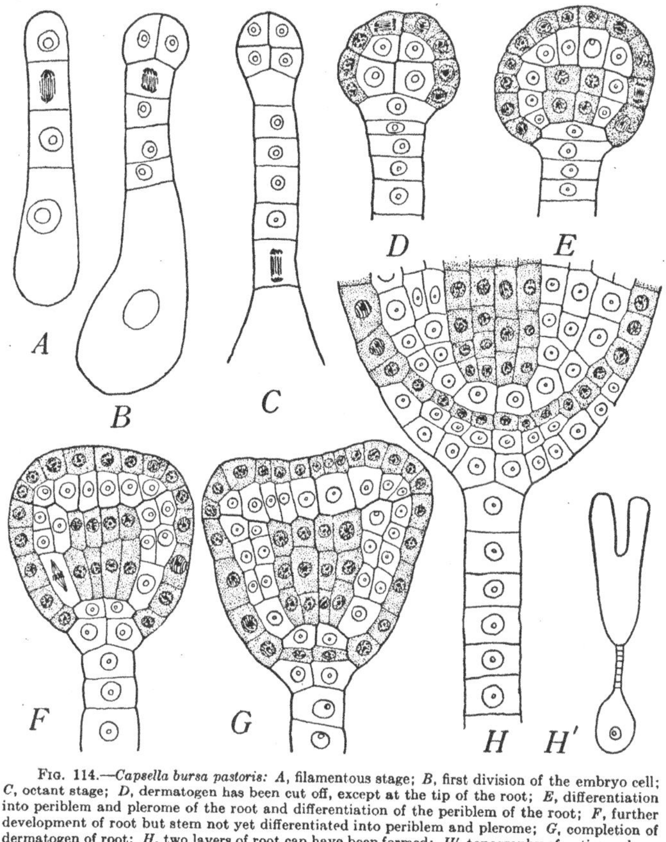

FIG. 114.—*Capsella bursa pastoris:* *A*, filamentous stage; *B*, first division of the embryo cell; *C*, octant stage; *D*, dermatogen has been cut off, except at the tip of the root; *E*, differentiation into periblem and plerome of the root and differentiation of the periblem of the root; *F*, further development of root but stem not yet differentiated into periblem and plerome; *G*, completion of dermatogen of root; *H*, two layers of root cap have been formed; *H'*, topography of entire embryo at stage shown in *H*. Only dermatogen and plerome shaded. *A–H*, ×580; *H'* ×75.

becomes too hard to cut in paraffin. Fix pieces about 3 mm. wide cut perpendicular to the face of the cheese-shaped seed. Do not try to cut the whole pod.

Sagittaria has been used quite extensively. It is easily obtained, the whole head can be cut with ease, even after the cotyledon and stem-tip are clearly differentiated, and the endosperm is instructive; but the embryo is curved, like that of Capsella, and good views are rather rare.

Zea Mays, especially the sweet corn, is a good type to illustrate the peculiar embryo of the grasses. Directions have been given on page 294.

In many forms good preparations of late stages may be secured by soaking the seeds in water until the embryo bursts the seed coat. Young seedlings furnish valuable material for a study of vascular anatomy.

Parthenogenesis.—Many embryos are developed without fertilization. *Taraxacum*, the common dandelion, is an omnipresent example. Other widely distributed illustrations are *Hieracium* and practically all species of the *Eualchemilla* section of the genus *Alchemilla*. Parthenogenetic forms show various irregularities in the mitoses leading up to the formation of the egg.

CHAPTER XXVI

USING THE MICROSCOPE

The investigator who desires to see all that his microscope is capable of showing must study the optics of his instrument. The fundamental principles are presented in any good textbook of physics. Excellent practical hints are given in two booklets published by the leading American optical companies. These booklets tell the beginner how to set up the microscope, how to keep it in order, and give directions concerning illumination, dry and immersion objectives, mirror, condenser, diaphragm, and various other things (Fig. 115). They were doubtless written for advertising purposes, but since they advertise by giving directions for securing the best results with the microscope, the information is very reliable. The Spencer Lens Company of Buffalo, New York, and the Bausch & Lomb Optical Company, of Rochester, New York, furnish these booklets free of charge.

In the histological laboratory where preparations are being made the microscope is in constant danger. A cheap microscope with a 16-mm. objective and one ocular, such an instrument as can be got new for $30 or less, can be used for examining preparations while they are wet with alcohols, oils, or other reagents. If it is necessary to use a better instrument for such work, cover the stage with a piece of glass—an old lantern slide is of about the right size—and be extremely careful not to get reagents upon the brass portions.

MICROMETRY

Everyone who expects to become at all proficient in the use of the microscope should learn to measure microscopic objects and should learn to form some estimate even without measuring, just as one guesses at the size of larger objects. In any measurement one should note the tube length, which is usually 160 mm. Since the use of the nosepiece is universal, it is convenient to have the length measure 160 mm. when the tube is pushed in. Some companies still make the tube so short that it must be pulled out about 15 mm. to reach the length of 160 mm., even when the nosepiece is in place. Where there is no revolving nosepiece, the draw-tube is simply pulled out until the length is 160 mm. (Fig. 116). Where a nosepiece is used, its height should be measured, and the draw-tube

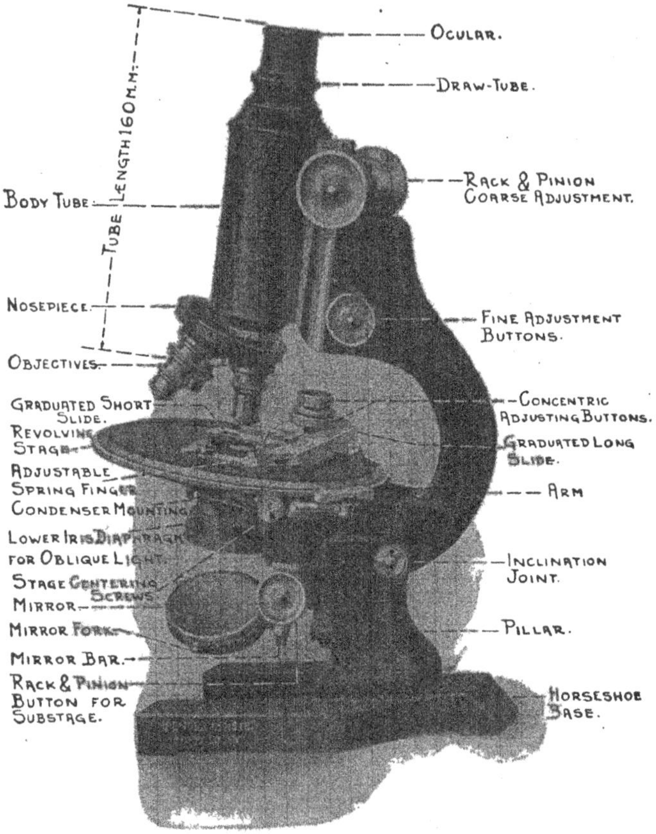

FIG. 115.—A modern microscope

should be pushed in a distance equal to the length of the nosepiece. There are in general use two practical methods of measuring microscopic objects, one by means of the ocular micrometer, and the other by means of camera lucida sketches.

Measuring with the Ocular Micrometer.—A stage micrometer and an ocular micrometer are necessary. A stage micrometer should be ruled in tenths and one-hundredths of a millimeter. It does not matter what the spacing in the ocular micrometer may be, except that the lines must be at equal distances from one another. As a matter of fact, the ocular micrometer is generally ruled in tenths of a millimeter, but this ruling is more or less magnified by the lens of the ocular.

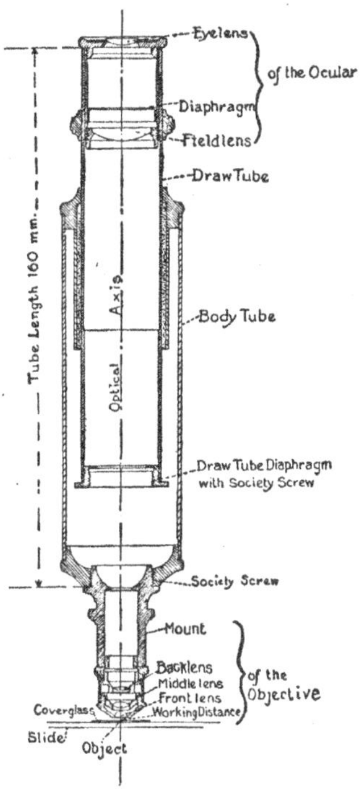

FIG. 116.—Tube length

Place the stage micrometer upon the stage and the ocular micrometer in the tube, and arrange the two sets of rulings so that the first line in the ocular micrometer will coincide with the first line of the stage micrometer, and then find the value of one space in the ocular micrometer. The method of finding this value is shown in the following case in which the tube length was 160 mm., the ocular a Zeiss ocular micrometer 2, and the objective a Leitz 3. In the ocular micrometer, ninety-eight spaces covered just fifteen of the larger spaces of the stage micrometer. Since the stage micrometer is ruled in tenths and one-hundredths of a millimeter, the fifteen spaces equal 1.5 mm., or 1,500 μ.[1] Then ninety-eight spaces of the ocular micrometer equal 1,500 μ; and one space in the ocular equals $\frac{1}{98}$ of 1,500 μ, or 15.3 μ. This value being determined, there is no further use for the stage micrometer. To measure the diameter of a pollen grain put the

[1] One millimeter = 1,000 μ. The Greek letter μ is an abbreviation for μικρόν, or micron.

preparation on the stage, using the same objective and ocular micrometer, and note how many spaces a pollen grain covers. If the pollen grain covers five spaces, its diameter is five times 15.3 μ, or 76.5 μ. In the same way, the value of a space in the ocular when used with the other objectives should be determined. The values for three

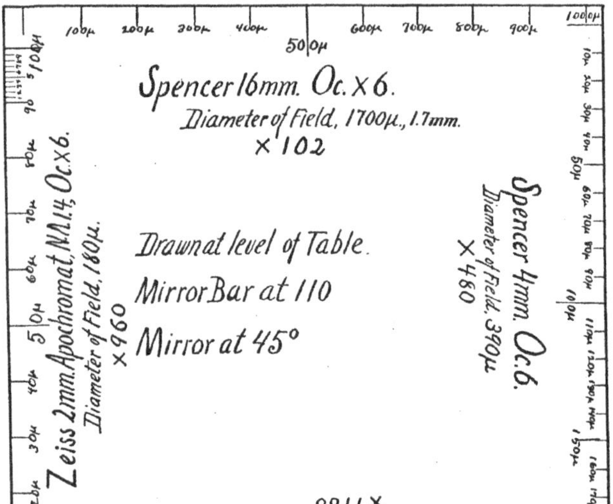

FIG. 117.—Card scale for practical use

or four objectives may be written upon an ordinary slide label and pasted upon the base of the microscope for convenient reference.

This method is the best one for measuring spores and for most measurements in taxonomy.

Measuring by Means of Camera Lucida Sketches.—This method is of great importance in research work, because various details can be measured with far greater rapidity than by the other method. Upon a piece of cardboard, about as thick as a postal card, draw a series of scales like those shown in Figure 117.

Make a scale for each objective. It is not necessary to make scales for all the oculars, but only for the one in most constant use. It is absolutely necessary to note the tube length, length of the bar of the camera mirror and inclination of the camera mirror, and the level at which the scale is made. A variation in any of these details will change the scale.

In using the stage micrometer, place the cardboard on the table, and with the aid of the camera lucida sketch the rulings of the micrometer. In Figure 117 note, for example, the scale drawn with Spencer 16 mm. objective, ocular $\times 6$. The spaces are drawn from the tenths of a millimeter rulings of the stage micrometer. Therefore each space on the card represents one-tenth of a millimeter or $100\,\mu$, and the ten spaces shown on the card represent 1 mm., or $1{,}000\,\mu$. By measuring with a metric rule the ten spaces upon the card, it is found that the scale is 102 mm. in length. The magnification of any drawing made with the same ocular and objective, under the same conditions, will therefore be 102 diameters. This does not mean that the magnifying power is 102 diameters, for the magnification of this combination is much less. A scale drawn at the level of the stage would show more nearly the magnifying power of the combination, but would still give too large a figure. The exact size of any object which has been sketched with this combination can now be measured by applying the cardboard scale, just as one would measure gross objects with a rule.

The diameter of the field with this combination is $1{,}700\,\mu$. By knowing the diameter of the field with the various combinations, one can guess approximately the size of objects.

Other combinations are made in the same way. An excellent check on the accuracy of the computations is to measure the same object by means of the ocular micrometer and by the scale card. If the results are the same, the computations are correct.

In making sketches, it is a good plan to add the data which would be needed at any time in making measurements; e.g., Spencer objective 16 mm., ocular $\times 6$, table, 110, 45°, would show that the sketch was made at the level of the table, with the mirror bar at 110, and the camera mirror at an angle of 45.°

ARTIFICIAL LIGHT

During a considerable part of the year daylight is often insufficient for successful work with the microscope. Numerous contrivances for artificial illumination have been devised, some of them fairly good, but most of them thoroughly unsatisfactory. More than two hundred years ago Hooke used a device for artificial illumination which probably suggested the apparatus used by the late Professor Strasburger at Bonn. The apparatus in use in our own laboratory is only a slightly modified form of that used in the Bonn laboratory.

The apparatus consists, essentially, of a hollow sphere filled with liquid. A fairly good and practical light can be got with an ordinary lamp by allowing the light to pass through a wash bottle filled with a weak solution of ammonia copper sulphate. A piece of dark paper with a circular hole in it serves as a diaphragm, and at the same time protects the eyes from the direct light of the lamp.

At present, we are using a white 50-watt, 115-volt, nitrogen Mazda bulb, with a shade to protect the eyes. This not only furnishes a strong light, without any glare, but throws a good light on the pencil, which is an important consideration in drawing with a camera lucida.

Optical companies are now making excellent lights for microscopes. These lights furnish good illumination and most of them have the effect of good daylight.

If laboratory tables are small, seating only one student, there should be a plug to attach the table to some convenient outlet; and also another outlet on the table for the microscope lamp. If the table is large, seating four or more students, there should be an outlet on the table for each student, and a single plug by which the whole table may be connected with a convenient outlet.

For elementary classes, which are not likely to use higher powers than a 4 mm. objective with an ocular magnifying five or six times, individual lamps are not necessary in a well-lighted laboratory. Half-a-dozen strong lights, of the overhead type, with white shades, serve very well for a class of twenty-five or thirty students.

CHAPTER XXVII

LABELING AND CATALOGUING PREPARATIONS

THE LABEL

We should say that the *first* thing to write upon a label is the genus and species of the plant; the next thing would be the name of the organ or tissue, and then might be added the date of collection, e.g., *Marchanita polymorpha*, young archegonia, January 10, 1915. The date of making the preparation is of no value unless the student is testing the permanence of stains or something of that sort. It is hardly worth while to write upon the label the names of the stains used, for the student will soon learn to recognize the principal stains. A hasty sketch on the label will often indicate any exceptionally interesting feature in the preparation. To facilitate finding such a feature, it is a good plan to mark the particular section or sections with ink, the marking being always on the underside of the slide so as not to cause any inconvenience if an immersion lens should be used.

CATALOGUING PREPARATIONS

As a collection grows, the student will need some device for locating readily any particular preparation. Some have their slides numbered and catalogued, but all devices of this sort are too cumbrous and slow for the practical worker in the laboratory. After thirty years' experience with a collection which now numbers more than thirty thousand preparations, we recommend the following system:

Four wooden slide boxes of the usual type will do for a beginning; they should be labeled: THALLOPHYTES, BRYOPHYTES, PTERIDOPHYTES, and SPERMATOPHYTES. As the collection grows and new boxes are needed, the classification can be made more definite, e.g., there should be a box labeled BRYOPHYTES *Hepaticae* and one labeled BRYOPHYTES *Musci*. As the liverwort collection grows, three boxes will be necessary, and should be labeled BRYOPHYTES *Hepaticae Marchantiales*, BRYOPHYTES *Hepaticae* Jungermanniales, and BRYOPHYTES *Hepaticae* Anthocerotales. It will readily be seen that the process can be continued almost indefinitely, and that new slides may be at any time dropped into their proper places. A rather complete label gradually built up in this way is shown in Figure 118.

The beginner will often find that the mere placing of a slide in the proper box and the box in its proper place on the shelf will refresh or

BRYOPHYTES

HEPATICAE

Jungermanniales

Porella platyphyllum

Archegonia

Fig. 118.—Label for slide box

increase his knowledge of classification. While this system is almost ideal for the careful worker, especially if he has some knowledge of classification, it is the worst possible method for a careless student since a slide in the wrong box is almost hopelessly lost if the collection is large enough to need thorough classifying.

CHAPTER XXVIII

A CLASS LIST OF PREPARATIONS

Where a regular course in histology is conducted, it is a good plan to give each student at the outset a complete list of the preparations which he is expected to make. In a three months' course a fairly representative collection of preparations can be made. The availability of material determines what a list shall be. Besides gaining an introduction to the use of the microscope and its accessories, a class meeting ten hours a week for twelve weeks should be able to do as much work as is outlined below.

In making the mounts, the order indicated in the list should not be followed. Begin with temporary mounts, and then study, in succession, freehand sections (the glycerin method), the Venetian turpentine method, the paraffin method (the celloidin method), and special methods. A large proportion of the time should be devoted to the paraffin method.

It is neither possible nor desirable that each student should in every case go through all the processes from collecting material to labeling. Some of the material may be in 85 per cent alcohol, some in formalin, some in glycerin, some in Venetian turpentine, and some in paraffin. One student may imbed in paraffin enough of the *Anemone* for the whole class; another may imbed the *Lilium* stamens; and by such a division of labor a great variety of preparations may be secured without a corresponding demand upon the time of the individual.

LIST OF PREPARATIONS

THALLOPHYTES

SCHIZOPHYTES

MYXOMYCETES

1. *Trichia varia.*—Paraffin sections, 5 μ. Safranin, gentian-violet, orange.

SCHIZOMYCETES

2. *Bacteria.*—*Coccus, Bacillus,* and *Spirillum* forms. Stain on cover-glass or slide.

3. *Bacillus anthracis.*—In liver of mouse. Paraffin sections, 5 μ. Stain in gentian-violet, Gram's method.
4. *Oscillatoria.*—Fix in special chromo-acetic-osmic acid and stain in iron-alum haematoxylin to show nuclei. Venetian turpentine method.
5. *Tolypothrix.*—Use the Venetian turpentine method. Should show heterocysts, hormogonia, and false branching.
6. *Nostoc.*—Venetian turpentine method.
7. *Wasserblüthe.*—The principal forms in this material are:
   a) *Coelosphaerium Kützingianum.*—Colonies in the form of hollow spheres.
   b) *Anabaena gigantea.*—Filaments straight. Preparations should show vegetative cells, heterocysts, hormogonia, and spores.
   c) *Anabaena flos-aquae.*—Filaments curved. Stain on the slide and mount in balsam. If material is abundant, stain in iron-alum haematoxylin and mount in Venetian turpentine.
8. *Gloeotrichia.*—Smear on the slide, stain in safranin and gentian-violet, and mount in balsam; or use the Venetian turpentine method, staining in Magdala red and anilin blue and crushing under the cover-glass.

ALGAE

CHLOROPHYCEAE

9. *Volvox.*—Use the Venetian turpentine method. If paraffin material is available, cut 5 μ in thickness and stain in safranin, gentian-violet, orange.
10. *Scenedesmus.*—Let a drop containing the material dry upon the slide stain, and mount in balsam.
11. *Hydrodictyon.*—Use the Venetian turpentine method.

    Each preparation should contain pieces of old and of young nets, and also at least one young net developing within an older segment. The greatest care must be taken not to injure the older segments while arranging the mount.
12. *Ulothrix.*—Use the Venetian turpentine method. Each mount should show various stages in the development of spores and gametes.
13. *Oedogonium.*—Stain in Magdala red and anilin blue and mount in Venetian turpentine.
14. *Coleochaete.*—Stain in Delafield's haematoxylin and mount in balsam.
15. *Cladophora.*—Stain some in iron-haematoxylin and some in Magdala red and anilin blue. Mount both together in Venetian turpentine.
16. *Diatoms.*—Make mounts of the frustules and also stained preparations showing the cell contents.
17. *Desmids.*—Make mounts of available forms. Use the Venetian turpentine method if material is sufficiently abundant.

18. *Zygnema.*—Stain in iron-haematoxylin, and mount in Venetian turpentine.
19. *Spirogyra.*—Stain in Magdala red and anilin blue, and mount in Venetian turpentine.
20. *Vaucheria.*—Stain in iron-alum haematoxylin, and mount in Venetian turpentine.
21. *Chara.*—Cut paraffin sections of the apical cell, oögonia, and antheridia.

PHAEOPHYCEAE

22. *Ectocarpus.*—Stain some in iron-haematoxylin and some in Magdala red and anilin blue. Mount both together in Venetian turpentine.
23. *Cutleria.*—Sections of oögonia, antheridia, and sporangia. Cut 10 μ thick and stain in iron-haematoxylin with about 7 minutes in safranin.
24. *Fucus vesiculosus.*—Antheridial conceptacle with paraphyses and antheridia; oögonial conceptacle with oögonia. Cut 10 μ thick and stain in iron-haematoxylin with about 5 minutes in safranin.

RHODOPHYCEAE

25. *Nemalion.*—Stain some in iron-haematoxylin and some in eosin. Each preparation should show trichogyne, carpogonium, and cystocarp. You cannot mount it in Venetian turpentine; use glycerin or glycerin jelly.
26. *Polysiphonia.*—Stain in iron-haematoxylin or Magdala red and anilin blue. Mount whole in Venetian turpentine. Each mount should show tetraspores, antheridia, and cystocarps. If material is in paraffin, cut sections about 7 μ thick.

FUNGI

PHYCOMYCETES

27. *Rhizopus nigricans.*—Stain young sporangia in eosin, dilute Delafield's haematoxylin, or in Magdala red and anilin blue. Some zygosporic material should be stained strongly in eosin: some should be stained in iron-alum haematoxylin, and in reducing the stain, some should be taken out early to show the coenocytic character of the mycelium, and some should be drawn farther to show the structure of the zygospores and suspensors. Venetian turpentine.
28. *Saprolegnia.*—Stain some in Magdala red and anilin blue, and some in iron-alum haematoxylin. Each mount should show sporangia and oögonia. Venetian turpentine.
29. *Albugo candida.*—Select white blisters which have not yet broken open. Paraffin 8 μ. Iron-alum haematoxylin and orange. Oögonia and antheridia, 5 μ, same stain.
30. *Albugo bliti* on *Amarantus retroflexus.*—Cut out small portions of leaves in which the oögonia can be seen in abundance. Paraffin, 5 μ.

ASCOMYCETES

31. *Peziza.*—Paraffin sections of young apothecia, 5 μ or less; sections of older apothecia, 10 or 15 μ. Safranin, gentian-violet, orange.
32. *Aspergillus, Eurotium.*—Stain in eosin and mount in Venetian turpentine.
33. *Penicillium.*—Treat like *Aspergillus.*
34. *Erysiphe commune* on *Polygonum aviculare.*—Strip the fungus from the leaf. Paraffin, 5 μ or less. Safranin, gentian-violet, orange.
35. *Uncinula necator* on *Ampelopsis quinquefolia.*—Stain in Magdala red and anilin blue. Mount whole in Venetian turpentine and break the perithecia under the cover.
36. *Xylaria.*—Paraffin sections of younger stages. Delafield's haematoxylin and erythrosin. Be sure that some section in each mount shows the opening of a perithecium.

LICHENS

37. *Physcia stellaris.*—Cut in paraffin, 5 μ. Stain in cyanin and erythrosin.

BASIDIOMYCETES

38. *Puccinia graminis.*—Aecidium stage on barberry leaf. Uredospore and teleutospore stage on oats. Cut 3 μ and stain in iron-haematoxylin.
39. *Coprinus micaceus.*—Paraffin. Transverse sections of gills showing trama, paraphyses, basidia, and spores. To show the basidium with four spores, the sections should be 15 μ thick. For development of the spores, cut 5 μ or less. Safranin, gentian-violet, orange. *Boletus, Hydnum,* and *Polyporus* are treated in the same manner.

BRYOPHYTES

HEPATICAE

40. *Riccia natans.*—Paraffin, 10 or 15 μ. Delafield's haematoxylin. Archegonia, antheridia, and sporophytes imbedded in the gametophyte.
41. *Marchantia polymorpha.*—Paraffin, 5 or 10 μ. Archegonia, antheridia, and sporophytes.
42. *Anthoceros laevis.*—Paraffin, 5 or 10 μ. Longitudinal and transverse sections of sporophyte. Safranin, gentian-violet, orange.
43. *Pellia epiphylla.*—Paraffin, 5 or 10 μ. Longitudinal sections of sporophyte attached to gametophyte. Safranin, gentian-violet, orange.
44. *Porella platyphylla.*—Paraffin, 10 μ. Delafield's haematoxylin. Archegonia, antheridia, sporophyte, and apical cell.

MUSCI

45. *Sphagnum.*—Leaf buds. Cut 5 μ and stain in safranin and anilin blue.
46. *Sphagnum.*—Capsule. Paraffin. Delafield's haematoxylin and erythrosin.

47. *Funaria hygrometrica.*—Paraffin. Longitudinal and transverse sections of young capsules. Delafield's haematoxylin.
48. *Funaria hygrometrica* or any favorable form. Protonema. Place the well-cleaned material directly into 50 per cent glycerin and allow it to concentrate. Mount in glycerin or glycerin jelly.
49. *Bryum.*—Paraffin. Antheridia, 10 μ; archegonia, 15 to 20 μ; capsule, 10 μ.

PTERIDOPHYTES

LYCOPODIALES

50. *Lycopodium lucidulum.*—Transverse section of stem. Paraffin sections. Safranin and Delafield's haematoxylin.
51. *Lycopodium inundatum.*—Paraffin. Longitudinal sections of strobilus.
52. *Selaginella.*—Paraffin. Longitudinal sections of rather mature strobili. Safranin, gentian-violet, orange.
53. *Isoetes echinospora.*—Transverse section of stem. Paraffin. Safranin and Delafield's haematoxylin.
54. *Isoetes echinospora.*—Paraffin. Longitudinal sections of microsporangia and megasporangia. Safranin, gentian-violet, orange.

EQUISETALES

55. *Equisetum arvense.*—Prothallia in Venetian turpentine. Stem-tips in paraffin. Transverse section of stem freehand or in celloidin.

FILICALES

56. *Botrychium virginianum.*—Paraffin. Stain rhizome, stipes, and root in safranin and Delafield's haematoxylin. Stain sporangia in iron-haematoxylin.
57. *Protostele.*—Use *Gleichenia.* Cut freehand and stain in safranin and anilin blue.
58. *Solenostele (amphiphloic siphonostele).*—Use *Adiantum* or *Dicksonia punctilobula.*
59. *Ectophloic siphonostele.*—Use *Osmunda cinnamomea.*
60. *Polystele.*—Use *Pteris aquilina* or any species of *Polypodium.*
61. *Sporangia.*—For development, use *Pteris, Aspidium, Cyrtomium,* or try any available species. For mitosis, *Osmunda* is exceptionally good.
62. *Antheridia and archegonia.*—Mount whole in Venetian turpentine. Iron-alum haematoxylin. Sections should be 5 to 10 μ thick. Stain in iron-haematoxylin and orange.
63. *Embryo.*—*Pteris* and *Adiantum* are good. Cut longitudinal vertical sections 10 μ thick.

SPERMATOPHYTES

GYMNOSPERMS

CYCADALES

64. *Zamia.*—Freehand sections of stem. Safranin and light green. Transverse sections of microsporophyll, 5 or 10 μ. Longitudinal sections of entire ovule, 10 to 15 μ; stain in safranin, gentian-violet, orange. Longitudinal sections of nucellus with pollen tubes, 10 μ. Iron-haematoxylin and orange.

GINKGOALES

65. *Ginkgo biloba.*—Longitudinal sections of endosperm showing archegonia or young embryos. Paraffin 10 μ.
Sections of microsporangia with nearly mature pollen, 5 μ.

CONIFERALES

66. *Pinus Laricio.*—Transverse sections of needles and young stem. Freehand. Safranin and light green.
67. *Pinus Strobus.*—Freehand sections of well-seasoned wood. Safranin and Delafield's haematoxylin.
68. *Pinus Laricio.*—Paraffin. Longitudinal section of mature staminate strobilus. Safranin, gentian-violet, orange.
69. *Abies balsamea* or *Pinus Laricio.*—Pollen at shedding stage shaken out and imbedded in paraffin; 5 μ. Stain in safranin, gentian-violet, orange.
70. *Pinus Laricio.*—Paraffin. Ovule with archegonia. Safranin, gentian-violet, orange.
71. *Pinus sylvestris* or *P. Laricio.*—Paraffin. Embryos. Cyanin and erythrosin, or safranin and anilin blue.

GNETALES

72. Transverse section of stem of *Ephedra.* Freehand.
73. Longitudinal section of the ovule of *Ephedra.*

ANGIOSPERMS

DICOTYLS

74. *Pelargonium.*—Transverse sections of stem to show phellogen and intrafascicular cambium. Freehand. Endarch siphonostele.
75. *Tilia americana.*—Celloidin or freehand. Transverse sections of small stems 3 mm. to 6 mm. in diameter. Safranin and Delafield's haematoxylin. Endarch siphonostele with annual rings.
76. *Sambucus nigra.*—Transverse section of primary root to show origin of secondary structures.
77. *Cucurbita.*—Longitudinal section of stem to show sieve tubes.

78. *Capsella bursa-pastoris.*—Paraffin. Floral development, 5 μ. Embryos, 5 to 10 μ. Stain both in Delafield's haematoxylin without any contrast stain.
79. *Taraxacum officinale.*—Paraffin. Floral development, 5 μ. Embryo sac, 10 to 15 μ.
80. *Ranunculus.*—Longitudinal sections of young flowers to show megaspore mother-cells and megaspores.
81. *Silphium.*—Longitudinal sections of the ovule at the fertilization period. Longitudinal sections of staminate flowers just before the shedding of pollen.
82. *Anemone patens.*—Paraffin. Embryo sac.

MONOCOTYLS

83. *Clintonia.*—Transverse section of stem to show siphonostele. Paraffin. Safranin and anilin blue.
84. *Acorus calamus.*—Transverse sections of rhizome, freehand or paraffin, to show amphivasal bundles.
85. *Zea Mays.*—Transverse section of stem to show scattered bundles; also good for companion cells. Freehand. Safranin and anilin blue.
86. *Tradescantia virginica.*—Longitudinal sections of root-tip. Paraffin, 5 and 10 μ. Stain for mitosis.
87. *Smilax herbacea.*—Transverse section of adult root. Freehand. Shows exarch, radial structure, and a highly developed endodermis. Safranin and Delafield's haematoxylin.
88. *Lilium.*—Transverse section of leaf. Freehand. Transverse section of ovaries in various stages from megaspore mother-cell to fertilization; transverse sections of anthers to show microspore mother-cells and reduction of chromosomes; also later stages with nearly mature pollen. Paraffin 5 to 10 μ.
89. *Iris.*—Section of young seeds to show embryo with cotyledon and stem-tip.
90. *Sagittaria.*—Longitudinal sections of ovulate flowers of various stages to show development of the embryo and endosperm.
91. *Zea Mays.*—Longitudinal and transverse sections of embryo (sweet corn, roasting-ear condition) to show structure of root and beginning of protoxylem.

CHAPTER XXIX

FORMULAS FOR REAGENTS

FIXING AGENTS

Absolute Alcohol.—Used alone without any mixtures.

Carnoy's Fluid.—

Absolute alcohol	2 parts
Chloroform	3 parts
Glacial acetic acid	1 part

Farmer's Fluid.—

Absolute alcohol	3 parts
Glacial acetic acid	1 part

Formalin Alcohol (Lynds Jones's formula).—

70 per cent alcohol	100 c.c.
Commercial formalin	2 c.c.

Formalin Alcohol (Dr. Land's formula).—

50 per cent alcohol	100 c.c.
Commercial formalin	6 c.c.

Formalin Acetic Alcohol.—

50 per cent alcohol	90 c.c.
Commercial formalin	5 c.c.
Glacial acetic acid	5 c.c.

Formalin Acetic Acid.—

Commercial formalin	10 c.c.
Glacial acetic acid	5 c.c.
Water	85 c.c.

Formalin.—

Commercial formalin	3 to 10 c.c.
Water	100 c.c.

Stock Chromo-Acetic Solution.—

Chromic acid	1 g.
Glacial acetic acid	1 c.c.
Water	100 c.c.

Schaffner's Chromo-Acetic Solution.—

Chromic acid	0.3 g.
Glacial acetic acid	0.7 c.c.
Water	99.0 c.c.

Chromo-Acetic Solution (for marine algae).—

Chromic acid	1.0 g.
Glacial acetic acid	0.4 c.c.
Sea-water	400.0 c.c.

Material must be washed in sea-water.

Strong Chromo-Acetic Solution.—

Chromic acid	1 g.
Glacial acetic acid	3 c.c.
Water	100 c.c.

Licent's Formula.—

1 per cent chromic acid	80 c.c.
Glacial acetic acid	5 c.c.
Formalin	15 c.c.

Flemming's Fluid (weaker solution).—

A	1 per cent chromic acid (in water)	25 c.c.
	1 per cent glacial acetic acid (in water)	10 c.c.
	Water	55 c.c.
B.	1 per cent osmic acid (in water)	10 c.c.

Keep the mixture A made up, and add B as the reagent is needed for use, since it does not keep well.

Flemming's Fluid (stronger solution).—

1 per cent chromic acid	45 c.c.
2 per cent osmic acid	12 c.c.
Glacial acetic acid	3 c.c.

Special Chromo-Acetic-Osmic Solution.—

Chromic acid	1 g.
Glacial acetic acid	3 c.c.
1 per cent osmic acid	1 c.c.
Water	100 c.c.

Good for algae (fresh-water), filamentous fungi, and many others; but not for root-tips, stem-tips, and many other things. The name, *special*, was selected because the formula is not general in its application.

Merkel's Fluid.—

1.4 per cent solution of chromic acid	25 c.c.
1.4 per cent solution of platinic chloride	25 c.c.

Benda's Fluid.—

1 per cent chromic acid	16 c.c.
2 per cent osmic acid	4 c.c.
Glacial acetic acid	2 drops

Hermann's Fluid.—

1 per cent platinic chloride	15 parts
Glacial acetic acid	1 part
2 per cent osmic acid	4 or 2 parts

Picric Acid.—

Picric acid	1 g.
Water or 70 per cent alcohol	100 c.c.

Bouin's Fluid.—

Commercial formalin	25 c.c.
Picric acid (saturated solution in water)	75 c.c.
Glacial acetic acid	5 c.c.

Corrosive Sublimate and Acetic Acid.—

Corrosive sublimate	3 g.
Glacial acetic acid	5 c.c.
70 per cent alcohol (or water)	100 c.c.

Corrosive Sublimate, Formalin, Acetic Acid.—

Corrosive sublimate	4 g.
Formalin	5 c.c.
Glacial acetic acid	5 c.c.
Water (or 50 per cent alcohol)	100 c.c.

For material to be mounted in Venetian turpentine, use the aqueous solution; for imbedding, use the alcoholic. Both this and the preceding solution should be used hot (85° C.).

Bensley's Formula (for mitochondria).—

2½ per cent corrosive sublimate in water	4 parts
2 per cent osmic acid	1 part

Corrosive Sublimate, Acetic Acid, and Picric Acid.—

Corrosive sublimate	5 g.
Glacial acetic acid	5 c.c.
Picric acid	1 g.
50 per cent alcohol	100 c.c.

Corrosive Sublimate and Picric Acid (Jeffrey's formula).—

Corrosive sublimate, saturated solution in 30 per cent alcohol	3 parts
Picric acid, saturated in 30 per cent alcohol	1 part

Iodine.—

To a saturated solution of potassium iodide in distilled water, add iodine to saturation. Filter and dilute with distilled water to a rich brown color. For fixing, dilute still farther to a light-brown color. Good for unicellular and filamentous algae. See directions in chapter ii.

Gram's Potassium Iodide Solution (for staining bacteria).—

Iodine	1 g.
Potassium iodide	2 g.
Water	300 c.c.

Iodine (solution for starch test).—

Dissolve 1 g. potassium iodide in 100 c.c. of water and add 0.3 g. iodine.

Gilson's Fluid.—

95 per cent alcohol	42 c.c.
Water	60 c.c.
Glacial acetic acid	18 c.c.
Concentrated nitric acid	2 c.c.
Corrosive sublimate (saturated solution in water)	11 c.c.

Bensley's Formula (for canal system).—

1. Bichromate of potash	$2\frac{1}{2}$ g.
2. Corrosive sublimate	5 g.
3. Water	90 c.c.
4. Formalin (neutral)	10 c.c.

Make a solution of 1, 2, 3, and then add the neutral formalin.

Osmic Acid (stock solution).—

Osmic acid	1 c.c.
Distilled water	1 c.c.

The bottle in which the solution is to be kept, and also the glass tube in which the acid is sold, must be thoroughly cleaned. Break off the end of the tube, and drop both tube and acid into the distilled water, or simply drop the tube into the bottle and shake the bottle until the tube breaks.

Osmic Acid.—

Five or six drops of the stock solution to 50 c.c. of water is good for unicellular and colonial algae. In many cases 1 or 2 c.c. to 100 c.c. of water is better.

STAINS

Delafield's Haematoxylin.—

To 100 c.c. of a saturated solution of ammonia alum add, drop by drop, a solution of 1 g. of haematoxylin dissolved in 6 c.c. of absolute alcohol. Expose to air and light for one week. Filter. Add 25 c.c. of glycerin and 25 c.c. of methyl alcohol. Allow to stand until the color is sufficiently dark. Filter and keep in a tightly stoppered bottle.[1]

The solution should stand for at least 2 months before it is ready for using.

Erlich's Haematoxylin.—

Distilled water	50 c.c.
Absolute alcohol	50 c.c.
Glycerin	50 c.c.
Glacial acetic acid	5 c.c.
Haematoxylin	1 g.
Alum in excess..	

Keep it in a dark place until the color becomes a deep red. If well stoppered, it will keep indefinitely.

Boehmer's Haematoxylin.—

A	Haematoxylin	1 g.
	Absolute alcohol	12 c.c.
B	Alum	1 g.
	Distilled water	240 c.c.

The solution A must ripen for 2 months. When wanted for use, add about 10 drops of A to 10 c.c. of B. Stain 10 to 20 minutes. Wash in water and proceed as usual.

Mayer's Haem-Alum.—Haematoxylin, 1 g., dissolved with heat in 50 c.c. of 95 per cent alcohol and added to a solution of 50 g. of

[1] Stirling and Lee.

alum in a liter of distilled water. Allow the mixture to cool and settle; filter; add a crystal of thymol to preserve from mold (Lee).

It is ready for use as soon as made up. Unless attacked by mold, it keeps indefinitely.

Haidenhain's Iron-Haematoxylin.—This stain was introduced by Haidenhain in 1892 and has gained a well-deserved popularity with those engaged in cytological work. Two solutions are used, and they are never mixed:

A. 1½ to 4 per cent aqueous solution of ammonia sulphate of iron. Use the ferric (violet) crystals, not the ferrous (green) crystals.

B. ½ per cent solution of haematoxylin in distilled water.

The crystals of haematoxylin will dissolve in the distilled water in about 10 days; the stain reaches its greatest efficiency in about 6 weeks. About 3 months from the time it is made up, it begins to deteriorate. A stain made by dissolving the crystals in strong alcohol and then diluting with water so as to get a practically aqueous solution is not so good.

Greenacher's Borax Carmine.—

Carmine	3 g.
Borax	4 g.
Distilled water	100 c.c.

Dissolve the borax in water and add the carmine, which is quickly dissolved with the aid of gentle heat. Add 100 c.c. of 70 per cent alcohol and filter (Stirling).

Alum Carmine.—A 4 per cent aqueous solution of ammonia alum is boiled 20 minutes with 1 per cent of powdered carmine. Filter after it cools (Lee).

Alum Cochineal.—

Powdered cochineal	50 g.
Alum	5 g.
Distilled water	500 c.c.

Dissolve the alum in water, add the cochineal, and boil; evaporate down to two-thirds of the original volume and filter. Add a few drops of carbolic acid to prevent mold (Stirling).

Picro-Carmine.—

Picro-carmine (picro-carminate of ammonia).....	1 g.
Water..	100 c.c.

Myer's Carmalum.—

Carminic acid..................................	1 g.
Alum...	10 g.
Distilled water................................	200 c.c.

Dissolve with heat; decant or filter, and add a crystal of thymol to avoid mold.

This is the stain recommended for *Volvox.*

Carmalum (Alum Lake).—

Carmalum.......................................	1 g.
Water..	100 c.c.
Ammonia..	1 c.c.

Filter, if there is any precipitate.

Aceto-Carmine.—

Heat a 45 per cent aqueous solution of glacial acetic acid to the boiling-point, with an excess of powdered carmine. Cool and filter.

Iron Aceto-Carmine.—

Add a trace of ferric hydrate, dissolve 45 per cent acetic acid, to a quantity of acetic carmine until the liquid becomes bluish red, but no precipitate forms. Then add an equal amount of ordinary acetic-carmine.

Eosin.—

Eosin..	1 g.
Water, or 70 per cent alcohol..................	100 c.c.

General Formula for Anilins.—Make a 3 per cent solution of anilin oil in distilled water; shake well and frequently for a day; add enough alcohol to make the whole mixture about 20 per cent alcohol; add 1 g. of cyanin, erythrosin, safranin, gentian-violet, etc., to each 100 c.c. of this solution.

Cyanin.—This general formula is not at all successful with Grüber's cyanin, but gives satisfactory results with an immensely cheaper cyanin, sold by H. A. Metz & Company, 122 Hudson Street, New York.

Anilin Blue.—

Anilin blue	1 g.
85 or 90 per cent alcohol	100 c.c.

For staining before mounting in Venetian turpentine, this stain should be made up in strong alcohol, even if the dry stain is intended for aqueous solution.

Iodine Green.—

Iodine green	1 g.
70 per cent alcohol	100 c.c.

Methyl Green.—

Methyl green	1 g.
Glacial acetic acid	1 c.c.
Water	100 c.c.

If the preparation is to be mounted in balsam, a slight trace of acetic acid and also a trace of methyl green should be added to the absolute alcohol used for dehydrating.

For staining vascular bundles, the acid may be omitted, even from the formula.

Light Green.—

Light green	1 g.
Clove oil	100 c.c.

or

Light green	1 g.
Clove oil	75 c.c.
Absolute alcohol	25 c.c.

Fuchsin.—

Fuchsin	1 g.
95 per cent alcohol	100 c.c.
Water	100 c.c.

Acid Fuchsin.—

Acid fuchsin	1 g.
Water	100 c.c.

Use this formula when staining woody tissues in methyl green and acid fuchsin.

Ziehl's Carbol Fuchsin.—

Fuchsin	1 g.
Carbolic-acid crystals	5 g.
95 per cent alcohol	10 c.c.
Water	100 c.c.

Fuchsin and Iodine Green Mixtures.—Two solutions are kept separate, since they do not retain their efficiency long after they are mixed.

A	Fuchsin (acid)	0.1 g.
	Distilled water	50.0 c.c.
B	Iodine green	0.1 g.
	Distilled water	50.0 c.c.
C	Absolute alcohol	100.0 c.c.
	Glacial acetic acid	1.0 c.c.
	Iodine	0.1 g.

Stain in equal parts of A and B. Transfer from the stain directly to solution C, and from C to xylol.

Another Formula.—

A	Acid fuchsin	0.5 g.
	Water	100.0 c.c.
B	Iodine green	0.5 g.
	Water	100.0 c.c.

Mix a pipette full of A with a pipette full of B; stain 2 to 8 minutes; dehydrate rapidly and mount in balsam.

Magdala Red.—

Magdala red	1 g.
85 or 90 per cent alcohol	100 c.c.

Use formula when staining in Magdala red and anilin blue, before mounting in Venetian turpentine. Phloxine is the American equivalent.

Safranin.—

Safranin	1 g.
95 per cent alcohol	50 c.c.
Water	50 c.c.

Safranin (another formula).—

Dissolve 1 g. of alcohol-soluble safranin in 100 c.c. of absolute alcohol; dissolve 1 g. of water-soluble safranin in 100 c.c. of distilled water. Mix equal parts of the two solutions.

Gentian-Violet.—

Gentian-violet	1 g.
95 per cent alcohol	20 c.c.
Water	80 c.c.
Anilin oil	3 c.c.

A 1 per cent solution in water keeps better.

A 1 per cent solution in clove oil is worth a thorough trial.

Pyoktanin.—This is sold by E. Merck, in Darmstadt, Germany. Dissolve 1 g. of pyoctanin in 30 c.c. of water.

Orange G.—

Orange G	1 g.
Water	100 c.c.

For most purposes a 1 per cent solution in clove oil is preferable. It is easier to get a solution by dissolving 1 g. of orange in 100 c.c. of absolute alcohol; then add 100 c.c. of clove oil. Let the absolute alcohol evaporate until there is left about 100 c.c. of the solution.

Gold Orange.—

Gold orange	1 g.
Clove oil	100 c.c.

Bismarck Brown.—

Bismarck brown	2 g.
70 per cent alcohol	100 c.c.

Nigrosin.—

Nigrosin	1 g.
Water	100 c.c.

Gram's Solution.—

Iodine	1 g.
Iodide of potassium	2 g.
Water	300 c.c.

MISCELLANEOUS

Mayer's Albumen Fixative.—

White of egg (active principle)	50 c.c.
Glycerin (to keep it from drying up)	50 c.c.
Salicylate of soda (to keep out bacteria, etc.)	1 g.

Shake well and filter.

Land's Gum Fixative.—

Gum arabic	1 g.
Potassium bichromate	1 g.
Water	98 c.c.

Dissolve the gum in water and add the bichromate of potash; or dissolve the gum in half the quantity of water and the bichromate of potash in the other half, and mix just before using. Le Page's liquid glue may be used instead of the gum arabic.

Land's Tropical Developer.—

Hydrokinon	8 g.
Metol	3 g.
Sodium Sulphite (dry) (60 g. if crystals are used)	30 g.
Sodium carbonate (dry) (90 g. if crystals are used)	30 g.
Potassium bromide	2 g.
Water	1,000 c.c.

Land's Contrast Developer.—

Hydrokinon	20 g.
Sodium sulphite (dry)	60 g.
Sodium carbonate (dry)	140 g.
Potassium bromide	12 g.
Water	1,000 c.c.

Schultze's Maceration Fluid.—

The ingredients are nitric acid and potassium chlorate. They are mixed only as the reagent is applied. See chapter on "Special Methods" (chap. xi).

Jeffrey's Maceration Fluid.—

Nitric acid (10 per cent in water)	50 c.c.
Chromic acid (10 per cent in water)	50 c.c.

For directions, see chapter xii.

Fehling's Solution.—

A	Cupric sulphate	3 g.
	Water	100 c.c.
B	Sodium potassium tartrate (Rochelle salts)	16 g.
	Water	100 c.c.
C	Caustic soda	12 g.
	Water	100 c.c.

Keep it in three bottles labeled A, B, and C. When needed for use, add 10 c.c. of water to 5 c.c. from each of the three bottles.

Millon's Reagent.—

Mercury	1 c.c.
Concentrated nitric acid	9 c.c.
Water	10 c.c.

Cuprammonia.—

Prepare by pouring 15 per cent ammonia water upon copper turnings or filings. Let it stand in an open bottle.

Phloroglucin.—

Use a 5 per cent solution in water or alcohol.

Celloidin.—

To make a 2 per cent solution, add one tablet of Schering's celloidin and enough ether-alcohol (equal parts absolute alcohol and ether) to make the whole weigh 2,000 g.

Where only a small quantity is needed, shave off 2 g. of celloidin and add 100 c.c. of ether alcohol.

Eycleshymer's Clearing Fluid.—

Mix equal parts of bergamot oil, cedar oil, and carbolic acid.

Cellulose Acetate.—

Cellulose acetate	12 g.
Pure acetone	100 c.c.

Mrs. Williamson recommends the cellulose acetate sold by Cellon, Ltd., 22 Cork Street, London, W. I., England.

Glycerine Jelly.—

One part (by weight) of finest French gelatin is left for 2 hours in 6 parts (by weight) of distilled water. Add 7 parts of glycerin and for every 100 g.

of the mixture add 1 g. of concentrated carbolic acid. Warm the whole mixture for 15 minutes, stirring all the time, until all the flakes produced by the carbolic acid have disappeared. Filter, while still warm, through a fine-meshed cheese-cloth.

Venetian Turpentine.—

To make a 10 per cent solution, add 90 c.c. of absolute alcohol to 10 c.c. of thick Venetian turpentine. Stir it with a glass rod. Guess at the amount of turpentine, for it is not easy to clean things which have contained Venetian turpentine.

The following need no formulas: acetic acid, hydrochloric acid, nitric acid, sulphuric acid, carbolic acid, hydrofluoric acid, acetone, chloroform, ether, xylol, cedar oil, clove oil, bergamot oil, turpentine, glycerin, paraffin, balsam.

BIBLIOGRAPHY

It is not our purpose to give a complete list of literature which might be useful, but merely to call attention to some papers and books which will extend the student's knowledge in various directions. Some of the papers are cited because they will aid in collecting material and in studying the preparations rather than for any specific technical methods; for we assume that preparations are made to study and not merely for the pleasure of making a good mount.

It will be worth while to watch the reports which are being published in *Science* by the Committee on the Standardization of Biological Stains. The reports which have been published as this book goes to press are given below under the name of H. J. Conn, the chairman of the Committee. His principal collaborators are L. W. Sharp, F. B. Mallory, S. L. Kornhauser, and J. A. Ambler.

BAILEY, I. W., Microtechnique for woody structures. Bot. Gaz. **49:**57–58. 1910.

BELLING, JOHN, On counting chromosomes in pollen mother cells. Amer. Nat. **55:**573–574. 1921.

BENEDICT, H. M., An imbedding medium for brittle or woody tissues. Bot. Gaz. **52:**232. 1911.

BLACKMAN, V. H., Congo red as a stain for Uredineae. New Phyt. **4:**173–174. 1905.

BOYCE, J. S., The imbedding and staining of diseased wood. New Phyt. **8:**432–436. 1918.

BRYAN, GEORGE S., The archegonium of *Sphagnum subsecundum.* Bot. Gaz. **59:**40–56. 1915.

———, The archegonium of *Catharinea angustata.* Bot. Gaz. **64:**1–20. 1917.

BUGNON, M. P., Sur une nouvelle méthode de coloration élective des membranes végétales lignifiées. C. R. Acad. Sci. Paris **168:**62–64. 1919.

CHURCH, MARGARET, Celloidin paraffin method. Sci. N.S. **47:**640. 1918.

CONN, H. J., An investigation of American stains. Jour. Bact. **7:**127–148. 1922.

———, Dye solubility in relation to staining solutions. Sci. N.S. **57:**638–639. 1923.

——— (Chairman) and Commission on Standardization of Biological Stains have published the following reports in Science:

The standardization of biological stains. January 13, 1922.

The production of biological stains in America. **53**:289–290. 1921.
The present supply of biological stains. **56**:562–563. 1922.
Collaborators in the standardization of biological stains. **56**:594–596. 1922.
Preliminary report on American biological stains. **56**:August 11, 1922.
American eosins. **56**:689–690. 1922.
The preparation of staining solutions. **57**:15–16. 1923.
Safranin and methyl green. **57**:304–305. 1923.
Dye solubility in relation to staining solutions. **57**:41–42. 1923.
Standardized nomenclature of biological stains. **57**:743–746. 1923.
Certified methylene blue. **58**:41–42. 1923.

COUPIN, H., Sur le montage de quelques préparations microscopiques. Rev. Gén. Bot. **31**:109–114. 1919.

COWDRY, E. V., Microchemical constituents of protoplasm. Carnegie Inst. Wash. Publ. 271. 1918.

CROCKER, E. C., An experimental study of the significance of "lignin" color reactions. Jour. Indust. and Eng. Chem. **13**:625–627. 1921.

DAVIS, W. H., Staining germinating spores. Phytopath. **12**:492–494. 1922.

DEGENER, OTTO, Four new stains of *Lycopodium* prothallia. Bot. Gaz. **77**:89–95. 1924.

DE ZEEUW, R., The value of double infiltration in botanical microtechnique. Papers Mich. Acad. Sci. **1**:83–84. 1923.

DICKSON, B. T., The differential staining of plant pathogen and host. Sci. N.S. **52**:63–64. 1920.

DOWSON, W. T., A new method of paraffin infiltration. Ann. Bot. **36**:577–578. 1922.

DURAND, E. J., The differential staining of intercellular mycelium. Phytopath. **1**:129–130. 1911.

FARMER, J. B., and SHOVE, DOROTHY, On the structure and development of the somatic and heterotype chromosomes of *Tradescantia virginica*. Quart. Jour. Mic. Sci. **48**:559–569. 1905.

GERRY, E., and DIEMER, M. E., Stains for the mycelium of molds and other fungi. Sci. N.S. **54**:629–630. 1921.

HILL, J. B., A method for dehydration of histological material. Bot. Gaz. **51**:255–256. 1916.

HUBERT, E. E., A staining method for hyphae of wood-inhabiting fungi. Phytopath. **12**:440–441. 1922.

KARSTON, G., Über embryonales Wachstun und seine Tagesperiode. Zeitschrift für Botanik **7**:1–34. 1915.

KELLICOTT, W. E., The daily periodicity of cell division and of elongation in the root of *Allium*. Bull. Torrey Bot. Club. **31**:529–550. 1904.

KLEINENBERG, Haematoxylin method. Quar. Jour. Micr. Sci. **19**:208. 1879.

KORNHAUSER, S. I., Celloidin paraffin method. Sci. **44**: 57–58. 1916.

LAND, W. J. G., Microtechnical methods. Bot. Gaz. **59**: 397–401. 1915.

———, A method of controlling the temperature of the paraffin block and microtome knife. Bot. Gaz. **57**: 520–523. 1914.

LANGDON, LADEMA MARY, Sectioning hard woody tissues. Bot. Gaz. **70**: 82–84. 1920.

LEE, H. N., The staining of wood fibers for permanent microscopic mounts. Bot. Gaz. **62**: 318–319. 1916.

MANN, ALBERT, The preparation of unbroken pollen mother cells and other cells for studies in mitoses. Sci. **36**: 151–153. 1912.

METZ, C. W., A simple method for handling small objects in making microscopic preparations. Anat. Record **21**: 373–374. 1921.

MOLLENDORF, W. VON, Vitale Färbungen an tierischen Zellen. Ergebn. Physiol. **18**: 141–306. 1920. (Bibliography.)

NEMEC, B., Gentian method for starch. Ber. Deu. Bot. Ges. **24**: 528–531. 1906.

NIEUWLAND, J. A., The mounting of algae (preserving the green color). Bot. Gaz. **47**: 237. 1909.

OSTERHOUT, W. J. V., Cell studies: I. Spindle formation in *Agave*. Proc. Cal. Acad. Sci. Bot. (3d series) **2**: 255–284. 1902.

PFEIFFER, NORMA E, Monograph of the Isoetaceae. Annals of the Missouri Botanical Garden. **9**: 79–232. 1922.

PICKETT, F. L., Preparation of whole pollen mother cells. Sci. **36**: 479–480. 1912.

———, A study of the stability of staining solutions. Trans. Amer. Mic. Soc. **42**: 129–132. 1923.

PLOWMAN, A. B., Celloidin method with hard tissues. Bot. Gaz. **37**: 456–461. 1904.

RUCKES, H., Washing microscopic organisms. Sci N.S. **48**: 44–45. 1918.

SAMPSON, H. C., Chemical changes accompanying abscission in *Coleus Blumei*. Bot. Gaz. **66**: 32–53. 1918. (Summarizes microchemical tests.)

SANDS, HAROLD C., The structure of the chromosomes in *Tradescantia virginica*. Amer. Jour. of Bot. **10**: 343–360. 1923.

SCHAEDE, R., Über das Verhalten von Pflanzenzellen gegenüber Anilinfarbstoffen. Jahrb. Wiss. Bot. **62**: 65–91. 1923.

SEIFRIZ, W., A method for inducing protoplasmic streaming. New Phytol. **21**: 107–112. 1922.

SHARP, L. W., An introduction to cytology. New York. McGraw-Hill Book Co., Inc.

SINNOTT, E. W., and BAILEY, I. W., Some technical aids for the anatomical study of decaying wood. Phytopath. **14**: 403. 1914.

SMITH, G. M., The preservation of fresh water algae. Plant World **16**: 219–230. 1913.

SPESSARD, E. A., Prothallia of *Lycopodium* in America. Bot. Gaz. **63**:66–76. 1917; **65**:362. 1918; **74**:392–413. 1922.

STEIL, W. N., Method for staining antherozoids of ferns. Bot. Gaz. **65**:562–563. 1918.

STEVENS, F. L., The fungi that cause disease. New York: Macmillan Co. 1919.

STEWART, A. B., The mounting of celloidin sections in series. Sci. **42**:872. 1915.

STOKEY, A. G., and STARR, A. M., *Lycopodium* prothallia in western Massachusetts. Bot. Gaz. **77**:80–88. 1924.

STRASBURGER, E., Das botanische Prakticum, siebente Auflage bearbeitet von Max Koernicke. Jena, Germany: Gustav Fischer, 1923.

SZOMBATHY, S., Neue Methode zum Aufkleben von Paraffinschnitten. Zeitschrift. Wiss. Mikr. **34**:334–336. 1918. Review in Bot. Gaz. **67**:373–374. 1919.

TAYLOR, W. R., A method for demonstrating the sheath structure of a desmid. Trans. Amer. Micr. Soc. **40**:94–95. 1921.

TISON, A., Méthode nouvelle de coloration des tissus subéreux. Compt. Rend. Assoc. Franc. Avanc. Sci. **28**:Pt. II, 454–456. 1899. (See Phytopath. **9**:483–496. 1919.)

TOBLER-WOLFF, G., Zur Methodik der mikroskopischen Pflanzenfaseruntersuchung. Zeitschr. Wiss. Mikr. **32**:129–136. 1916.

VAUGHAN, R. E., A method for the differential staining of fungus and host cells. Ann. Mo. Bot. Garden **1**:241–242. 1914.

WALKER, ELDA, The gametophytes of *Equisetum laevigatum*. Bot. Gaz. **71**:378–391. 1921.

WILLIAMSON, H. S., A new method of preparing sections of hard vegetable structures. Ann. Bot. January, 1921.

WOODS, A. F., Method of preserving the green color of plants for exhibition purposes. Bot. Gaz. **24**:206. 1897.

———, Preserving the green color of algae, protonema, etc. *Quar. Jour. Mic. Sci.* **14**:225–228. 1921.

INDEX

INDEX

[The references are to pages. Italic figures indicate illustrations.

PRINTED IN THE U.S.A.

Spotted something that isn't quite right?

Let us know and we'll fix it for future reprints. And as a thank you for your diligence, we'll send you a free Sothis book of your choice.

publisher@sothis.press

This page intentionally left blank

Made in the USA
Monee, IL
15 November 2026

68605195R00215